T0076256

METHODS IN MOLECULAR BIOLOGY™

For further volumes:
http://www.springer.com/series/7651

Microbial Carotenoids From Fungi

Methods and Protocols

Edited by

José-Luis Barredo

Gadea Biopharma, Parque Tecnológico de León, León, Spain

Editor
José-Luis Barredo
Gadea Biopharma
Parque Tecnológico de León
León, Spain

ISSN 1064-3745 ISSN 1940-6029 (electronic)
ISBN 978-1-61779-917-4 ISBN 978-1-61779-918-1 (ebook)
DOI 10.1007/978-1-61779-918-1
Springer New York Heidelberg Dordrecht London

Library of Congress Control Number: 2012939488

Printed on acid-free paper

Humana Press is a brand of Springer
Springer is part of Springer Science+Business Media (www.springer.com)

Preface

Carotenoids are a family of yellow to orange-red terpenoid pigments synthesized by photosynthetic organisms and by many bacteria and fungi. They have beneficial health effects, protecting against oxidative damage, and may be responsible for the colors associated with plants and animals. Carotenoids are also desirable commercial products used as colorants, feed supplements, and nutraceuticals in the food, medical, and cosmetic industries. Only a few of the more than 600 identified carotenoids are produced industrially, with β-carotene (a popular additive for butter, ice cream, orange juice, candies, etc.) being the most prominent.

Commercial production of natural carotenoids from microorganisms is a new approach more eco-friendly than synthetic manufacture by chemical procedures. Despite the availability of a variety of natural and synthetic carotenoids, there is currently renewed interest in microbial sources. Due to its increasing importance, industrial biotechnological methods of carotenoids production have been developed with the algae *Dunaliella salina* and *Haematococcus pluvialis*, the fungus *Blakeslea trispora*, and the heterobasidiomycetous yeast *Xanthophyllomyces dendrorhous*.

This book is intended to provide practical experimental laboratory procedures for a wide range of carotenoids producing microorganisms. Although not an exhaustive treatise, it provides a detailed "step-by-step" description of the most recent developments in applied biotechnological processes useful for screening and selection of carotenoids producing fungi, and construction of new carotenoid biosynthetic pathways. The detailed protocols are cross-referenced in the Notes section, providing special details, little problems, troubleshooting, and safety comments that may not normally appear in journal articles and can be particularly useful for those not familiar with specific techniques.

The two lead chapters of this volume represent overviews on the advancement of biotechnology and on microbial carotenoids. The subsequent chapters show comprehensive experimental methods for the manipulation and metabolic engineering of the β-carotene producing fungi *Blakeslea trispora* and *Mucor circinelloides*. Furthermore, methods for lycopene production with the yeast *Yarrowia lipolytica*, and peroxisome targeting of lycopene pathway enzymes in *Pichia pastoris* are shown. This book also comprises several chapters on the manipulation of the heterobasidiomycetous yeast *X. dendrorhous*, which produces astaxanthin, a red xanthophyll with large importance in the aquaculture, pharmaceutical, and food industries. Additionally, the book includes a DNA assembler method for construction of zeaxanthin-producing strains of *Saccharomyces cerevisiae*, production of neurosporaxanthin by *Neurospora* and *Fusarium*, and production of torularhodin, torulene, and β-carotene by *Rhodotorula* yeasts.

This book has been written by outstanding experts in this field and provides a reference source for laboratory and industrial professionals, as well as for graduate students in a number of biological disciplines (biotechnology, microbiology, genetics, molecular biology, etc.).

I am indebted to the authors that, in spite of their professional activities, agreed to participate in this book, to Dr. J. Walker, Series Editor, for his encouragement and advice while reviewing the manuscripts, and to the rest of the staff of The Humana Press, Inc., for their assistance in assembling this volume and their efforts in keeping this project on schedule. Last but not least, I warmly acknowledge my wife Natalia and our children Diego, José Luis, Álvaro, and Gonzalo for their patience and support.

León, Spain ***José-Luis Barredo***

Contents

Contributors

JOSÉ L. ADRIO • *Neuron BioIndustrial Division, Neuron BPh S.A., Granada, Spain; Parque Tecnologico de Ciencias de la Salud, Granada, Spain*
MARÍA ISABEL ALVAREZ • *Área de Genética, Departamento de Microbiología y Genética, University of Salamanca, Salamanca, Spain*
JAVIER AVALOS • *Departamento de Genética, Facultad de Biología, Universidad de Sevilla, Sevilla, Spain*
MARCELO BAEZA • *Centro de Biotecnología y Dpto. Cs. Ecológicas, Facultad de Ciencias, Universidad de Chile, Santiago, Chile*
JOSÉ-LUIS BARREDO • *Gadea Biopharma, Parque Tecnológico de León, León, Spain*
PAUL S. BERNSTEIN • *Department of Ophthalmology and Visual Sciences, Moran Eye Center, University of Utah School of Medicine, Salt Lake City, UT, USA*
PRAKASH BHOSALE • *Department of Ophthalmology and Visual Sciences, Moran Eye Center, University of Utah School of Medicine, Salt Lake City, UT, USA*
BYRON F. BREHM-STECHER • *Department of Bacteriology, Food Research Institute, University of Wisconsin-Madison, Madison, USA*
JOANNA BULTMAN • *Department of Bacteriology, University of Wisconsin-Madison, Madison, WI, USA*
ENRIQUE CERDÁ-OLMEDO • *Departamento de Genética, Facultad de Biología, Universidad de Sevilla, Sevilla, Spain*
VÍCTOR CIFUENTES • *Centro de Biotecnología y Dpto. Cs. Ecológicas, Facultad de Ciencias, Universidad de Chile, Santiago, Chile*
ÁRPÁD CSERNETICS • *Department of Microbiology, Faculty of Science and Informatics, University of Szeged, Szeged, Hungary*
JUAN-LUIS DE LA FUENTE • *Gadea Biopharma, Parque Tecnológico de León, León, Spain*
ARNOLD L. DEMAIN • *Charles A. Dana Research Institute for Scientists Emeriti (R.I.S.E.), Drew University, Madison, NJ, USA*
ALMA ROSA DOMÍNGUEZ-BOCANEGRA • *Department of Biotechnology and Bioengineering, CINVESTAV, Mexico DF, Mexico*
ARTURO P. ESLAVA • *Área de Genética, Departamento de Microbiología y Genética, University of Salamanca, Salamanca, Spain*
ALEJANDRO F. ESTRADA • *Departamento de Genética, Facultad de Biología, Universidad de Sevilla, Sevilla, Spain*
JACK W. FELL • *Rosenstiel School of Marine and Atmospheric Science, University of Miami, Key Biscayne, FL, USA*
MARÍA FERNÁNDEZ-LOBATO • *Centro de Biología Molecular Severo Ochoa, Dpto. Biología Molecular (UAM-CSIC), Universidad Autónoma de Madrid, Spain, Madrid, Spain*
VICTORIANO GARRE • *Departamento de Genética y Microbiología (Unidad Asociada al IQFR-CSIC), Facultad de Biología, Campus de Espinardo, Universidad de Murcia, Murcia, Spain*

Enrique A. Iturriaga • *Área de Genética, Departamento de Microbiología y Genética, University of Salamanca, Salamanca, Spain*
Eric A. Johnson • *Department of Bacteriology, University of Wisconsin, Madison, WI, USA*
Pyung Cheon Lee • *Department of Molecular Science and Technology and Department of Biotechnology, Ajou University, Suwon, South Korea*
Diego Libkind • *Laboratorio de Microbiología Aplicada y Biotecnología Bariloche, INIBIOMA (CONICET-Universidad del COMAHUE), San Carlos de Bariloche, Río Negro, Argentina*
Guangyun Lin • *Department of Bacteriology, University of Wisconsin-Madison, Madison, WI, USA*
Sergio López-García • *Departamento de Genética y Microbiología (Unidad Asociada al IQFR-CSIC), Facultad de Biología, Campus de Espinardo, Universidad de Murcia, Murcia, Spain*
Yunzi Luo • *Department of Chemical and Biomolecular Engineering, University of Illinois at Urbana-Champaign, Urbana, IL, USA*
Bina J. Mehta • *Departamento de Genética, Facultad de Biología, Universidad de Sevilla, Sevilla, Spain*
Martín Moliné • *Laboratorio de Microbiología Aplicada y Biotecnología Bariloche, INIBIOMA (CONICET-Universidad del COMAHUE), San Carlos de Bariloche, Río Negro, Argentina*
Eusebio Navarro • *Departamento de Genética y Microbiología (Unidad Asociada al IQFR-CSIC), Facultad de Biología, Campus de Espinardo, Universidad de Murcia, Murcia, Spain*
Mauricio Niklitschek • *Centro de Biotecnología y Dpto. Cs. Ecológicas, Facultad de Ciencias, Universidad de Chile, Santiago, Chile*
Ildikó Nyilasi • *Department of Microbiology, Faculty of Science and Informatics, University of Szeged, Szeged, Hungary*
Tamás Papp • *Department of Microbiology, Faculty of Science and Informatics, University of Szeged, Szeged, Hungary*
Alfonso Prado-Cabrero • *Departamento de Genética, Facultad de Biología, Universidad de Sevilla, Sevilla, Spain*
Marta Rodríguez-Sáiz • *Gadea Biopharma, Parque Tecnológico de León, León, Spain*
Rosa M. Ruiz-Vázquez • *Departamento de Genética y Microbiología (Unidad Asociada al IQFR-CSIC), Facultad de Biología, Campus de Espinardo, Universidad de Murcia, Murcia, Spain*
C. Schimek • *General Microbiology and Microbial Genetics, Institute of Microbiology, Friedrich-Schiller-University Jena, Jena, Germany*
Zengyi Shao • *Department of Chemical and Biomolecular Engineering, Institute for Genomic Biology, University of Illinois at Urbana-Champaign, Urbana, IL, USA*
Pamela L. Sharpe • *E.I. DuPont de Nemours, Inc., Wilmington, DE, USA*
Celia Tognetti • *Laboratorio de Microbiología Aplicada y Biotecnología Bariloche, INIBIOMA (CONICET-Universidad del COMAHUE), San Carlos de Bariloche, Río Negro, Argentina*
Santiago Torres-Martínez • *Departamento de Genética y Microbiología (Unidad Asociada al IQFR-CSIC), Facultad de Biología, Campus de Espinardo, Universidad de Murcia, Murcia, Spain*

PREEJITH VACHALI • *Department of Ophthalmology and Visual Sciences, Moran Eye Center, University of Utah School of Medicine, Salt Lake City, UT, USA*
CSABA VÁGVÖLGYI • *Department of Microbiology, Faculty of Science and Informatics, University of Szeged, Szeged, Hungary*
MARIA VAN BROOCK • *Laboratorio de Microbiología Aplicada y Biotecnología Bariloche, INIBIOMA (CONICET-Universidad del COMAHUE), San Carlos de Bariloche, Río Negro, Argentina*
ANA VILA • *Departamento de Genética y Microbiología (Unidad Asociada al IQFR-CSIC), Facultad de Biología, Campus de Espinardo, Universidad de Murcia, Murcia, Spain*
J. WÖSTEMEYER • *General Microbiology and Microbial Genetics, Institute of Microbiology, Friedrich-Schiller-University Jena, Jena, Germany*
RICK W. YE • *E328/148B, DuPont Experimental Station, Wilmington, DE, USA*
HUIMIN ZHAO • *Department of Chemical and Biomolecular Engineering, Institute for Genomic Biology, University of Illinois at Urbana-Champaign, Urbana, IL, USA; Departments of Chemistry, Biochemistry, and Bioengineering, University of Illinois at Urbana-Champaign, Urbana, IL, USA*
QUINN ZHU • *E328/150B, DuPont Experimental Station, Wilmington, DE, USA*

Chapter 1

Essential Role of Genetics in the Advancement of Biotechnology

Arnold L. Demain and José L. Adrio

Abstract

Microorganisms are one of the greatest sources of metabolic and enzymatic diversity. In recent years, emerging recombinant DNA and genomic techniques have facilitated the development of new efficient expression systems, modification of biosynthetic pathways leading to new metabolites by metabolic engineering, and enhancement of catalytic properties of enzymes by directed evolution. Complete sequencing of industrially important microbial genomes is taking place very rapidly and there are already hundreds of genomes sequenced. Functional genomics and proteomics are major tools used in the search for new molecules and development of higher-producing strains.

Key words: Primary metabolites, Secondary metabolites, Bioconversions, Enzymes, Hosts, Biopharmaceuticals, Metabolic engineering, Genetic engineering, Agriculture, Polymers

1. Introduction

Microorganisms have advantages in the production of compounds as compared to isolation from plants and animals or synthesis by chemists. These are (1) rapid uptake of nutrients that supports high rates of metabolism and biosynthesis, (2) capability of carrying out a wide variety of reactions, (3) facility to adapt to a large array of different environments, (4) ease of genetic manipulation, both in vivo and in vitro, to increase production, to modify structures and activities, and to make entirely new products, (5) simplicity of screening procedures, and (6) a wide diversity.

Microbial products are very diverse, ranging from very large molecules, such as proteins, nucleic acids, carbohydrate polymers, or even cells, to small molecules that are usually divided into primary metabolites, i.e., those essential for vegetative growth, and secondary metabolites, i.e., those nonessential for growth.

José-Luis Barredo (ed.), *Microbial Carotenoids From Fungi: Methods and Protocols*, Methods in Molecular Biology, vol. 898, DOI 10.1007/978-1-61779-918-1_1, © Springer Science+Business Media New York 2012

1.1. Primary Metabolites

These are microbial products made during the exponential phase of growth whose synthesis is an integral part of the normal growth process. They are intermediates or end products of the pathways, are building blocks for essential macromolecules (e.g., amino acids and nucleotides), or are converted into coenzymes (e.g., vitamins). Other primary metabolites (e.g., organic acids and ethanol) result from catabolic metabolism; they are not used for building cellular constituents but their production, which is related to energy production and substrate utilization, is essential for growth. Industrially, the most important primary metabolites are amino acids, nucleotides, vitamins, alcohols, and organic acids.

Production of a particular primary metabolite by deregulated organisms may inevitably be limited by the inherent capacity of the particular organism to make the appropriate biosynthetic enzymes. Recent approaches utilize the techniques of modern genetic engineering to correct such deficiencies and develop strains overproducing primary metabolites. There are two ways to accomplish this (1) to increase the number of copies of structural genes coding for these enzymes and (2) to increase the frequency of transcription.

Novel genetic technologies are important for the development of overproducers. Ongoing genome-sequencing projects involving hundreds of genomes, the availability of sequences corresponding to model organisms, new DNA microarray and proteomic tools, as well as the new techniques for mutagenesis and recombination described below, will no doubt accelerate strain improvement programs.

Genome-based strain reconstruction achieves the construction of a superior strain which only contains mutations crucial to hyperproduction but not other unknown mutations which accumulate by brute-force mutagenesis and screening (1).

The directed improvement of product formation or cellular properties via modification of specific biochemical reactions or introduction of new ones with the use of recombinant DNA technology is known as metabolic engineering (2, 3). Analytical methods are combined to quantify fluxes and to control them with molecular biological techniques to implement suggested genetic modifications. The overall flux through a metabolic pathway depends on several steps, not just a single rate-limiting reaction. Amino acid production is one of the fields with many examples of this approach (4). Other processes improved by this technique include vitamins, organic acids, ethanol, 1,3-propanediol, and carotenoids.

Reverse (inverse) metabolic engineering is another approach that involves choosing a strain which has a favorable cellular phenotype, evaluating and determinating the genetic and/or environmental factors that confer that phenotype, and finally transferring that phenotype to a second strain via direct modifications of the identified genetic and/or environmental factors (5, 6).

Molecular breeding techniques are based on mimicking natural recombination by in vitro homologous recombination (7). DNA shuffling not only recombines DNA fragments but also introduces point mutations at a very low controlled rate (8). Unlike site-directed mutagenesis, this method of pooling and recombining parts of similar genes from different species or strains has yielded remarkable improvements in a very short amount of time (9). Whole genome shuffling is another technique that combines the advantage of multiparental crossing allowed by DNA shuffling with the recombination of entire genomes through recursive protoplast fusion (10, 11).

Systems biology is an integrated, systemic approach to the analysis and optimization of cellular processes by introducing a variety of perturbations and measuring the system response (12). Altered phenotypes are created by molecular biological techniques or by altering environments. Further characterization of the phenotype leading to maximal product formation is analyzed and quantified through the use of genome-wide high-throughput *omics* data and genome-scale computational analysis.

1.2. Amino Acids

Many of the techniques mentioned above have made a great impact on the production of amino acids. Different strategies including (1) amplification of a rate-limiting enzyme of pathway, (2) amplification of the first enzyme after a branchpoint, (3) cloning of a gene encoding an enzyme with more or less feedback regulation, (4) introduction of a gene encoding an enzyme with a functional or energetic advantage as a replacement for the normal enzyme, (5) amplification of the first enzyme leading from central metabolism to increase carbon flow into the pathway, and (6) isolation of a transport mutant decreasing amino acid uptake and intracellular feedback control while improving excretion.

Among the amino acids, L-glutamate and L-lysine, mostly used as feed and food additives, represent the largest products in this category. Produced by fermentation are 1.5 million tons of L-glutamate and 850,000 tons of L-lysine HCl. The total amino acid market was about 4.5 billion dollars in 2004 (13). High titers of amino acids are shown in Table 1.

1.2.1. L-Glutamate

Glutamate was the first amino acid to be produced by fermentation because of its use as a flavoring agent (monosodium glutamate, MSG). The process employs various species of the genera *Corynebacterium* and *Brevibacterium*. Molar yields of glutamate from sugar are 50–60% and broth concentrations have reached 150 g/L. Glutamic acid overproduction is feedback regulated. However, mutant strains with modifications in the cell membrane are able to pump glutamate out of the cell, thus allowing its

Table 1
High amino acids levels produced by fermentation

Amino acid	Titer (g/L)
L-Alanine	114
L-Arginine	96
L-Glutamate	150
L-Glutamine	49
L-Histidine	42
L-Hydroxyproline	41
L-Isoleucine	40
L-Leucine	34
L-Lysine HCl	170
L-Methionine	25
L-Phenylalanine	51
L-Proline	108
L-Serine	65
L-Threonine	100
L-Tryptophan	60
L-Tyrosine	55
L-Valine	105

biosynthesis to proceed unabated. Introduction of the *Vitreoscilla* hemoglobin gene *vgb* into *Corynebacterium glutamicum* increased cell growth and glutamic acid and glutamine production via increased oxygen uptake (14). Workers at Ajinomoto Co., Inc. increased glutamate production from glucose by 9% by suppressing CO_2 liberation in the pyruvate dehydrogenase reaction (15). They did this by cloning the *xfp* gene-encoding phosphoketolase from *Bifidobacterium animalis* into *C. glutamicum* and overexpressing it.

1.2.2. L-Lysine

Since the bulk of the cereals consumed in the world are deficient in L-lysine, this essential amino acid became an important industrial product. Although the lysine biosynthetic pathway is controlled very tightly in an organism such as *Escherichia coli* and in lysine-producing organisms (e.g., mutants of *C. glutamicum* and its relatives), there is only a single aspartate kinase, which is regulated via concerted feedback inhibition by threonine plus lysine. Metabolic engineering has been used in *C. glutamicum* to improve L-lysine production (4).

Comparative genome analysis between a wild-type strain and an L-lysine-producing *C. glutamicum* strain identified three mutations that increased L-lysine production when introduced into the wild-type strain. Introduction of the 6-phosphogluconate dehydrogenase *gnd* mutation increased yield by 15% (16). Further improvement was achieved by introducing the *mqo* mutation (malate:quinone oxidoreductase) resulting in an increased titer of 95 g/L in a fed-batch culture (17). Transcriptome analysis revealed that strain B-6, as compared to the wild type, is upregulated in both the pentose phosphate pathway genes and the amino acid biosynthetic genes and downregulated in TCA cycle genes. Lysine HCl titers have reached 170 g/L (18). Metabolic flux studies of wild-type *C. glutamicum* and improved lysine-producing mutants showed that yield increased from 1.2% to 24.9% relative to the glucose flux. Different approaches have been used to increase lysine production by *C. glutamicum* mutants including (1) deletion of different genes from the glycolytic pathway (19), (2) improving the availability of pyruvate by eliminating pyruvate dehydrogenase activity (20), (3) overexpression of pyruvate carboxylase or DAP dehydrogenase genes, and (4) overexpression of gene NCgl0855 encoding a methyltransferase or the *amtA-ocd-soxA* operon (21).

1.2.3. L-Threonine

Overproduction of L-threonine has been achieved in *Serratia marcescens* by transductional crosses which combined several feedback control mutations in a single organism. This resulted in titers up to 25 g/L (22). Combination in a single strain of another six regulatory mutations derived by resistance to amino acid analogs led to desensitization and derepression of the key enzymes of threonine synthesis. The resultant transductant produced 40 g/L of threonine (23). The use of recombinant DNA technology led to strains that made 63 g/L threonine (24). An *E. coli* strain was developed via mutation and genetic engineering and optimized by the inactivation of threonine dehydratase (TD) resulting in a process yielding 100 g/L of L-threonine in 36 h of fermentation (25).

In *C. glutamicum* ssp. *lactofermentum*, L-threonine production reached 58 g/L when a strain producing both threonine and lysine was transformed with a plasmid carrying its own *hom*, *thrB*, and *thrC* genes (26). A recombinant isoleucine auxotrophic strain of *E. coli* (carrying extra copies of the *thrABC* operon, an inactivated *tdh* gene and mutated to resist high concentrations of L-threonine and L-homoserine) produced 80 g/L L-threonine in 1.5 days with a yield of 50% (27). Cloning extra copies of threonine export genes into *E. coli* also increased threonine production (28).

Systems metabolic engineering was used to develop an L-threonine1-overproducing *E. coli* strain (29). Feedback inhibition of aspartokinase I and III (encoded by *thrA* and *lysC*, respectively) and transcriptional attenuation regulation (*thrL*) were removed. Deletion of *tdh* and mutation of *ilvA* avoid threonine degradation. The *metA* and *lysA* genes were deleted to make more precursors available for threonine biosynthesis. Further target genes to be engineered were identified by transcriptome profiling combined with in silico flux response analysis. The final engineered *E. coli* strain was able to reach a high yield of 0.393 g of threonine per gram of glucose and 82.4 g/L in a fed-batch culture.

1.2.4. L-Valine

Combined rational modification, transcriptome profiling, and systems-level in silico analysis was used to develop an *E. coli* strain for the production of this amino acid (30). The *ilvA*, *leuA*, and *panB* genes were deleted to make more precursors available for L-valine biosynthesis. This engineered strain, harboring a plasmid overexpressing the *ilvBN* genes, produced 1.3 g/L L-valine. Overexpession of the *lrp* and *ygaZH* genes (encoding a global regulator Lrp and L-valine exporter, respectively), and amplification of the *lrp*, *ygaZH*, and *lrp-ygaZH* genes enhanced production of L-valine by 21.6, 47.1, and 113%, respectively. Further improvement was achieved by using in silico gene knockout simulation, which identified the *aceF*, *mdh*, and *pfkA* genes as knockout targets. The VAMF strain (Val Δ*aceF* Δ*mdh* Δ*pfkA*) was able to produce 7.5 g/L L-valine from 20 g/L glucose in batch culture, resulting in a high yield of 0.378 g of L-valine per gram of glucose. Production by mutant strain VAL1 of *C. glutamicum* amounted to 105 g/L (31). The mutant was constructed by overexpressing biosynthetic enzymes via a plasmid, eliminating *ilvA* encoding threonine dehydratase, and deleting two genes encoding enzymes of pantothenate biosynthesis. The culture was grown with limiting concentrations of isoleucine and pantothenate.

1.2.5. L-Isoleucine

Isoleucine processes have been devised in various bacteria such as *S. marcescens*, *C. glutamicum* ssp. *flavum*, and *C. glutamicum*. In *S. marcescens*, resistance to isoleucine hydroxamate and α-aminobutyric acid led to derepressed L-threonine deaminase (TD) and acetohydroxyacid synthase (AHAS) and production of 12 g/L of isoleucine (32). Further work involving transductional crosses into a threonine-overproducer yielded isoleucine at 25 g/L (22).

Feedback regulation in *C. glutamicum* was eliminated (33) by replacing the native threonine dehydratase gene *ilvA* with the feedback resistant gene from *E. coli*. By introducing additional copies of genes encoding branched amino acid biosynthetic enzymes, lysine- or threonine-producing strains were converted into L-isoleucine producers with improved titers (34–36). Amplification of the wild-type threonine dehydratase gene *ilvA* in a threonine-producing strain of *Corynebacterium lactofermentum* led to isoleucine production (37).

A threonine-overproducing strain of *C. glutamicum* was sequentially mutated to resistance to thiaisoleucine, azaleucine, and aminobutyric acid; it produced 10 g/L of isoleucine (38). Metabolic engineering studies involving overexpression of biosynthetic genes were useful in improving isoleucine production by this species. Colon et al. (37) obtained an isoleucine-producing strain by cloning multiple copies of *hom* (encoding HDI), and wild-type *ilvA* (encoding TD) into a lysine-overproducer, and by increasing HK (encoded by *thrB*); a titer of 15 g/L isoleucine was obtained. Independently, Morbach et al. (39) cloned three copies of the feedback-resistant HD gene (*hom*) and multicopies of the deregulated TD gene (*ilvA*) in a deregulated lysine producer of *C. glutamicum*, yielding an isoleucine producer (13 g/L) with no threonine production and reduced lysine production. Application of a closed loop control fed-batch strategy raised production to 18 g/L (40). Further metabolic engineering work involving amplification of feedback inhibition-insensitive biosynthetic enzymes converted lysine overproducers and threonine overproducers into *C. glutamicum* strains yielding 30 g/L of isoleucine (41).

C. glutamicum ssp. *flavum* studies employed resistance to α-amino-β-hydroxyvaleric acid and the resultant mutant produced 11 g/L (42). D-Ethionine resistance was used by Ikeda et al. (43) to yield a mutant producing 33 g/L in a fermentation continuously fed with acetic acid.

1.2.6. L-Alanine

Lee et al. (44) introduced into an *E. coli* double mutant (lacking genes encoding a protein of the pyruvate dehydrogenase complex [*aceF*] and lactate dehydrogenase [*ldhA*] a plasmid containing the *Bacilus sphaericus* alanine dehydrogenase gene (*alaD*). The strain produced L-alanine in 27 h with a yield on glucose of 0.63 g/g and a maximum productivity of 2 g/L/h. Further work has raised titer to 114 g/L.

1.2.7. L-Proline

Proline-hyperproducing strains of bacteria, exhibiting reduced proline-mediated feedback inhibition of γ-glutamyl kinase (GK) activity (a result of single-base pair substitutions in the bacterial *proB* gene coding region), have been isolated based on their resistance to toxic proline analogs (L-azetidine-2-carboxylic acid and 3,4-dehydro-DL-proline), compounds which inhibit GK activity while not interfering with protein synthesis. Cloning of the three genes of proline biosynthesis in *E. coli* on multicopy plasmids and selection of mutants of such plasmid-containing strains to resistance to 3,4-dehydroproline led to a process producing 20 g/L proline (45).

A mutant of *S. marcescens* resistant to 3,4-dehydroproline, thiazolidine-4-carboxylate, and azetidine-2-carboxylate and unable to utilize proline produced 50–55 g/L L-proline (46). Cloning of a gene bearing the dehydroproline-resistance locus on a plasmid yielded a recombinant strain of *S. marcescens* producing 75 g/L (47). Further development work increased the production to over 100 g/L (48).

A sulfaguanidine-resistant mutant of *C. glutamicum* ssp. *flavum* produced 35 g/L proline (49). When a glutamate-producing strain of *C. glutamicum* was grown under modified conditions, it made 48 g/L (50). A strain of *Corynebacterium acetoacidophilum* produced 108 g/L proline when grown in the presence of glutamate (51).

1.2.8. L-Hydroxyproline

Introduction of the proline 4-hydroxylase gene from *Dactylosporangium* sp. into recombinant *E. coli* producing L-proline at 1.2 g/L led to a new strain producing 25 g/L of hydroxyproline (*trans*-4-hydroxy-L-proline) (52). When proline was added, the hydroxyproline titer reached 41 g/L, with a yield of 87% based on the amount of proline added.

1.3. Nucleotides

Nucleotide fermentations became commercially important due to the activity of two purine ribonucleoside 5′-monophosphates, namely guanylic acid (GMP) and inosinic acid (IMP), as enhancers of flavor. Titers of IMP and GMP have reached 30 g/L (53). The techniques used to achieve such production are similar to those used for amino acid fermentations. In Japan, 2,500 tons of GMP and IMP are produced annually with a market of $350 million. GMP can also be made by bioconversion of xanthylic acid (XMP). Genetic modification of *Corynebacterium ammoniagenes* involving trasnsketolase, an enzyme of the nonoxidative branch of the pentose phosphate pathway, resulted in the accumulation of 39 g/L of XMP (54).

1.4. Vitamins

Riboflavin. Production of riboflavin (vitamin B_2) reached over 6,000 tons per year and a titer of 20 g/L by overproducers such as the yeast-like molds, *Eremothecium ashbyii* and *Ashbya gossypii*. A bacterial process using *C. ammoniagenes* (previously *Brevibacterium ammoniagenes*) was developed by cloning and overexpressing the organism's own riboflavin biosynthetic genes (55) and its own promoter sequences. The resulting culture produced 15 g/L riboflavin in 3 days. Genetic engineering of a *Bacillus subtilis* strain, already containing purine analog-resistance mutations, led to improved production of riboflavin (56). The industrial strain of *B. subtilis* was produced by (1) making purine analog-resistance mutations to increase guanosine triphosphate (GTP; a precursor) production and (2) using a riboflavin analog (roseflavin)-resistance mutation in *ribC* that deregulated the entire pathway (57). Resultant production was over 25 g/L. A genome-wide transcript expression analysis (58) was successfully used to discover new targets for further improvement of the fungus *A. gossypii* (59). The authors identified 53 genes of known function, some of which could clearly be related to riboflavin production.

Biotin has traditionally been made by chemical synthesis but recombinant microbes have approached a competitive economic position. Cloning of a biotin operon (*bioABFCD*) on a multicopy plasmid allowed *E. coli* to produce 10,000 times more biotin than did the wild-type strain (60). Sequential mutation of *S. marcescens* to resistance to the biotin antimetabolite acedomycin (= actithiazic acid) led to mutant strain SB412 which produced 20 mg/L of biotin (61). Further improvements were made by mutating selected strains to ethionine resistance (strain ET2, 25 mg/L), then mutating ET2 to S-2-aminoethylcysteine resistance (strain ETA23, 33 mg/L) and finally cloning in the resistant *bio* operon yielding a strain able to produce 500 mg/L in a fed-batch fermention along with 600 mg/L of biotin vitamers. Later advances led to production by recombinant *S. marcescens* of 600 mg/L of biotin (62). A process using an *E. coli* mutant resistant to β-hydroxynorvaline (a threonine antimetabolite) yielding 970 mg/L has been patented (63). Biotin production was further increased to over 1 g/L by the use of a *B. subtilis* strain resistant to 5-(2-thenyl) pentanoic acid (a biotin analog) and overexpressing several *bio* genes.

Vitamin C (ascorbic acid) has traditionally been made in a five-step chemical process by first converting glucose to 2-keto-L-gulonic acid (2-KGA) with a yield of 50% and then converting the 2-KGA by acid or base to ascorbic acid. Annual production is 110,000 tons generating revenue of over 600 million dollars. A novel process for vitamin C synthesis involved the use of a genetically engineered *Erwinia herbicola* strain containing a gene from *Corynebacterium* sp. The engineered organism converted glucose into 1 g/L of 2-KGA (64). A better process was devised independently which converted 40 g/L of glucose into 20 g/L of 2-KGA (65). This process involved cloning and expressing the gene encoding 2, 5-diketo-D-gluconate reductase from *Corynebacterium* sp. into *Erwinia citreus*. Another process used a recombinant strain of *Gluconobacter oxydans* containing genes encoding L-sorbose dehydrogenase and L-sorbosone dehydrogenase from *G. oxydans* T-100. The new strain was an improved producer of 2-KGA (66). Further mutation to suppress the L-idonate pathway and to improve the promoter led to the production of 130 g/L of 2-KGA from 150 g/L of sorbitol.

Vitamin B_{12} production depends on avoidance of feedback repression by vitamin B_{12}. The vitamin is industrially produced by *Propionibacterium shermanii* or *Pseudomonas denitrificans*. The early stage of the *P. shermanii* fermentation is conducted under anaerobic conditions in the absence of the precursor 5,6-dimethylbenzimidazole. These conditions prevent vitamin B_{12} synthesis and allow for the accumulation of the intermediate, cobinamide. Then, the culture is aerated and dimethylbenzimidazole is added, converting cobinamide to the vitamin. In the *P. denitrificans* fermentation, the entire process is carried out under low levels of oxygen. A high level of oxygen results in an oxidizing intracellular environment which represses formation of the early enzymes of the pathway. Production of vitamin B_{12} has reached levels of 150 mg/L, 10 tons per year, and a world market of $71 million.

Other vitamins. Recombinant *E. coli*, transformed with genes encoding pantothenic acid (vitamin B_5) biosynthesis, and resistant to salicylic and/or other acids produce 65 g/L of D-pantothenic acid from glucose using alanine as precursor (67). A total of 7,000 tons per year are made chemically and microbiologically. Thiamine (vitamin B1) is produced synthetically at 4,000 tons per year. Pyridoxine (vitamin B6) is made chemically at 2,500 tons per year.

1.5. Carotenoids

Carotenoid production processes have been extensively studied (68), but they have had difficulty to economically challenge chemical methods. Of over 600 microbial carotenoids, only β-carotene and astaxanthin are produced industrially by fermentation (69). A semi-industrial β-carotene process was developed using mated cultures of *Blakeslea trispora* plus and minus strains. β-Carotene was produced at 1 g/L in the early 1960s (70). By the addition of carotogenic chemicals and antioxidants, the titer was raised to over 3 g/L (71). Processes in development include those yielding β-carotene, lycopene, zeaxanthin, and astaxanthin. Some have been improved by metabolic engineering and directed evolution (72–74). Metabolic engineering of *E. coli* has led to strains forming 0.2 g/L of lycopene (75). Lutein, a xanthophyll carotenoid with antioxidant properties, had sales as a food colorant of $150 million in the USA (76). It is thought that this carotenoid prevents age-related macular degeneration and cataracts. It is made from petals of marigold, but microalgae are a potential new source.

Chemical production of *trans*-astaxanthin has a selling price of $2,000/kg (77) and a market of over $100 million per year. It is mainly used for pigmentation of salmonids raised in aquaculture, a multibillion dollar industry (78). Astaxanthin can be made by the yeast *Phaffia rhodozyma* (*Xanthophyllomyces dendrorhous*) and the micralga *Haematococcus pluvialis*. Genetically improved strains of *P. rhodozyma* produce 10 mg/g cells in industrial fermentors. Recent improvements in astaxanthin production have been published by de la Fuente et al. (79) and Rodríguez-Sáiz et al. (80) to get maximal astaxathin titers of 420 mg/L when *X. dendrorhous* was fermented under continuous white light.

1.6. Organic Acids

Microbial production of organic acids is an excellent approach for obtaining building-block chemicals from renewable carbon sources (81). Production of some organic acids started decades ago and titers have been improved by classical mutation and screening/selection techniques as well as by metabolic engineering (82).

1.6.1. Citric Acid

Citric acid has a market of $2 billion. About 1 million tons per year are produced annually by *Aspergillus niger* and yeasts. The commercial process employs *A. niger* in media deficient in iron and manganese. Other factors contributing to high citric acid production are a high intracellular concentration of fructose 2,6-biphosphate, inhibition of isocitrate dehydrogenase, and low pH (1.7–2.0). In approximately 4–7 days, the major portion (80%) of the sugar (glucose or sucrose) provided is converted into citric acid. *A. niger* titers have reached over 200 g/L (83).

Alternative processes have been developed with *Candida* yeasts, especially from hydrocarbons. Such yeasts are able to convert *n*-paraffins to citric and isocitric acids in extremely high yields (150–170% on a weight basis). Titers as high as 225 g/L have been reached (84).

1.6.2. Acetic Acid

Titers of acetic acid reached 53 g/L with genetically engineered *E. coli* (85) and 83 g/L with a *Clostridium thermoaceticum* mutant (86). Cloning of the aldehyde dehydrogenase gene from *Acetobacter polyoxogenes* on a plasmid vector into *Acetobacter aceti* subsp. *xylinum* increased the rate of acetic acid production by over 100% (1.8–4 g/L/h) and the titer by 40% (68–97 g/L) (87).

1.6.3. Lactic Acid

Whole genome shuffling was used to improve the acid tolerance of a commercial lactic acid-producing *Lactobacillus* sp. (88). Further approaches using this recursive protoplast fusion technique yielded strains of *Lactobacillus rhamnosus* ATCC 11443 with improved glucose tolerance (160–200 g/L glucose), while simultaneously enhancing L-lactic acid production by 71% as compared to the wild type. Shuffling of a mutant strain of *Lactobacillus delbrueckii* NCIM 2025 and *Bacillus amyloliquefaciens* ATCC 23842 produced a fusant that could utilize liquefied cassava bagasse starch directly to yield a titer of 40 g/L of lactic acid with a 96% conversion of starch to lactic acid (89).

Although lactobacilli make more lactic acid than *Rhizopus oryzae*, they produce mixed isomers. The fungus, however, produces L-(+) lactic acid exclusively. The yield is about 60–80% of added glucose, the remainder going to ethanol. By increasing lactic dehydrogenase levels via plasmid transformation with *ldhA*, more lactate could be made from pyruvate and production was increased to 78 g/L, whereas the undesirable coproduct ethanol was reduced from 10.6 to 8.7 g/L (90). A recombinant *Saccharomyces cerevisiae* strain containing six copies of bovine L-lactate dehydrogenase produced 122 g/L from sugar cane with an optical purity of 99.9% or higher (91). Expression of the same bovine enzyme and a deletion of the pyruvate decarboxylase gene in *Kluyveromyces lactis* produced 109 g/L (92).

A recombinant *E. coli* strain was constructed that produced optically active pure D-lactic acid from glucose at virtually the theoretical maximum yield, e.g., two molecules from one molecule of glucose (93). D-Lactic acid has also been produced at 61 g/L by a recombinant strain of *S. cerevisiae* containing the D-lactic dehydrogenase gene from *Leuconostoc mesenteroides* (94).

1.6.4. Succinic Acid

Sanchez et al. (95) used metabolic engineering to create an *E. coli* strain which had three deactivated genes of the central metabolic pathway, i.e., *adhE*, *ldhA*, and *act-pta*, and an inactivated *iclR* gene which resulted in activation of the glyoxylate pathway. The strain produced 40 g/L of succinate. Metabolic engineering of *Mannheimia succiniciproducens* led to a strain producing 52 g/L of succinic acid at a yield of 1.16 mol/mol glucose and a productivity of 1.8 g/L/h in fed-batch culture (96). A metabolically engineered succinate-producing strain of *E. coli* yielded 58 g/L succinate in a 59 h fed-batch fermentation under aerobic conditions (97). The average succinate yield was 0.94 mol/mol of glucose, the average productivity was 1.08 g/L/h, and the average specific activity was 90 mg/g/h. A titer of 99 g/L with a productivity of 1.3 g/L/h has been reached with recombinant *E. coli* (98).

1.6.5. Other Organic Acids

Metabolic engineering of *Clostridium tyrobutyricum* created a fermentation strain yielding 80 g/L butyric acid and a yield on glucose of 0.45 g/g (99). *S. cerevisiae* normally produces 2 g/L of malic acid from fumaric acid. However, a recombinant strain containing a cloned fumarase gene was able to produce 125 g/L with a yield of almost 90% (100). Microbial fermentation titers of some other organic acids are 135 g/L pyruvic acid, 107 g/L fumaric acid, 90 g/L shikimic acid, 69 g/L dehydroshikimic acid, 85 g/L itaconic acid, 504 g/L gluconic acid, 106 g/L propionic acid, 68 g/L oxalic acid, and 136 g/L glyceric acid. An oxidative bioconversion of saturated and unsaturated linear aliphatic 12–22 carbon substrates to their terminal dicarboxylic acids was developed by gene disruption and gene amplification (101). Product concentrations reached 200 g/L and problematic side reactions such as unsaturation, hydroxylation and chain-shortening did not occur.

1.7. Alcohols

Ethanol. Fermentation of sugars by *S. cerevisiae* in the case of hexoses, and *Kluyveromyces fragilis* or *Candida* species with lactose or a pentose, results in the production of ethanol. Under optimum conditions, approximately 120 g/L ethanol can be obtained. Such a high concentration slows down growth and the fermentation ceases.

A *S. cerevisiae* fusant library obtained by genome shuffling was screened for growth at 35, 40, 45, 50, and 55°C on agar plates containing different concentrations of ethanol (102). After three rounds of genome shuffling, a strain was obtained which was able to grow on plates up to 55°C, completely utilized 20% (w/v) glucose at 45–48°C, produced 99 g/L ethanol, and tolerated 25% (v/v) ethanol stress.

In silico metabolic models have been used to overcome the redox imbalance in *S. cerevisiae* engineered with the *Xyl1* and *Xyl2* genes from *Pichia stipitis* (103, 104). Overexpression of both genes led to an accumulation of NADH and a shortage of NADPH. Deletion of $NADP^+$-dependent glutamate dehydrogenase (GGH1) and overexpression of NAD^+-dependent GDH2 led to an increase in ethanol production using xylose as fermentation substrate.

An in silico genome-scale gene insertion strategy was used to improve ethanol production and decrease the production of by-products glycerol and xylitol (105). Introduction of glyceraldehyde-3-phosphate dehydrogenase in *S. cerevisiae* led to a 58% reduction in glycerol, a 33% reduction in xylitol, and a 24% increase in ethanol production.

When biomass is used as a carbon source for ethanol production, its breakdown results in liberation of acetic acid. The acid interferes with ethanol production. Tolerance of *Candida krusei* GL560 to acetic acid was improved by genome shuffling (106). A mutant, S4-3, which was isolated and selected after four rounds, had a higher viability in different media containing acetic acid than did the parent strain GL560. The mutant also improved its multiple stress tolerance to ethanol, H_2O_2, heat, and freeze-thawing.

E. coli was converted into an ethanol producer (43 g/L) by cloning the alcohol dehydrogenase II and pyruvate decarboxylase genes from *Zymomonas mobilis* (107). By cloning and expressing the same two genes in *Klebsiella oxytoca*, the recombinant was able to convert crystalline cellulose to ethanol in high yield when fungal cellulase was added (108). Maximum theoretical yield was 81–86% and a titer of 47 g/L of ethanol was produced from 100 g/L of cellulose.

Recombinant strains of *E. coli*, *Zymomonas*, and *Saccharomyces* can convert corn fiber hydrolysate to 21–35 g/L ethanol with yields of 0.41–0.50 ethanol per gram of sugar consumed (109). For a recombinant *E. coli* strain making 35 g/L, time was 55 h and yield was 0.46 g ethanol per gram of available sugar, which is 90% of the attainable maximum.

1,3-Propanediol. A strain of *Clostridium butyricum* converts glycerol to 1,3-propanediol (PDO) at a yield of 0.55 g/g glycerol consumed (110). A major metabolic engineering feat was carried out in *E. coli* leading to a culture growing on glucose and producing PDO at 135 g/L, with a yield of 51% and a rate of 3.5 g/L/h (111). To do this, eight new genes were introduced to convert dihydroxyacetone phosphate (DHAP) into PDO. Production was further improved by modifying 18 *E. coli* genes, including regulatory genes. PDO is the monomer used to chemically synthesize industrial polymers such as polyurethanes and the polyester fiber Sorono™ by DuPont. This bioplastic is polytrimethylene terephthalate (3GT polyester) made by reacting terephthalic acid with PDO (112). PDO is also used as a polyglycol-like lubricant and as a solvent.

D-*Mannitol* is a naturally occurring polyol, widely used in the food, chemical, and pharmaceutical industries. A whole cell bioconversion of D-fructose to D-mannitol was developed by metabolic engineering of *E. coli* (113). The *mdh* gene encoding mannitol dehydrogenase from *Leuconostoc pseudomesenteroides* and the *fdh* gene encoding formate dehydrogenase from *Mycobacterium vaccae* were coexpressed in *E. coli* along with the *glf* gene encoding the glucose facilitator protein of *Z. mobilis.* The process yielded 75–91 g/L of D-mannitol, a specific productivity of 3.1–4.1 g/g/h and a molar yield of 84–92% with no by-products. An improved bioconversion process was developed with a recombinant *E. coli* strain in the presence of added glucose isomerase yielding 145 g/L of D-mannitol from 180 g/L glucose (114). Supplementation of the medium used for mannitol production by *Candida magnoliae* with Ca^{2+} and Cu^{2+} increased production up to 223 g/L (115).

Sorbitol, also called D-glucitol, is 60% as sweet as sucrose and is used in the food, pharmaceutical, and other industries. Its worldwide production is estimated to be higher than 500,000 tons per year, and it is made chemically by catalytic hydrogenation of D-glucose or syrup with a 50:50 mixture of glucose and fructose. It is also produced by extraction from seaweed as a by-product of alginate and iodine manufacture. However, excellent microbial processes have been developed (116). Toluenized (permeabilized) cells of *Z. mobilis* produce 290 g/L of sorbitol and 283 g/L of gluconic acid from a glucose and fructose mixture in 16 h with yields near 95% for both products (117). Other leading organisms are recombinant *C. glutamicum* at 285 g/L, *Lactobacillus intermedius* at 227 g/L, *C. magnoliae* at 223 g/L, and many others producing between 100 and 200 g/L. Metabolic engineering of *Lactobacillus plantarum* for high sorbitol production was successfully achieved by a simple two-step strategy overexpressing the two sorbitol-6-phosphate dehydrogenase genes (*srlD1* and *srlD2*) identified in the genome sequence (118).

n-Butanol is a good alternative fuel additive as it has two more carbons than ethanol which results in an energy content about 40% higher. Also, automobile engines do not require modification until the percentage of butanol reaches over 40% of the total automobile fuel. In contrast, modification is required when ethanol is added to gasoline at levels exceeding 15%. Butanol can be obtained from the acetone–butanol–ethanol fermentation of *Clostridium beijerinkii* or *Clostridium acetobutylicum.* Butanol-resistant mutants showed increased production of butanol and acetone (119). Biochemical engineering modifications were able to increase total acetone, butanol, and ethanol production (ABE) to 69 g/L (120). A mutant in the presence of added acetate was able to produce almost 21 g/L butanol and 10 g/L of acetone from glucose (121).

Because butanol's octane number is lower than that of ethanol and the octane number increases with methyl branching and double bonds, other higher alcohols are also being considered as biofuels, e.g., branched C_4 and C_5 alcohols. They are also desirable because of their higher energy density, lower vapor pressure, and lower hygroscopicity as compared to ethanol (122). They include isopropanol, 1-propanol, 1-butanol (*n*-butanol), isobutanol (2-methyl-propanol), 3-methyl-1-butanol, 2-methyl-1-butanol, isopentanol (3-methyl-1 butanol), and isopentenol (3-methyl-3-buten-1-ol). A novel screening method based on overcoming the toxicity associated with the accumulation of prenyl diphosphate was used to screen a library of 19,000 clones harboring fragments of *Bacillus* genomic DNA (123). Two genes, *yhfR* and *nudF*, coding for proteins capable of overcoming the toxicity associated with accumulating IPP and DMAPP were isolated. Both protein products have an affinity for IPP and DMAPP, converting them into isopentenol. *Clostridium beijerinckii* (*Clostridium butylicum*) produces 20 g/L of 1-butanol and 2 g/L of isopropanol as part of a mixed product. Recombinant *E. coli* can produce 4.9 g/L of isopropanol. Recently, a new strategy for the production of these alcohols has been reported (124). This approach is based on the diversion of 2-keto acid intermediates from the endogenous amino acid pathway to alcohol biosynthesis especially that of isobutanol. As a result, engineered *E. coli* can produce 22 g/L of isobutanol in 110 h with a yield of 86% of the theoretical maximum.

Other alcohols. The noncariogenic, noncaloric, and diabetic-safe sweetener erythritol has 70–80% of the sweetness of sucrose. It can be produced by *Aureobasidium* sp. (165 g/L), an acetate-negative mutant of *Yarrowia lipolytica* (170 g/L), a *C. magnoliae* osmophilic mutant (187 g/L), the osmophile *Trichosporon* sp. (188 g/L), *Torula* sp. (200 g/L), and the yeast *Pseudozyma tsukubaensis* (245 g/L) (125).

Xylitol is a naturally occurring sweetener with anti-cariogenic properties which is used for some diabetes patients. It can be produced by chemical reduction of D-xylose or by fermentation. Xylitol production at 150 g/L was obtained with *Candida guilliermondii* 2,581 at pH 6.0 and shaking at 60 rpm (126).

2. Secondary Metabolites

As a group that includes antibiotics, pesticides, pigments, toxins, pheromones, enzyme inhibitors, immunomodulating agents, receptor antagonists and agonists, pesticides, antitumor agents, immunosuppressants, cholesterol-lowering agents, plant protectants, and animal and plant growth factors, these metabolites have tremendous economic importance. This remarkable group of compounds is produced by certain restricted taxonomic groups of organisms and is usually formed as mixtures of closely related members of a chemical family.

2.1. Antibiotics

Antibiotics are the most well-known secondary metabolites and have a tremendous importance. The most well known are the β-lactams, tetracyclines, aminoglycosides, chloramphenicol, macrolides, and other polyketides, polyenes, glycopeptides, among others. They have been crucial in the increase in average life expectancy in the USA from 47 years in 1900 to 74 for males and 80 for women in 2000 (127). More than 350 agents have reached the market as antimicrobials. The global market for finished antibiotics has reached $35 billion.

2.1.1. β-Lactams

The β-lactams are the most important class of antibiotics in terms of use. Included are the penicillins, cephalosporins, cephamycins, clavulanic acid, and the carbapenems. Many of the current penicillins and cephalosporins are semisynthetic. All of the above are of great importance in chemotherapy of bacterial infections.

β-Lactamases of pathogenic bacteria are the major cause of resistance development and there are over 450 such enzymes. However, β-lactams are still very useful due to the discovery of β-lactamase inhibitors. Although clavulanic acid is a β-lactam compound, it has only low antibacterial activity, but is used widely as an inhibitor of β-lactamase, in combination with β-lactam antibiotics. Conventional strain improvement by protoplast fusion of auxotrophic strains yielded a fusant producing 30-fold more clavulanic acid than the wild type (128). Inactivation of two G-3-P dehydrogenases, encoded by *gap1* and *gap2* by targeted gene disruption, doubled clavulanic acid production (129). Also, increased dosage of biosynthetic genes *ceas* and *cs2* (130) or overexpression of positive regulatory genes increased production two- to three-fold (131, 132).

Yield improvements have been achieved through different strategies. Thus, protoplast fusion between strains of *Penicillium chrysogenum* yielded a higher-producing strain of penicillin G (133). Metabolic engineering was used to replace the normal promoter with the ethanol dehydrogenase promoter (134), increasing penicillin production up to 30-fold.

Protoplast fusion was also been carried out with strains of *Acremonium chrysogenum* to obtain a strain that produced 40% more cephalosporin C than the parent (135). Production was also improved by cloning multiple copies of cyclase (136) or the *pcbC* and the *cefEF* genes (137).

Two improved cephamycin C-producing strains from *Nocardia lactamdurans* were fused to obtain cultures which produced 10–15% more antibiotic (138). Overexpression of *lat*, encoding lysine-aminotransferase also led to an overproducing strain (139). High-level expression of *ccaR*, a positive regulatory gene in

Streptomyces clavuligerus (140) led to a two- to threefold increase in antibiotic production.

Chemical methods had traditionally been used to produce 7-aminocephalosporanic acid (7-ACA) and 7-aminodeacetoxycepalosporanic acid (7-ADCA), chemical precursors of semisynthetic cephalosporins. These processes are being replaced by safer microbiological processes. Transformation of *P. chrysogenum* with bacterial *cefD* and *cefE* genes allowed the production of deacetoxycephalosporin C (DAOC) (141), another key intermediate in the commercial production of semisynthetic cephalosporins. Also, cloning of *cefE* from *S. clavuligerus* or *cefEF* and *cefG* from *Acremonium chrysogenum* into *P. chrysogenum* fed with adipic acid as side-chain precursor (142) resulted in the formation of several adipyl-6/7-intermediates. Enzymatic removal of the adipoyl side chain led to the production of 7-ADCA.

Disruption and one-step replacement of the *cefEF* gene of an industrial strain of *A. chrysogenum* yielded strains accumulating up to 20 g/L of penicillin N. Cloning and expression of the *cefE* gene from *S. clavuligerus* into those high-producing strains yielded recombinant strains producing high titers of DAOC (143). An *E. coli* strain containing the D-amino acid oxidase gene from *Trigonopsis variabilis* and the glutaryl-7-aminocephalosporanic acid acylase gene from *Pseudomonas* sp. was able to convert cephalosporin C directly to 7-ACA (144).

Natural carbapenems, such as thienamycin, are made by *Streptomyces cattleya*, *Erwinia carotovora* subsp. *carotovora*, *Serratia* sp., and *Photorhabdus* luminescens (145). Carbapenems are resistant to attack by most β-lactamases. The commercial carbapenems are made synthetically and include imipenem, meropenem, and ertapenem. Thienamycin is one of the most potent and broadest in spectrum of all antibiotics known today. Although a β-lactam, it is not a member of the penicillins or cephalosporins. It is active against aerobic and anaerobic bacteria, both Gram positive and Gram negative, including *Pseudomonas*. This novel structure was isolated in Spain from a new soil species which was named *S. cattleya* (146). Interestingly, this culture also produces penicillin N and cephamycin C. Also used to combat β-lactamase-containing pathogens are new carbapenems, such as the recently approved doripenem (S-4661) which has broad spectrum activity against resistant bacteria including *Pseudomonas aeruginosa*.

2.1.2. Other Antibiotics

The biosynthesis of polyketide macrolides has been subjected to genetic engineering (147). This group of compounds includes antibiotics such as erythromycin, oleandomycin, pikromycin, tylosin, and amphotericin B. Reverse metabolic engineering increased erythromycin production by *Aeromicrobium erythreum* (148). The technique is also known as inverse metabolic engineering and as combinatorial engineering. Tylosin production was increased up to

60% in *Streptomyces fradiae* by transposing a second copy of *tylF*, encoding for macrocin *O*-methyltransferase, into a neutral site on its chromosome (149).

Genetic engineering of the nystatin biosynthetic pathway yielded polyenes with high antifungal activity which are less toxic than amphotericin B (150).

The production of antibiotics in heterologous hosts via combinatorial biosynthesis is becoming very popular in antibiotic production and discovery (151). New derivatives of antibiotics have been obtained after the biosynthetic paths were elucidated and the biosynthetic genes isolated (152). Over 200 new polyketides have been made by combinatorial biosynthesis (153, 154). The discovery of new antibiotics has also been achieved by genetic recombination between producers of different or even the same antibiotics (155–157).

Combinatorial biosynthesis been used to construct macrolides with new sugar moieties (158, 159). Methymycin and pikromycin, produced by a gene cluster of *Streptomyces venezuelae* and normally containing the sugar desosamine, were modified by cloning of a gene from the calicheamicin producer, *Micromonospora echinospora* spp. *calichensis*. The gene encodes TDP-glycero-hexulose aminotransferase. Transfer of a 12.6 kb DNA fragment from the tetracenomycin C-producing *Streptomyces glaucescens* to *Streptomyces lividans* resulted in tetracenomycin C production by the latter (160). The fragment contains 12 genes of biosynthesis and resistance. Novel hybrid tetracenomycins were produced by introducing a 25 kb cosmid from the elloramycin biosynthetic pathway of *Streptomyces olivaceus* into the polyketide synthase (PKS)-deleted mutant of the urdamycin producer, *S. fradiae* and into the mithramycin producer, *Streptomyces argillaceus* (161). The cosmid contains a glycosyltransferase gene whose enzyme has broad substrate specificity and thus produces hybrid products containing different D- and L-sugars.

For more than 35 years, vancomycin and teicoplanin were the only antibiotics active against multidrug-resistant Gram-positive bacteria. Their use became severely limited by an increase in multidrug resistance. One group of narrow-spectrum compounds are the streptogramins which are synergistic pairs of antibiotics made by a single microbial strain. The pairs are constituted by a (Group A) polyunsaturated macrolactone containing an unusual oxazole ring and a dienylamide fragment and a (Group B) cyclic hexadepsipeptide possessing a 3-hydroxypicolinoyl exocyclic fragment. Such streptogramins include virginiamycin and pristinamycin (162). Pristinamycin, made by *Streptomyces pristinaespiralis*, is a mixture of a cyclodepsipeptide (pristinamycin I) and a polyunsaturated macrolactone (pristinamycin II). Increasing resistance to pristinamycin in the pristinamycin producer *S. pristinaespiralis* was combined with genome shuffling to increase pristinamycin production ninefold (163).

An important new strategy to improve the discovery of new antibiotics is genome mining which has come about due to advances in microbial genomics (164). Mining of whole genome sequences and genome scanning allows the rapid identification of more than 450 clusters of genes in antibiotic-producing cultures encoding biosynthesis of new bioactive products and the prediction of structure based on gene sequences (165, 166). These efforts include mining of whole genome sequences, genome scanning, heterologous expression, and discovery of novel chemistry. Genomics will also provide a huge group of new targets against which natural products can be screened (167).

2.2. Antitumor Agents

Microorganisms have played a crucial role in identifying compounds with therapeutic benefit against cancer (168). Most of the important compounds used for chemotherapy of tumors are microbially produced antibiotics mainly made by actinomycetes. Among the most well known are actinomycin D (dactinomycin), anthracyclines (including daunorubicin, doxorubicin, epirubicin, pirirubicin, idarubicin, valrubicin, and amrubicin), glycopeptolides (bleomycin and phleomycin), mitomycin C, anthracenones (mithramycin, streptozotocin, and pentostatin), the enediyne calicheamycin attached to a monoclonal antibody (Mylotarg®) and recently, the epothilones.

Novel anthracyclines have been produced by metabolic engineering, i.e., cloning genes from antitumor-producing species into other producing or nonproducing strains, or by blocking deoxysugar biosynthesis. A new anthracycline, 11-hydroxyaclacinomycin A, was produced by cloning the doxorubicin resistance gene and the aklavinone 11-hydroxylase gene *dnrF* from the doxorubicin producer, *Streptomyces peucetius* subsp. *caesius*, into the aclacinomycin A producer (169). The hybrid molecule showed greater activity against leukemia and melanoma than aclacinomycin A. Another hybrid molecule produced was 2″-amino-11-hydroxyaclacinomycin Y, which was highly active against tumors (170). Additional new anthracyclines have been made by introducing DNA from *Streptomyces purpurascens* into *Streptomyces galilaeus*, both of which normally produce known anthracyclines (171). Novel anthracyclines were produced by cloning DNA from the nogalomycin producer, *Streptomyces nogalater*, into *S. lividans* and into an aclacinomycin-negative mutant of *S. galilaeus* (172). Cloning of the actI, actIV and actVII genes from *Streptomyces coelicolor* into the 2-hydroxyaklavinone producer, *S. galilaeus* 31,671 yielded novel hybrid metabolites, desoxyerythrolaccin and 1-*O*-methyl-desoxyerythrolaccin (173). Similar studies yielded the novel metabolite aloesaponarin II (174).

Epirubicin (4′-epidoxorubicin) is a semisynthetic anthracycline with less cardiotoxicity than doxorubicin (175). Genetic engineering of a blocked *S. peucetius* strain provided a new method to produce it (176). The gene introduced was *avrE* of the avermectin-producing *Streptomyces avermitilis* or the *eryBIV* genes of the erythromycin producer, *Saccharopolyspora erythrea*. These genes and the blocked gene in the recipient are involved in deoxysugar biosynthesis.

Taxol®, a diterpene alkaloid, is approved for breast and ovarian cancer and acts by blocking depolymerization of microtubules. In addition, taxol promotes tubulin polymerization and inhibits rapidly dividing mammalian cancer cells (177). Taxol was originally isolated from the bark of the Pacific yew tree (*Taxus brevifolia*), but it took six trees of 100 years of age to treat one cancer patient (178). It is now produced by plant cell culture or by semisynthesis from taxoids made by *Taxus* species. Early genetic engineering of *S. cerevisiae* yielded no taxadiene (the taxol precursor) because too little of the intermediate, geranylgeranyl diphosphate, was formed. When the *Taxus canadensis* geranylgeranyl diphosphate synthase gene was introduced, 1 mg/L of taxadiene was obtained (179). More recent metabolic engineering studies (180) yielded a *S. cerevisiae* strain producing over 8 mg/L taxadiene and 33 mg/L geranylgeraniol. Taxol has sales of $1.6 billion per year.

Epothilones are an important group of new anticancer agents produced by the myxobacterium, *Sorangium cellulosum* (181). Rounds of classical mutation and screening followed by recursive protoplast fusion resulting in fusants able to produce 130-fold more epothilone B compared to the starting strain. Epothilones have a mode of action similar to Taxol and, very importantly, are active against Taxol-resistant tumors.

2.3. Cholesterol-Lowering Agents

The largest segment of the pharmaceutical business is that of cholesterol-lowering drugs, amounting to about 30% of global sales. The first member of the fungal statins, i.e., compactin, was discovered in cultures of *Penicillium brevicompactum* (182) and *Penicillium citrinum* (183). A few years later, the more active methylated form of compactin known as lovastatin (monacolin K, mevinolin, Mevacor™) was isolated from broths of *Monascus ruber* and *Aspergillus terreus* (184, 185). Simvastatin (Zocor™), a semisynthetic derivative of lovastatin, reached a market of over $7 billion. Pravastatin, a product of compactin bioconversion, attained sales of $5 billion. The synthetic statin, atorvastatin (Lipitor™), became the world's leading drug at $12 billion per year.

Recently, association analysis, a strategy that integrates transcriptional and metabolic profiles, led to an improvement in lovastatin production of over 50% (186). Improvement was achieved by increasing the dosage of lovastatin biosynthetic genes and of regulatory genes for secondary metabolism.

Microbially produced polyethers such as monensin, lasalocid, and salinomycin dominate the coccidiostat market.

2.4. Antihelmintics

The avermectins, a family of secondary metabolites having both antihelmintic and insecticidal activities, produced by *S. avermitilis*, have a market of over 1 billion dollars per year. Despite their macrolide structure, avermectins lack antibiotic activity, do not inhibit protein synthesis nor are they ionophores; instead, they interfere with neurotransmission in many invertebrates.

Although the Merck Laboratories had earlier developed a commercially useful synthetic product, thiobenzole, they had enough foresight to also examine microbial broths for antihelmintic activity and found a nontoxic fermentation broth which killed the intestinal nematode, *Nematospiroides dubius*, in mice. The *S. avermitilis* culture, which was isolated by Omura and coworkers at the Kitasato Institute in Japan (187), produced a family of secondary metabolites having both antihelmintic and insecticidal activities. These were discovered by Merck scientists and named "avermectins." They are disaccharide derivatives of macrocyclic lactones with exceptional activity against parasites, i.e., at least ten times higher than any synthetic antihelmintic agent known. They have activity against both nematode and arthropod parasites in sheep, cattle, dogs, horses, and swine.

Incorporation of multiple copies of *afsR2*, a global regulatory gene, from *S. lividans* into wild-type *S. avermitilis* increased avermectin production by 2.3-fold (188). Transposon mutagenesis was used to eliminate production of the troublesome toxic oligomycin in *S. avermitilis* (189). DNA shuffling of the *ave C* gene of *S. avermitilis* gave an improved ratio of the undesirable CHC-B2 to the desirable CHC-B1 of 0.07:1. This was an improvement of 21-fold over the ratio with the starting strain (190).

A semisynthetic derivative, 22, 23-dihydroavermectin B1 (Ivermectin) is 1,000 times more active than thiobenzole and is a commercial veterinary product. Ivermectin is made by hydrogenation at C22–C23 of avermectin B1a and B1b with rhodium chloride acting as catalyst. By genetic engineering of *S. avermitilis*, in which certain PKS genes were replaced by genes from the PKS of *S. venezuelae* (the pikromycin producer), ivermectin could be made directly by fermentation, thus avoiding semisynthesis (191).

A new avermectin, called Doramectin (=cyclohexyl avermectin B1), was developed at Pfizer by the technique of mutational biosynthesis (192). Indeed, it was the first commercially successful example of mutational biosynthesis.

2.5. Immunosuppressive Agents

Cyclosporin A was originally discovered as a narrow spectrum antifungal peptide produced by the mold, *Tolypocladium niveum* (previously *Tolypocladium inflatum*). Discovery of immunosuppressive activity led to its use in heart, liver, and kidney transplants and to the overwhelming success of the organ transplant field. Sales of cyclosporin A have reached $1.5 billion. Although cyclosporin A had been the only product on the market for many years, two other products, produced by actinomycetes, provided new opportunities. These are rapamycin (=sirolimus) (193) and the independently discovered FK-506 (=tacrolimus) (194). They are both narrow spectrum polyketide macrolide antifungal agents, which are 100-fold more potent that cyclosporin as immunosuppressants and less toxic. FK-506 and rapamycin have been used clinically for many years. FK-506 had a market of $2 billion in 2007. Mutants developed by increasing resistance to FK-506 produce higher titers (195).

Genome shuffling using mutants of the rapamycin producer, *S. hygroscopicus*, as well as interspecies fusion of protoplasts of *Streptomyces hygroscopicus* D7-804 and *Streptomyces erythreus* ZJU325, generated improved rapamycin-producing strains (196). Two genes of the rapamycin biosynthetic cluster in *S. hygroscopicus*, i.e., *rap G* and *rap H*, encode positive regulatory proteins for rapamycin production (197). Overexpression of either gene increased rapamycin formation, whereas their deletions eliminated rapamycin biosynthesis. They act by affecting the promoter of the operon.

2.6. Bioinsecticides

The insecticidal bacterium, *Bacillus thuringiensis* (BT), owes its activity to its crystal protein produced during sporulation. Crystals plus spores had been applied to plants for years to protect them against lepidopteran insects. BT preparations are highly potent, some 300 times more active on a molar basis than synthetic pyrethroids and 80,000 times more active than organophosphate insecticides. In 1993, BT represented 90% of the biopesticide market and had annual sales of $125 million.

A very important insecticide is Spinosad (Naturalyte®) produced by *Saccharopolyspora spinosa* and used for protection of crops and feedstock animals. Spinosad is a mixture of two tetracyclic macrolides containing forosamine and tri-*O*-methyl rhamnose with different levels of methylation on the polyketide moiety. The two components are spinosyns A and D which differ by a methyl group at position 6 of the polyketide. Spinosad is an excellent nontoxic agricultural insecticide. Genome shuffling has been used for strain improvement (198).

3. Recombinant Proteins: Biopharmaceuticals

Biopharmaceutical proteins can be categorized into four major groups: (1) protein therapeutics with enzymatic activity (e.g., insulin), (2) protein vaccines, (3) protein therapeutics with special targeting activity (e.g., monoclonal antibodies), and (4) protein diagnostics (e.g., biomarkers) (199). Biologics accounted for over

$80 billion in sales in 2008. Six of these therapeutic proteins were among the best selling drugs in the USA in that year. Monoclonal antibodies and Fc-fusion proteins made up 43% of this market value. The projected market was expected to reach $94 billion in 2010.

By means of genetic engineering, desired proteins are massively generated to meet the copious demands of industry (200). Protein quality, functionality, production speed, and yield are the most important factors to consider when choosing the right expression system for recombinant protein production.

Non-glycosylated proteins are usually made in *E. coli* or yeasts and they constitute 55% of the therapeutic protein market (39% by *E. coli*, 1% by other bacteria and 15% by yeasts) (201). On the other hand, N-glycosylated proteins are usually made in mammalian cells which mimic human glycosylation. Chinese hamster ovary (CHO) cells provide about 35% of the therapeutic protein market but the process is very expensive and the glycoproteins made are not exactly of the human type. Although yeasts, molds, and insect cells are generally unable to provide mammalian glycosylation, the methylotrophic yeast, *Pichia pastoris*, has been genetically engineered to produce a human type of glycosylation (202).

Directed evolution of proteins has been reviewed by Yuan et al. (203). Strategies include DNA shuffling, whole genome shuffling, heteroduplex, random chimeragenesis of transient templates, assembly of designed oligonucleotides, mutagenic and unidirectional reassembly, exon shuffling, Y-ligation-based block shuffling, nonhomologous recombination, and the combining of rational design with directed evolution.

3.1. Bacteria

Bacterial systems are used to make somatostatin, insulin, bovine growth hormone for veterinary applications, α–1 antitrypsin, interleukin-2, tumor necrosis factor, β-interferon, and γ-interferon. *E. coli* is one of the earliest and most widely used hosts for the production of heterologous proteins (204). As early as 1993, recombinant processes of *E. coli* were responsible for almost $5 billion worth of products, i.e., insulin, human growth hormone, interferons, and G-CSF. Advantages of *E. coli* include rapid growth, rapid expression, ease of culture and genome modifications, low cost, and high product yields (205). It is used for massive production of many commercialized proteins. This system is excellent for functional expression of non-glycosylated proteins.

E. coli genetics are far better understood than those of any other microorganism. Recent progress in the fundamental understanding of transcription, translation, and protein folding in *E. coli*, together with the availability of improved genetic tools, are making this bacterium more valuable than ever for the expression of complex eukaryotic proteins. Its genome can be quickly and precisely modified with ease, promotor control is not difficult, and plasmid copy number can be readily altered. This system also features alteration of metabolic carbon flow, avoidance of incorporation of amino acid analogs, formation of intracellular disulfide bonds, and reproducible performance with computer control. *E. coli* can accumulate recombinant proteins up to 80% of its dry weight and survives a variety of environmental conditions. Recombinant protein production in *E. coli* can be increased by mutations which eliminate acetate production (206). Avecia Biologics achieved a titer of 14 g/L of recombinant protein using *E. coli* (207).

The value of transcriptome analysis in process improvement was shown by Choi et al. (208). They analyzed an *E. coli* process yielding human insulin-like growth factor 1 fusion protein (IGF-If) in a high-density culture. Of 200 or so genes whose expression was down-regulated after induction, the ones involved in biosynthesis of amino acids or nucleotides were studied. Amplification of two of these, *prsA* (encoding PRPP synthetase) and *glpF* (encoding the glycerol transporter), raised product formation from 1.8 to 4.3 g/L.

Bacilli have yields as high as 3 g/L. The organisms used are usually *Bacillus megaterium*, *B. subtilis*, and *Bacillus brevis*. *Staphylococcus carnosus* can produce 2 g/L of secreted mammalian protein.

An improved Gram-negative host for recombinant protein production has been developed using *Ralstonia eutropha* (209). The system appears superior to *E. coli* with respect to inclusion body formation. Organophosphohydrolase, a protein prone to inclusion body formation with a production of less than 100 mg/L in *E. coli*, was produced at 10 g/L in *R. eutropha*. Another useful bacterium is *Pseudomonas fluorescens* MB101 developed by Dowpharma (210). This system has produced 4 g/L of TNF-α.

Table 2
Advantages of yeast expression systems

Advantages
Stable strains
High density of growth
Durability
High production titers and yields
Protein glycosylation
Reasonable cost
Product processing similar to that of mammalian cells
Suitable for isotopically labeled proteins
Suitable for S–S-rich proteins
Protein folding assisted
Chemically defined media supporting rapid growth

3.2. Yeasts

Yeasts, the single-celled eukaryotic fungal organisms, are often used to produce recombinant proteins that are not produced well in *E. coli* because of problems dealing with folding or the need for glycosylation. The major advantages of yeast expression systems are listed in Table 2. The yeast strains are genetically well characterized and are known to perform many posttranslational modifications. They are easier and less expensive to work with than insect or mammalian cells and are easily adapted to fermentation processes.

The two most utilized yeast strains are *S. cerevisiae* and the methylotrophic yeast *P. pastoris.* Glucose oxidase from *A. niger* is produced by *S. cerevisiae* at 9 g/L. Recombinant products on the market which are made in *S. cerevisiae* are insulin, hepatitis B surface antigen, urate oxidase, glucagons, granulocyte macrophage colony-stimulating factor (GM-CSF), hirudin, and platelet-derived growth factor.

P. pastoris has the desirable qualities of dense growth and methanol-induced expression and secretion of recombinant proteins. It is used for the commercial production of non-glycosylated human serum albumin and glycosylated vaccines. Strains have been developed which are capable of human type N-glycosylation and such products are already in clinical testing. High recombinant protein yields can be obtained with *P. pastoris*, e.g., 10 g/L of tumor necrosis factor, (211) 14.8 g/L of gelatin, 15 g/L of mouse collagen (212), and *E. coli* phytase and *Candida parapsilosis* lipase/acetyltransferase at 6 g/L. Bacterial proteins such as intracellular tetanus fragment C were produced as 27% of protein with a titer of 12 g/L (213). Production of serum albumin in *S. cerevisiae* amounted to 0.15 g/L, whereas in *P. pastoris* the titer was 10 g/L (214). Indeed, claims have been made that *P. pastoris* can make 20–30 g/L of recombinant proteins (215).

Heterologous gene expression in another methylotroph *Hansenula polymorpha* yielded 13.5 g/L of phytase, and other proteins were made at levels over 10 g/L. Among the advantages of methylotrophic yeasts over *S. cerevisiae* as a cloning host are the following: (1) higher protein productivity, (2) avoidance of hyperglycosylation, (3) growth in reasonably strong methanol solutions that would kill most other microorganisms, (4) a system that is cheap to set up and maintain, and (5) integration of multicopies of foreign DNA into chromosomal DNA yielding stable transformants (216).

3.3. Filamentous Fungi (Molds)

Filamentous fungi are attractive hosts for recombinant DNA technology because of their ability to secrete high levels of bioactive proteins with posttranslational processing such as glycosylation. *A. niger* excretes 25 g/L of native glucoamylase (83, 217). Foreign genes can be incorporated via plasmids into chromosomes of the filamentous fungi where they integrate stably into the chromosome as tandem repeats providing superior long-term genetic stability. An excellent comprehensive review of heterologous expression in *Aspergillus* has been published by Lubertozzi and Keasling (218). Like mammalian cells, fungi possess the cellular machinery for translation of proteins, protein folding, and post-translational modification. Like bacteria, they are easy to culture. Genetic development of aspergilli has included (1) transformation systems, (2) expression constructs, (3) targeted integration and copy number manipulation, (4) promoters, (5) improved gene design, (6) engineering of proteases, secretion, and glycosylation, and (7) tools for tagging, targeting, and silencing of genes.

A 1,000-fold increase in phytase production was achieved in *A. niger* by the use of recombinant technology (219). Recombinant *Aspergillus oryzae* can produce 2 g/L of human lactoferrin (218) and 3.3 g/L of *Mucor* renin (221). Production of human lactoferrin (222) by *Aspergillus awamori* via rDNA technology and classical strain improvement amounted to 2 g/L of extracellular protein (220). *A. niger* glucoamylase was made by *A. awamori* at 4.6 g/L.

The fungus *Chrysosporium lucknowense* has been genetically converted into a nonfilamentous, less viscous, low protease-producing strain that is capable of producing very high yields of heterologous proteins (223). Dyadic International, Inc., the company responsible for the development of the *C. lucknowense* system, claims protein production levels of up to 100 g/L of native extracellular protein.

3.4. Mammalian Cells

CHO cells constitute the preferred system for producing monoclonal antibodies and some other recombinant proteins. Other cell types include (1) various mouse myelomas such as NS0 murine myeloma cells (224), (2) baby hamster kidney (BHK) cells for production of cattle foot-and-mouth disease vaccine, (3) green monkey kidney cells for polio vaccine (225), and (4) human cell lines such as human embryonic kidney (HEK) cells. NSO is a non-secreting subclone of the NS-1 mouse melanoma cell line. By 2006, production of therapeutic proteins by mammalian systems reached $20 billion (226).

Animal-free, protein-free, and even chemically defined media with good support of production have been developed (227). Protein production by mammalian cells (CHO) went from 5–50 mg/L in 1985 to 50–500 mg/L in 1995 and to 5 g/L in 2005 (228). A number of mammalian processes are producing 3–5 g/L of recombinant protein (229) and in some cases, protein titers have reached 10 g/L in industry (230), including antibodies (231). A rather new system is that of a human cell line known as PER.C6 of Crucell Holland BV, which, in cooperation with DSM Biologics, was reported to produce 15 g/L (232) and then later, 27 g/L of a monoclonal antibody (233). Protein production of over 20 g/L has been achieved in serum-free medium, but the production of 2–3 g/L in such media is more usual.

3.5. Insect Cells

Insect cells are able to carry out more complex posttranslational modifications than can be accomplished with fungi. They also have the best machinery for the folding of mammalian proteins and are therefore quite suitable for making soluble proteins of mammalian origin (234). The most commonly used vector system for recombinant protein expression in insects is the baculovirus, especially the nuclear polyhedrosis virus (*Autographa californica*) which contains circular double-stranded DNA, is naturally pathogenic for lepidopteran cells, and can be grown easily in vitro. The usual host is the fall armyworm (*Spodoptera frugiperda*) in suspension culture. A larval culture can be used which is much cheaper than mammalian cell culture.

Baculovirus-assisted insect cell expression offers many advantages as follows: (1) Eukaryotic posttranslational modifications without complication, including phosphorylation, N- and O-glycosylation, correct signal peptide cleavage, proper proteolytic processing, acylation, palmitylation, myristylation, amidation, carboxymethylation, and prenylation (235, 236). (2) Proper protein folding and S–S bond formation, unlike the reducing environment of *E. coli* cytoplasm. (3) High expression levels. The virus contains a gene encoding the protein polyhedrin which is made at very high levels normally and is not necessary for virus replication. The gene to be cloned is placed under the strong control of the viral polyhedrin promoter, allowing expression of heterologous protein of up to 30% of cell protein. (4) Easy scale up with high-density suspension culture. (5) Safety, expression vectors are prepared from the baculovirus which can attack invertebrates but not vertebrates or plants. (6) Lack of limit on protein size. (7) Efficient cleavage of signal peptides. (8) Simultaneous expression of multiple genes (237).

Production of recombinant proteins with the baculovirus expression vector system in insect cells reached 600 mg/L in 1988 (238). Recent information indicates that the baculovirus insect cell system can produce 11 g/L of recombinant protein (239). Recombinant insect cell cultures have yielded over 200 proteins encoded by genes from viruses, bacteria, fungi, plants, and animals (240).

4. Enzymes

Over the years, high titers of enzymes were obtained using "brute force" mutagenesis and random screening of microorganisms. Recombinant DNA technology acted as a boon for the enzyme industry in the following ways (241): (1) plant and animal enzymes could be made by microbial fermentations, e.g., chymosin; (2) enzymes from organisms difficult to grow or handle genetically were now produced by industrial organisms such as species of

Aspergillus and *Trichoderma*, and *K. lactis*, *S. cerevisiae*, *Y. lipolytica*, and *Bacillus licheniformis* (e.g., thermophilic lipase was produced by *A. oryzae* and *Thermoanaerobacter* cyclodextrin glycosyl transferase by *Bacillus*); (3) enzyme productivity was increased by the use of multiple gene copies, strong promoters, and efficient signal sequences; (4) production of a useful enzyme from a pathogenic or toxin-producing species could now be done in a safe host; and (5) protein engineering was employed to improve the stability, activity, and/or specificity of an enzyme.

Genes encoding many microbial enzymes have been cloned and the enzymes expressed at levels hundreds of times higher than those naturally produced. Over 60% of the enzymes used in different applications including detergent, food and starch processing industry are recombinant proteins (242). Recombinant molds are one of the main sources of enzymes for industrial applications. Yields as high as 4.6 g/L have been reached for several hosts including *A. niger*, *A. oryzae*, *A. awamori*, *C. lucknowense*, and *A. chrysogenum*.

Plant phytase (243), produced in recombinant *A. niger* was used as a feed for 50% of all pigs in Holland. A 1,000-fold increase in phytase production was achieved in *A. niger* by the use of recombinant technology (219). Mammalian chymosin was cloned and produced by *A. niger* or *E. coli* and recombinant chymosin was approved in the USA, its price was half that of natural calf chymosin.

Three fungal recombinant lipases are currently used in the food industry. They are from *Rhizomucor miehi*, *Thermomyces lanuginosus*, and *Fusarium oxysporum* and are produced in *A. oryzae*. They are used for laundry cleaning, inter-esterification of lipids and esterification of glucosides, producing glycolipids which have applications as biodegradable nonionic surfactants for detergents, skin care products, contact lenses and as food emulsifiers. Washing powders have been improved in activity and low temperature operation has been achieved by the application of recombinant DNA technology and site-directed mutagenesis of genes encoding proteases and lipases (244, 245). The first commercial recombinant lipase used in a detergent was from *Humicola lanuginose*. The gene was cloned into the *A. oryzae* genome.

A multicopy plasmid of *B. subtilis* was used to increase by 2,500-fold the production of an α-amylase from *B. amyloliquefaciens* (246). An exoglucanase from *Cellulomonas fimi* was overproduced in *E. coli* to a level of over 20% of cell protein (247). The same host has also been used to clone the endo-β-glucanase components from *Thermomonospora* and *Clostridium thermocellum* as well as the cellobiohydrolase I gene of *Trichoderma reesei* (248). *P. pastoris* was engineered to produce and excrete *S. cerevisiae* invertase into the medium at 100 mg/L (249). Self-cloning of the xylanase gene in *S. lividans* resulted in sixfold overproduction of the enzyme (250).

The properties of many enzymes have been modified by random mutagenesis and screening of microorganisms over the years leading to changes in substrate specificity, feedback inhibition, kinetic parameters (V_{max}, K_m or K_i), pH optimum, thermostability, and carbon source inhibition. Based on this information, more rational techniques as site-directed mutagenesis were used to introduce single changes in amino acid sequences yielding similar types of changes in a large variety of enzymes. Modification of eight amino acids increased heat tolerance and temperature stability at 100°C of a protease from *Bacillus stearothermophilus* (251). Interestingly, all mutations were far from the active site of the enzyme.

Molecular breeding techniques (e.g., *DNA shuffling* and *DNA family shuffling*) are being currently used to generate enzymes with improved properties such as activity and stability at different pH values and temperatures (252), increased or modified enantioselectivity (253), altered substrate specificity (254), stability in organic solvents (255), novel substrate specificity and activity (256), increased biological activity of protein pharmaceuticals and biological molecules (257) as well as novel vaccines (258, 259). Two proteins from directed evolution work were already on the market by 2,000 (260). These were green fluorescent protein of Clontech (261) and Novo Nordisk's LipoPrime® lipase.

5. Closing Remarks

The fermentation industry developed slowly from the beginning of the twentieth century to the early 1970s using brute force mutagenesis followed by screening or selection. However, the birth of the era of recombinant DNA (rDNA) in 1971–1973 catalyzed a major change in the way useful processes could be developed. Production of primary metabolites was markedly improved by modern genetic techniques. Environmentally friendly fermentations replaced chemical synthesis to a great extent. Of great interest has been the application of rDNA technology to the production of secondary metabolites and to the elucidation of their biosynthetic pathways. Tools include transposition mutagenesis, targeted deletions, genetic recombination via combinatorial biosynthesis, transcriptome analysis, proteomics, metabolomics, metabolic engineering, etc. Genes encoding many enzymes have been cloned and the enzymes have been expressed at levels hundreds of times higher than those naturally produced. Random redesign techniques have generated enzymes with improved properties including activity, stability, increased or modified enantioselectivity, altered substrate specificity, etc. An entirely new field of industrial microbiology has arisen out of rDNA, i.e., the biopharmaceutical industry which is the most rapidly expanding segment of the biological industry, especially that of monoclonal antibodies. The best is yet to come from the fantastic combination of industrial microbiology and rDNA technology. Many of the new techniques are carried out by small companies and academic groups who could play a major role in rescuing us from the antibiotic crisis that we are now experiencing. Furthermore, we look forward to its role in eventually replacing the environmentally dangerous energy sources that we live with today, i.e., petroleum, coal, etc., with future biofuels made from agricultural and forest biomass.

References

1. Ohnishi J, Mitsuhashi S, Hayashi M, Ando S, Yokoi H, Ochiai K et al (2002) A novel methodology employing *Corynebacterium glutamicum* genome information to generate a new L-lysine-producing mutant. Appl Microbiol Biotechnol 58:217–223
2. Stephanopoulos G (1999) Metabolic fluxes and metabolic engineering. Metab Eng 1:1–11
3. Nielsen J (2001) Metabolic engineering. Appl Microbiol Biotechnol 55:263–283
4. Sahm H, Eggeling L, de Graaf AA (2000) Pathway analysis and metabolic engineering in *Corynebacterium glutamicum*. Biol Chem 381:899–910
5. Santos CSS, Stephanopoulos G (2008) Combinatorial engineering of microbes for optimizing cellular phenotype. Curr Opin Chem Biol 12:168–176
6. Bailey JE, Sburlati A, Hatzimanikatis V, Lee K, Renner WA, Tsai PS (1996) Inverse metabolic engineering: a strategy for directed genetic engineering of useful phenotypes. Biotechnol Bioeng 52:109–121
7. Ness JE, del Cardayre SB, Minshull J, Stemmer WP (2000) Molecular breeding: the natural approach to protein design. Adv Protein Chem 55:261–292
8. Zhao H, Arnold FH (1997) Optimization of DNA shuffling for high fidelity recombination. Nucleic Acids Res 25:1307–1308
9. Patten PA, Howard RJ, Stemmer WP (1997) Applications of DNA shuffling to pharmaceuticals and vaccines. Curr Opin Biotechnol 8:724–733
10. Zhang YX, Perry K, Vinci VA, Powell K, Stemmer WP, del Cardayre SB (2002) Genome shuffling leads to rapid phenotypic improvement in bacteria. Nature 415: 644–646
11. Hou L (2009) Novel methods of genome shuffling in *Saccharomyces cerevisiae*. Biotechnol Lett 31:671–677
12. Stephanopoulos G, Alper H, Moxley J (2004) Exploiting biological complexity for strain improvement through systems biology. Nat Biotechnol 22:1261–1267
13. Leuchtenberger W, Huthmacher K, Drauz K (2005) Biotechnological production of amino acids and derivatives: current status and prospects. Appl Microbiol Biotechnol 69:1–8
14. Liu Q, Zhang J, Wei X-X, Ouyang S-P, Wu Q, Chen G-Q (2008) Microbial production of L-glutamate and L-glutamine by recombinant *Corynebacterium glutamicum* harboring *Vitreoscilla* hemoglobin gene *vgb*. Appl Microbiol Biotechnol 77:1297–1304
15. Chinen A, Kozlov YI, Hara Y, Izui H, Yasueda H (2007) Innovative metabolic pathway design for efficient L-glutamate production by suppreassing CO_2 emission. J Biosci Bioeng 103:262–269
16. Ohnishi J, Katahira R, Mitsuhashi S, Kakita S, Ikeda M (2005) A novel gnd mutation leading to increased L-lysine production in *Corynebacterium glutamicum*. FEMS Microbiol Lett 242:265–274
17. Ikeda M, Ohnishi J, Hayashi M, Mitsuhashi S (2006) A genome-based approach to create a minimally mutated *Corynebacterium glutamicum* strain for efficient L-lysine production. J Ind Microbiol Biotechnol 33:610–615
18. Hiyashi M, Ohnishi J, Mitsuhashi S, Yonetani Y, Hashimoto S, Ikeda M (2006) Transcriptome analysis reveals global expression changes in an industrial L-lysine producer of *Corynebacterium glutamicum*. Biosci Biotechnol Biochem 70:546–550
19. Radmacher E, Eggeling L (2007) The three tricarboxylate synthase activities of *Corynebacterium glutamicum* and increase of L-lysine synthesis. Appl Microbiol Biotechnol 76:587–595
20. Blombach B, Schreiner ME, Moch M, Oldiges M, Eikmanns BJ (2007) Effect of pyruvate dehydrogenase complex deficiency on L-lysine production with *Corynebacterium glutamicum*. Appl Microbiol Biotechnol 76: 615–623
21. Sindelar G, Wendisch VF (2007) Improving lysine production by *Corynebacterium glutamicum* through DNA microarray-based identification of novel target genes. Appl Microbiol Biotechnol 76:677–689

22. Komatsubara S, Kisumi M, Chibata I (1979) Transductional construction of a threonine-producing strain of *Serratia marcescens*. Appl Environ Microbiol 38:1045–1051
23. Komatsubara S, Kisumi M, Chibata I (1983) Transductional construction of a threonine-hyperproducing strain of *Serratia marcescens*: lack of feedback controls of three aspartokinases and two homoserine dehydrogenases. Appl Environ Microbiol 45:1445–1452
24. Sugita T, Komatsubara S (1989) Construction of a threonine-hyperproducing strain of *Serratia marcescens* by amplifying the phosphoenolpyruvate carboxylase gene. Appl Microbiol Biotechnol 30:290–293
25. Debabov VG (2003) The threonine story. Adv Biochem Eng Biotechnol 79:113–136
26. Ishida M, Kawashima H, Sato K, Hashiguchi K, Ito H, Enei H et al (1994) Factors improving L-threonine production by a three L-threonine biosynthetic genes-amplified recombinant strain of *Brevibacterium lactofermentum*. Biosci Biotechnol Biochem 58:768–770
27. Eggeling L, Sahm H (1999) Amino acid production: principles of metabolic engineering. In: Lee SY, Papoutsakis ET (eds) Metabolic engineering. Marcel Dekker, New York, pp 153–176
28. Kruse D, Kraemer R, Eggeling L, Rieping M, Pfefferle W, Tchieu JH et al (2002) Influence of threonine exporters on threonine production in *Escherichia coli*. Appl Microbiol Biotechnol 59:205–210
29. Lee KH, Park JH, Kim TY, Kim HU, Lee SY (2007) Systems metabolic engineering of *Escherichia coli* for L-threonine production. Mol Syst Biol 3:149–157
30. Park JH, Lee KH, Kim TY, Lee SY (2007) Metabolic engineering of *Escherichia coli* for the production of L-valine based on transcriptome analysis and in silico gene knockout stimulation. Proc Natl Acad Sci USA 104:7797–7802
31. Lange C, Rittmann D, Wendisch VF, Bott M, Sahm H (2003) Global expression profiling and physiological characterization of *Corynebacterium glutamicum* grown in the presence of L-valine. Appl Environ Microbiol 69:2521–2532
32. Kisumi M, Komatsubara S, Chibata I (1977) Enhancement of isoleucine hydroxamate-mediated growth inhibition and improvement of isoleucine-producing strains of *Serratia marcescens*. Appl Environ Microbiol 34: 647–653
33. Guillouet S, Rodal AA, An G-H, Lessard PA, Sinskey AJ (1999) Expression of the *Escherichia coli* catabolic threonine dehydratase in *Corynebacterium glutamicum* and its effect on isoleucine production. Appl Environ Microbiol 65:3100–3107
34. Morbach S, Sahm H, Eggeling L (1996) L-Isoleucine production with *Corynebacterium glutamicum*: further flux increase and limitation of export. Appl Environ Microbiol 62: 4345–4351
35. Hashiguchi K, Takesada H, Suzuki E, Matsui H (1999) Construction of an L-isoleucine overproducing strain of *Escherichia coli* K-12. Biosci Biotechnol Biochem 63:672–679
36. Eggeling L, Morbach S, Sahm H (1977) The fruits of molecular physiology: engineering the L-isoleucine biosynthesis pathway in *Corynebacterium glutamicum*. J Biotechnol 56:167–182
37. Colon GE, Nguyen TT, Jetten MSM, Sinskey A, Stephanopoulos G (1995) Production of isoleucine by overexpression of *ilvA* in *Corynebacterium lactofermentum* threonine producer. Appl Microbiol Biotechnol 43: 482–488
38. Kase H, Nakayama K (1977) L-Isoleucine production by analog-resistant mutants derived from threonine-producing strain of *Corynebacterium glutamicum*. Agric Biol Chem 41:109–116
39. Morbach S, Sahm H, Eggeling L (1995) Use of feedback-resistant threonine dehydratases of *Corynebacterium glutamicum* to increase carbon flux towards L-isoleucine. Appl Environ Microbiol 61:4315–4320
40. Morbach S, Kelle R, Winkels S, Sahm H, Eggeling L (1996) Engineering the homoserine dehydrogenase and threonine dehydratase control points to analyse flux towards L-isoleucine in *Corynebacterium glutamicum*. Appl Microbiol Biotechnol 45:612–620
41. Sahm H, Eggeling L, Morbach S, Eikmanns B (1999) Construction of L-isoleucine-overproducing strains of *Corynebacterium glutamicum*. Naturwissenschaften 86:33–38
42. Shiio I, Sasaki A, Nakamori S, Sano K (1973) Production of L-isoleucine by AHV resistant mutants of *Brevibacterium flavum*. Agric Biol Chem 37:2053–2061
43. Ikeda S, Fujita I, Hirose Y (1976) Culture conditions of L-isoleucine fermentation from acetic acid. Agric Biol Chem 40:517–522
44. Lee M, Smith GM, Eiteman MA, Altman E (2004) Aerobic production of alanine by *Escherichia coli aceF IdhA* mutants expressing the *Bacillus sphaericus alaD* gene. Appl Microbiol Biotechnol 65:56–60
45. Bloom F, Smith CJ, Jessee J, Veileux B, Deutch AH (1984) The use of genetically engineered strains of *Escherichia coli* for the

overproduction of free amino acids: proline as a model system. In: Downey K, Voellmy RW (eds) Advances in gene technology: molecular genetics of plants and animals. Academic, New York, pp 383–394
46. Sugiura M, Takagi T, Kisumu M (1995) Proline production by regulatory mutants of *Serratia marcescens.* Appl Microbiol Biotechnol 21:213–239
47. Sugiura M, Imai Y, Takagi T, Kisumi M (1985) Improvement of a proline-producing strain of *Serratia marcescens* by subcloning of a mutant allele of the proline gene. J Biotechnol 3:47–58
48. Masuda M, Takamatu S, Nishimura N, Komatsubara S, Tosa T (1993) Improvement of culture conditions for L-proline production by a recombinant strain of *Serratia marcescens.* Appl Biochem Biotechnol 43:189–197
49. Tsuchida T, Kubota K, Yoshinaga F (1986) Improvement of L-proline production by sulfaguanidine resistant mutants derived from L-glutamic acid-producing bacteria. Agric Biol Chem 50:2201–2207
50. Nakanishi T, Yokote Y, Taketsugu Y (1973) Conversion of L-glutamic acid fermentation to a L-proline fermentation by *Corynebacterium glutamicum.* J Ferment Technol 51:742–749
51. Nakanishi T, Hirao T, Azuma T, Sakurai M, Hagino H (1987) Application of L-glutamate to L-proline fermentation by *Corynebacterium acetoacidophilum.* J Ferment Technol 65: 139–144
52. Shibasaki T, Hashimoto S, Mori H, Ozaki A (2000) Construction of a novel hydroxyproline-producing recombinant *Escherichia coli* by introducing a proline 4-hydroxylase gene. J Biosci Bioeng 90:522–525
53. Vandamme EJ (2007) Microbial gems: microorganisms without frontiers. SIM News 57:81–90
54. Kamada N, Yasuhara A, Takano Y, Nakano T (2001) Effect of transketolase modifications on carbon flow to the purine-nucleotide pathway in *Corynebacterium ammoniagenes.* Appl Microbiol Biotechnol 56:710–717
55. Koizumi S, Yonetani Y, Maruyama A, Teshiba S (2000) Production of riboflavin by metabolically engineered *Corynebacterium ammoniagenes.* Appl Microbiol Biotechnol 51: 674–679
56. Perkins JB, Pero J (1993) Biosynthesis of riboflavin, biotin, folic acid, and cobalamin. In: Sonenshein AL, Hoch JA, Losick R (eds) *Bacillus subtilis* and other gram positive bacteria: biochemistry, physiology and molecular genetics. American Society for Microbiology, Washington, DC, pp 319–334
57. Perkins JB, Sloma A, Hermann T, Theriault K, Zachgo E, Erdenberger T et al (1999) Genetic engineering of *Bacillus subtilis* for the commercial production of riboflavin. J Ind Microbiol Biotechnol 22:8–18
58. Brenner S, Johnson M, Bridgham J, Golda G, Lloyd DH, Johnson D et al (2000) Gene expression analysis by massively parallel signature sequencing (MPSS) on microbead arrays. Nat Biotechnol 18:630–634
59. Karos M, Vilarino C, Bollschweiler C, Revuelta JL (2004) A genome wide transcription analysis of a fungal rivoflavin overproducer. J Biotechnol 113:69–76
60. Levy-Schil S, Debussche L, Rigault S, Soubrier F, Bacchette F, Lagneaux D et al (1993) Biotin biosyntheric pathway in a recombinant strain of *Escherichia coli* overexpressing bio genes: evidence for a limiting step upstream from KAPA. Appl Microbiol Biotechnol 38:755–762
61. Sakurai N, Imai Y, Masuda M, Komatsubara S, Tosa T (1994) Improvement of a d-biotin-hyperproducing recombinant strain of *Serratia marcescens.* J Biotechnol 36:63–73
62. Masuda M, Takahashi K, Sakurai N, Yanagiya K, Komatsubara S, Tosa T (1995) Further improvement of D-biotin production by a recombinant strain of *Serratia marcescens.* Process Biochem 30:553–562
63. Matsui J, Ifuku O, Kanzaki N, Kawamoto T, Nakahama K (2001) Microorganism resistant to threonine analogue and production of biotin. US patent 6284500
64. Anderson S, Marks CB, Lazarus R, Miller J, Stafford K, Seymour J et al (1985) Production of 2-keto-L-gulonate: an intermediate in L-ascorbate synthesis by a genetically modified *Erwinia herbicola.* Science 230:144–149
65. Grindley JF, Payton MA, van de Pol H, Hardy KG (1988) Conversion of glucose to 2-keto-L-gulonate, an intermediate in L-ascorbate synthesis, by a recombinant strain of *Erwinia citreus.* Appl Environ Microbiol 54: 1770–1775
66. Saito Y, Ishii Y, Hayashi H, Imao Y, Akashi T, Yoshikawa K et al (1997) Cloning of genes coding for L-sorbose and L-sorbosone dehydrogenases from *Gluconobacter oxydans* and microbial production of 2-keto-gulonate, a precursor of L-ascorbic acid, in a recombinant *G. oxydans* strain. Appl Environ Microbiol 63:454–460
67. DeBaets S, Vandedrinck S, Vandamme EJ (2000) Vitamins and related biofactors, microbial production. In: Lederberg J (ed) Encyclopedia of microbiology, vol 4, 2nd edn. Academic, New York, pp 837–853

68. Johnson EA, Schroeder WA (1995) Microbial carotenoids. Adv Biochem Eng Biotechnol 53:119–178
69. López-Nieto MJ, Costa J, Peiro E, Méndez E, Rodríguez-Sáiz M, de la Fuente JL, Cabri W, Barredo JL (2004) Biotechnological lycopene production by mated fermentation of *Blakeslea trispora*. Appl Microbiol Biotechnol 66:153–159
70. Ciegler A (1965) Microbial carotenogenesis. Adv Appl Microbiol 7:1–34
71. Ninet L, Renaut J (1979) Carotenoids. In: Peppler HJ, Perlman D (eds) Microbial technology, vol 1, 2nd edn. Academic, New York, pp 529–544
72. Barkovich R, Liao JC (2001) Metabolic engineering of isoprenoids. Metab Eng 3: 27–39
73. Lee PC, Schmidt-Dannert C (2002) Metabolic engineering towards biotechnological production of carotenoids in microorganisms. Appl Microbiol Biotechnol 60:1–11
74. Tao L, Jackson RE, Cheng Q (2005) Directed evolution of copy number of a broad host range plasmid for metabolic engineering. Metab Eng 7:10–17
75. Alper H, Miyaoku K, Stephanopoulos G (2006) Characterization of lycopene-overproducing *E. coli* strains in high cell density fermentations. Appl Microbiol Biotechnol 72:968–974
76. Fernández-Sevilla JM, Acien Fernández FG, Molina Grima E (2010) Biotechnological production of lutein and its applications. Appl Microbiol Biotechnol 86:27–40
77. Margalith PZ (1999) Production of ketocarotenoids by microalgae. Appl Microbiol Biotechnol 51:431–438
78. Johnson EA (2003) *Phaffia rhodozyma:* a colorful odyssey. Int Microbiol 6:169–174
79. de la Fuente JL, Rodríguez-Sáiz M, Schleissner C, Díez B, Peiro E, Barredo JL (2010) High-titer production of astaxanthin by the semi-industrial fermentation of *Xanthophyllomyces dendrorhous*. J Biotechnol 148:144–146
80. Rodríguez-Sáiz M, de la Fuente JL, Barredo JL (2010) *Xanthophyllomyces dendrorhous* for the industrial production of astaxanthin. Appl Microbiol Biotechnol 88:645–658
81. Sauer M, Porro D, Mattanovich D, Branduardi P (2008) Microbial production of organic acids: expanding the markets. Trends Biotechnol 26:100–108
82. Sánchez S, Demain AL (2008) Metabolic regulation and overproduction of primary metabolites. Microb Biotechnol 1:283–319
83. Ward OP, Qin WM, Hanjoon JD, Singh EJYA (2006) Physiology and biotechnology of *Aspergillus*. Adv Appl Microbiol 58:1–75
84. Kubicek CP, Roehr M (1986) Citric acid fermentation. Crit Rev Biotechnol 3:331–373
85. Causey TB, Zhou S, Shanmugam KT, Ingram LO (2003) Engineering the metabolism of *Escherichia coli* W3110 for the conversion of sugar to redox-neutral and oxidized products: homoacetate production. Proc Natl Acad Sci USA 100:825–832
86. Parekh SR, Cheryan M (1994) High concentrations of acetate with a mutant strain of *C. thermoaceticum*. Biotechnol Lett 16: 139–142
87. Fukaya M, Tayama K, Tamaki T, Tagami H, Okumura H, Kawamura Y et al (1989) Cloning of the membrane-bound aldehyde dehydrogenase gene of *Acetobacter polyoxogenes* and improvement of acetic acid production by use of thecloned gene. Appl Environ Microbiol 55:171–176
88. Patnaik R, Louie S, Gavrilovic V, Perry K, Stemmer WPC, Ryan CM et al (2002) Genome shuffling of *Lactobacillus* for improved acid tolerance. Nat Biotechnol 20:707–712
89. John RP, Gangadharan D, Madhavan Nampoothiri K (2008) Genome shuffling of *Lactobacillus delbrueckii* mutant and *Bacillus amyloliquuefaciens* through protoplastic fusion for L-latic acid production from starchy wastes. Bioresour Technol 99:8008–8015
90. Skory CD (2004) Lactic acid production by *Rhizopus oryzae* transformants with modified lactate dehydrogenase activity. Appl Microbiol Biotechnol 64:237–242
91. Saito S, Ishida N, Onishi T, Tokuhiro K, Nagamori E, Kitamoto K et al (2005) Genetically engineered wine yeast produces a high concentration of L-lactic acid of extremely high optical purity. Appl Environ Microbiol 71:2789–2792
92. Porro D, Bianchi MM, Brambilla L, Menghini R, Bolzani D, Carrera V et al (1999) Replacement of a metabolic pathway for large scale production of lactic acid from engineered yeasts. Appl Environ Microbiol 65:4211–4215
93. Zhou S, Yomano LP, Shanmugam KT, Ingram LO (2003) Fermentation of 10% (w/v) sugar to D(-)-lactate by engineered *Escherichia coli* B. Biotechnol Lett 27:1891–1896
94. Ishida N, Suzuki T, Tokuhiro K, Nagamori E, Onishi T, Saitoh S et al (2006) D-Lactic acid production by metabolically engineered *Saccharomyces cerevisiae*. J Biosci Bioeng 101: 172–177

95. Sánchez AM, Bennett GN, San KY (2005) Novel pathway engineering design of the anaerobic central metabolic pathway in *Escherichia coli*. Metab Eng 7:229–239
96. Lee SJ, Song H, Lee SY (2006) Genome-based metabolic engineering of *Mannheimia succiniciproducens* for succinic acid production. Appl Environ Microbiol 72:1939–1948
97. Lin H, Bennett GN, San KY (2005) Fed-batch culture of a metabolically engineered *Escherichia coli* strain designed for high-level succinate production and yield under aerobic conditions. Biotechnol Bioeng 90:775–779
98. Vemuri GN, Eiteman MA, Altman E (2002) Succinate production in dual-phase *Escherichia coli* fermentations depends on the time of transition from aerobic to anaerobic conditions. J Ind Microbiol Biotechnol 28: 325–332
99. Liu X, Yang S-T (2005) Metabolic engineering of *Clostridium tyrobutyricum* for butyric acid fermentation. Proceedings of the 229th ACS National Meeting, San Diego, Abstract 70
100. Neufeld RJ, Peleg Y, Rokem JS, Pines O, Goldberg I (1991) L-Malic acid formation by immobilized *Saccharomyces cerevisiae* amplified for fumarase. Enzyme Microb Technol 13:991–996
101. Picataggio S, Rohver T, Deander K, Lanning D, Reynolds R, Mielenz J et al (1992) Metabolic engineering of *Candida tropicalis* for the production of long-chain dicarboxylic acids. Nat Biotechnol 10:894–898
102. Shi DJ, Wang CL, Wang KM (2009) Genome shuffling to improve thermotolerance, ethanol tolerance and ethanol productivity of *Saccharomyces cerevisiae*. J Ind Microbiol Biotechnol 36:139–147
103. Chu BCH, Lee H (2007) Genetic improvement of *Saccharomyces cerevisiae* for xylose fermentation. Biotechnol Adv 25:425–441
104. Jeffries TW (2006) Engineering yeast for xylose metabolism. Curr Opin Biotechnol 17:320–326
105. Bro C, Regenberg B, Forster J, Nielsen J (2006) In silico aided metabolic engineering of *Saccharomyces cerevisiae* for improved bioethanol production. Metab Eng 8:102–111
106. Wei P, Li Z, He P, Lin Y, Jiang N (2008) Genome shuffling in the ethanologenic yeast *Candida krusei* to improve acetic acid tolerance. Biotechnol Appl Biochem 49:113–120
107. Ingram LO, Conway T, Clark DP, Sewell GW, Preston JF (1987) Genetic engineering of ethanol production in *Escherichia coli*. Appl Environ Microbiol 53:2420–2425
108. Doran JB, Ingram LO (1993) Fermentation of crystalline cellulose to ethanol by *Klebsiella oxytoca* containing chromosomally integrated *Zymomonas mobilis* genes. Biotechnol Prog 9:533–538
109. Dien BS, Nichols NN, O'Bryan PJ, Bothast RJ (2000) Development of new ethanologenic *Escherichia coli* strains for fermentation of lignocellulosic biomass. Appl Biochem Biotechnol 84(86):181–196
110. Papanikolaou S, Ruiz-Sánchez P, Pariset B, Blanchard F, Fick M (2000) High production of 1,3-propanediol from industrial glycerol by a newly isolated *Clostridium butyricum* strain. J Biotechnol 77:191–208
111. Sanford K, Valle F, Ghirnikar R (2004) Pathway engineering through rational design. Tutorial: designing and building cell factories for biobased production. Genet Eng News 24:44–45
112. Nakamura C, Whited G (2003) Metabolic engineering for the microbial production of 1,3 propanediol. Curr Opin Biotechnol 14:454–459
113. Kaup B, Bringer-Meyer S, Sahm H (2004) Metabolic engineering of *Escherichia coli*: construction of an efficient biocatalyst for D-mannitolformationinawhole-cellbiotransformation. Appl Microbiol Biotechnol 64:333–339
114. Kaup B, Bringer-Meyer S, Sahm H (2005) D-Mannitol formation from D-glucose in a whiole-cell biotransformation with recombinant *Escherichia coli*. Appl Microbiol Biotechnol 69:397–403
115. Lee J-K, Oh D-K, Song H-Y, Kim I-W (2007) Ca2+ and Cu2+ supplementation increases mannitol production by *Candida magnoliae*. Biotechnol Lett 29:291–294
116. Song SH, Vieille C (2009) Recent advances in the biological production of mannitol. Appl Microbiol Biotechnol 84:55–62
117. Chun UH, Rogers PL (1988) The simultaneous production of sorbitol and gluconic acid by *Zymomonas mobilis*. Appl Microbiol Biotechnol 29:19–24
118. Ladero V, Ramos A, Wiersma A, Goffin P, Schanck A, Kleerbezem M (2007) High-level; production of the low-calorie sugar sorbitol by *Lactobacillus plantarum* through metyabolic engineering. Appl Environ Microbiol 73:1864–1872
119. Hermann M, Fayolle F, Marchal R, Podvin L, Sebald M, Vandecasteele J-P (1985) Isolation and characterization of butanol-resistant mutants of *Clostridium acetobutylicum*. Appl Environ Microbiol 50:1238–1243
120. Qureshi N, Maddox IS, Freidl A (1992) Application of continuous substrate feeding

to the ABE fermentation: relief of product inhibition using extraction, perstraction, stripping and pervaporation. Biotechnol Prog 8:382–390

121. Chen C-K, Blaschek HP (1999) Acetate enhances solvent production and prevents degeneration in *Clostridium beijerinckii* BA101. Appl Microbiol Biotechnol 52: 170–173

122. Connor MR, Liao JC (2009) Microbial production of advanced transportation fuels in non-natural hosts. Curr Opin Biotechnol 20:307–315

123. Whiters ST, Gottlieb SS, Lieu B, Newman JD, Keasling JD (2007) Identification of isopentenol biosynthetic genes from *Bacillus subtilis* by a screening method based on isoprenoid precursor toxicity. Appl Environ Microbiol 73:6277–6283

124. Atsumi S, Hanai T, Liao JC (2008) Non-fermentative pathways for synthesis of branched-chain higher alcohols as biofuels. Nature 451:86–89

125. Jeya M, Lee K-M, Tiwari MK, Kim J-S, Gunasekaran P, Kim S-Y, Kim I-W, Lee J-K (2009) Isolation of a novel high erythritol-producing *Pseudomonas tsukubaensis* and scale-up of erythritol fermentation to industrial level. Appl Microbiol Biotechnol 83:225–231

126. Zagustina NA, Rodionova NA, Mestechkina NM, Shcherbukhin VD, Bezborodov AM (2001) Xylitol production by a culture of *Candida guilliermondii* 2581. Appl Biochem Microbiol 37:489–492

127. Lederberg J (2000) Pathways of discovery: infectious history. Science 288:287–293

128. Choi KP, Kim KH, Kim JW (1997) Strain improvement of clavulanic acid producing *Streptomyces clavuligerus*. Proc 10th Internat Symp Biol Actinomycetes (ISBA), Beijing, Abstr 12, 9

129. Li R, Townsend CA (2006) Rational strain improvement for enhanced clavulanic acid production by genetic engineering of the glycolytic pathway in *Streptomyces clavuligerus*. Metab Eng 8:240–252

130. Pérez-Redondo RA, Rodríguez-García A, Martín JF, Liras P (1999) Deletion of the *pyc* gene blocks clavulanic acid biosynthesis except in glycerol-containing medium: evidence for two different genes in formation of the C3 unit. J Bacteriol 181:6922–6928

131. Paradkar AS, Aiodoo KA, Jensen SE (1998) A pathway-specific transcriptional activator regulates late steps of clavuklanic acid biosynthesis in *Streptomyces clavuligerus*. Mol Microbiol 27:831–843

132. Pérez-Llarena FJ, Liras P, Rodríguez-García A, Martín JF (1997) A regulatory gene (*ccaR*) required for cephamycin and clavulanic acid production in *Streptomyces clavuligerus*: amplification results in overproduction of both β-lactam compounds. J Bacteriol 179: 2053–2059

133. Lein J (1986) The Panlabs penicillin strain improvement program. In: Vanek Z, Hostalek Z (eds) Overproduction of microbial metabolites; strain improvement and process control strategies. Butterworth, Boston, pp 105–139

134. Kennedy J, Turner G (1996) γ-(L-α-Aminoadipyl)-L-cysteinyl-D-valine synthetase is a rate limiting enzyme for penicillin production in *Aspergillus nidulans*. Mol Gen Genet 253:189–197

135. Hamlyn PF, Ball C (1979) Recombination studies with *Cephalosporium acremonium*. In: Sebek OK, Laskin AI (eds) Genetics of industrial microorganisms. American Society for Microbiology, Washington, DC, pp 185–191

136. Skatrud PL, Fisher DL, Ingolia TD, Queener SW (1987) Improved transformation of *Cephalosporium acremonium*. In: Alacevic M, Hranueli D, Toman Z (eds) Genetics of industrial microorganisms, part B. Zagreb, Pliva, pp 111–119

137. Skatrud PL, Tietz AJ, Ingolia TD, Cantwell CA, Fisher DL, Chapman JL et al (1989) Use of recombinant DNA to improve production of cephalosporin C by *Cephalosporium acremonium*. Nat Biotechnol 7:477–485

138. Wesseling AC, Lago B (1981) Strain improvement by genetic recombination of cephamycin producers, *Nocardia lactamdurans* and *Streptomyces griseus*. Dev Ind Microbiol 22:641–651

139. Chary VK, de la Fuente JL, Leitao AL, Liras P, Martín JF (2000) Overexpression of the *lat* gene in *Nocardia lactamdurans* from strong heterologous promoters results in very high levels of lysine-6-aminotransferase and up to a two-fold increase in cephamycin C production. Appl Microbiol Biotechnol 53:282–288

140. Bignell DRD, Tahlan K, Colvin KR, Jensen SE, Leskiw BK (2005) Expression of *ccaR*, encoding the positive activator of cephamycin C and clavulanic acid production in *Streptomyces clavuligerus*, is dependent on *bldG*. Antimicrob Agents Chemother 49:1529–1541

141. Cantwell C, Beckmann R, Whiteman P, Queener SW, Abraham EP (1992) Isolation of deacetoxycephalosporin C from fermentation broths of *Penicillium chrysogenum* transformants: construction of a new fungal

biosynthetic pathway. Proc R Soc Lond (Biol) 248:283–289

142. Crawford L, Stepan AM, Mcada PC, Rambosek JA, Conder MJ, Vinci VA et al (1995) Production of cephalosporin intermediates by feeding adipic acid to recombinant *Penicillium chrysogenum* strains expressing ring expansion activity. Nat Biotechnol 13:58–62
143. Velasco J, Adrio JL, Moreno MA, Díez B, Soler G, Barredo JL (2000) Environmentally safe production of 7-aminodeacetoxycephalosporanic acid (7ADCA) using recombinant strains of *Acremonium chrysogenum*. Nat Biotechnol 18:857–861
144. Luo H, Yu H, Qiang L, Shen Z (2004) Cloning and co-expression of D-amino acid oxidase and glutaryl-7-aminocephalosporanic acid acylase genes in *Escherichia coli*. Enzyme Microb Technol 35:514–518
145. Coulthurst SJ, Barnard AM, Salmond GP (2005) Regulation and biosynthesis of carbapenem antibiotics in bacteria. Nat Rev Microbiol 3:295–306
146. Kahan JS, Kahan FM, Goegelman R, Currie SA, Jackson M, Stapley EO, Miller TW, Miller AK, Hendlin D, Mochales S, Hernandez S, Woodruff HB, Birnbaum J (1979) Thienamycin, a new β-lactam antibiotic. 1. Discovery, taxonomy, isolation and physical properties. J Antibiot 32:1–12
147. Park SR, Han AR, Ban Y-H, Yoo YJ, Kim EJ et al (2010) Genetic engineering of macrolide biosynthesis: past advances, current state and future prospects. Appl Microbiol Biotechnol 85:1227–1239
148. Reeves AR, Cernota WH, Brikun IA, Wesley RK, Weber JM (2004) Engineering precursor flow for increased erythromycin production in *Aeromicrobium erythreum*. Metab Eng 6:300–312
149. Solenberg PJ, Cantwell CA, Tietz AJ, McGilvray D, Queener SW, Baltz RH (1996) Transposition mutagenesis in *Streptomyces fradiae*: identification of a neutral site for the stable insertion of DNA by transposon exchange. Gene 16:67–72
150. Brautaset T, Sletta H, Nedal A, Borgos SEF, Degnes KF, Bakke I, Volokhan O, Sekurova ON, Treshalin ID, Mirchink EP, Dikiy A, Ellingsen TE, Zotchev SB (2008) Improved antifungal polyene macrolides via engineering of the nystatin biosynthetic genes in *Streptomyces noursei*. Chem Biol 15:1198–1206
151. Galm U, Shen B (2006) Expression of biosynthetic gene clusters in heterologous hosts for natural product production and combinatorial biosynthesis. Expet Opin Drug Discov 1:409–437
152. Méndez C, Salas JA (2003) On the generation of novel anticancer drugs by recombinant DNA technology: the use of combinatorial biosynthesis to produce novel drugs. Comb Chem High Throughput Screen 6:513–526
153. Rodríguez E, McDaniel R (2001) Combinatorial biosynthesis of antimicrobials and other natural products. Curr Opin Microbiol 4: 526–534
154. Trefzer A, Blanco G, Remsing L, Kunzel E, Rix U, Lipata F et al (2002) Rationally designed glycosylated premithramycins: hybrid aromatic polyketides using genes from three different biosynthetic pathways. J Am Chem Soc 124:6056–6062
155. Gomi S, Ikeda D, Nakamura H, Naganawa H, Yamashita F, Hotta K et al (1984) Isolation and structure of a new antibiotic, indolizomycin, produced by a strain SK2-52 obtained by interspecies fusion treatment. J Antibiot 37:1491–1494
156. Traxler P, Schupp T, Wehrli W (1982) 16, 17-dihydrorifamycin S and 16,17-dihydro-17-hydroxyrifamycin S, two novel rifamycins from a recombinant strain C5/42 of *Nocardia mediterranei*. J Antibiot 35:594–601
157. Okanishi M, Suzuki N, Furuta T (1996) Variety of hybrid characters among recombinants obtained by interspecific protoplast fusion in streptomycetes. Biosci Biotechnol Biochem 60:1233–1238
158. Zhou L, Ahlert J, Xue Y, Thorson JS, Sherman DH, Liu H-W (1999) Engineering a methymycin/pikromycin-calicheamicin hybrid: construction of two new macrolides carrying a designed sugar moiety. J Am Chem Soc 121:9881–9882
159. Méndez C, Salas JA (2001) Altering the glycosylation pattern of bioactive compounds. Trends Biotechnol 19:449–456
160. Decker H, Hutchinson CR (1993) Transcriptional analysis of the *Streptomyces glaucescens* tetracenomycin biosynthesis gene cluster. J Bacteriol 175:3887–3892
161. Wohlert S-E, Blanco G, Lombo F, Fernández E, Brana AF, Reich S, Udvarnoki G, Méndez C, Decker H, Frevert J et al (1998) Novel hybrid tetracenomycins through combinatorial biosynthesis using a glycosyltransferase encoded by the *elm* genes in cosmid 16 F4 which shows a very broad sugar substrate specificity. J Am Chem Soc 120: 10596–10601
162. Barriere JC, Berthaud N, Beyer D, Dutka-Malen S, Paris JM, Desnottes JF (1998)

Recent developments in streptogramin research. Curr Pharm Des 4:155–180
163. Xu B, Jin Z, Wang H, Jin Q, Jin X et al (2008) Evolution of *Streptomyces pristinaespiralis* for resistance and production of pristinamycin by genome shuffling. Appl Microbiol Biotechnol 80:261–267
164. Van Lanen SG, Shen B (2006) Microbial genomics for the improvement of natural product discovery. Curr Opin Microbiol 9:252–260
165. Jenke-Kodama H, Sandmann A, Müller R, Dittmann E (2005) Evolutionary implications of bacterial polyketide synthases. Mol Biol Evol 22:2027–2039
166. Zazopoulos E, Hwang K, Staffa A, Liu W, Bachmann BO, Nonaka K et al (2003) A genomics-guided approach for discovering and expressing cryptic metabolic pathways. Nat Biotechnol 21:187–190
167. Moir DT, Shaw KJ, Hare RS, Vovis GF (1999) Genomics and antimicrobial drug discovery. Antimicrob Agents Chemother 43:439–446
168. Newman DJ, Shapiro S (2008) Microbial prescreens for anticancer activity. SIM News 58:132–150
169. Hwang CK, Kim HS, Hong YS, Kim YH, Hong SK, Kim SJ, Lee JJ (1995) Expression of *Streptomyces peucetius* genes for doxorubicin resistance and aklavinone 11-hydroxylase in *Streptomyces galilaeus* ATCC 31133 and production of a hybrid aclacinomycin. Antimicrob Agents Chemother 39: 1616–1620
170. Kim HS, Hong YS, Kim YH, Yoo OJ, Lee JJ (1996) New anthracycline metabolites produced by the aklavinone 11-hydroxylase gene in *Streptomyces galilaeus* ATCC 3113. J Antibiot 49:355–360
171. Niemi J, Mäntäslä P (1995) Nucleotide sequences and expression of genes from *Streptomyces purpurascens* that cause the production of new anthracyclines. J Bacteriol 177:2942–2945
172. Ylihonko K, Hakala J, Kunnari T, Mäntsälä P (1996) Production of hybrid anthracycline antibiotics by heterologous expression of *Streptomyces nogalater* nogalamycin biosynthesis genes. Microbiology 142:1965–1972
173. Strohl WR, Bartel PL, Li Y, Connors NC, Woodman RH (1991) Expression of polyketide biosynthesis and regulatory genes in heterologous streptomycetes. J Ind Microbiol 7:163–174
174. Bartel PL, Zhu CB, Lampel JS, Dosch DC, Connors NC, Strohl WR, Beale JM Jr, Floss HG (1990) Biosynthesis of anthraquinones by interspecies cloning of actinorhodin genes in streptomycetes: clarification of actinorhodin gene functions. J Bacteriol 172: 4816–4826
175. Arcamone F, Penco S, Vigevani A, Redaelli S, Franchi G, Di Marco A, Casazza AM, Dasdia T, Formelli F, Necco A, Soranzo C (1975) Synthesis and antitumor properties of new glycosides of daunomycinone and adriamycinone. J Med Chem 18:703–707
176. Madduri K, Kennedy J, Rivola G, Inventi-Solari A, Filippini S, Zanuso G, Colombo AL, Gewain KM, Occi JL, MacNeil DJ, Hutchinson CR (1998) Production of the antitumor drug epirubicin (4′-epidoxorubicin) and its precursor by a genetically engineered strain of *Streptomyces peucetius.* Nat Biotechnol 16:69–74
177. Manfredi JJ, Horowitz SB (1984) Taxol: an antimitotic agent with a new mechanism of action. Pharmacol Ther 25:83–125
178. Horwitz SB (1994) Taxol (paclitaxel): mechanisms of action. Ann Oncol 5(Suppl 6): S3–S6
179. Dejong JM, Liu Y, Bollon AP, Long RM, Jennewein S, Williams D, Croteau RB (2005) Genetic engineering of taxol biosynthetic genes in *Saccharomyces cerevisiae.* Biotechnol Bioeng 93:212–224
180. Engels B, Dahm P, Jennewein S (2008) Metabolic engineering of taxadiene biosynthesis in yeast as a first step towards Taxol (Paclitaxel) production. Metab Eng 10: 201–206
181. Borzlleri RM, Vite GD (2002) Epothilones: new tubulin polymerization agents in preclinical and clinical development. Drugs Future 27:1149–1163
182. Brown AG, Smale TC, King TJ, Hasenkamp R, Thompson RH (1976) Crystal and molecular structure of compactin: a new antifungal metabolite from *Penicillium brevicompactum.* J Chem Soc Perkin Trans I (11):1165–1170
183. Endo A, Kuroda M, Tsujita Y (1976) ML-236B and ML-236 C, new inhibitors of cholesterolgenesis produced by *Penicillium citrinun.* J Antibiot 29:1346–1348
184. Endo A (1979) Monacolin K, a new hypocholesterolemic agent produced by *Monascus* species. J Antibiot 32:852–854
185. Alberts AW, Chen J, Kuron G, Hunt V, Huff J, Hoffman C et al (1980) Mevinolin: a highly potent competitive inhibitor of hydroxymethylglutaryl-coenzyme A reductase and a cholesterol-lowering agent. Proc Natl Acad Sci USA 77:3957–3961
186. Askenazi M, Driggers EM, Holtzman DA, Norman TC, Iverson S, Zimmer DP et al

(2003) Integrating transcriptional and metabolite profiles to direct the engineering of lovastatin-producing fungal strains. Nat Biotechnol 21:150–156

187. Stapley EO (1982) Avermectins, antiparasitic lactones produced by *Streptomyces avermitilis* isolated from a soil in Japan. In: Umezawa H, Demain AL, Hata R, Hutchinson CR (eds) Trends in antibiotic research. Japan Antibiotic Research Association, Tokyo, pp 154–170
188. Lee J-Y, Hwang Y-S, Kim S-S, Kim E-S, Choi C-Y (2000) Effect of a global regulatory gene, *afsR2*, from *Streptomyces lividans* on avermectin production in *Streptomyces avermitilis.* Biosci Bioeng 89:606–608
189. Ikeda H, Takada Y, Pang C-H, Tanaka H, Omura S (1993) Transposon mutagenesis by Tn4560 and applications with avermectin-producing *Streptomyces avermitilis.* J Bacteriol 175:2077–2082
190. Stutzman-Engwall K, Conlon S, Fedechko R, McArthur H, Pekrun K, Chen Y, Jenne S, La C, Trinh N, Kim S, Zhang Y-X, Fox R, Gustafsson C, Krebber A (2005) Semi-synthetic DNA shuffling of *aveC* leads to improved industrial scale production of doramectin by *Streptomyces avermitilis.* Metab Eng 7:27–37
191. Zhang X, Chen Z, Li M, Wen Y, Song Y, Li J (2006) Construction of ivermectin producer by domain swaps of avermectin polyketide synthase in *Streptomyces avermitilis.* Appl Microbiol Biotechnol 72:986–994
192. McArthur HIA (1998) The novel avermectin, Doramectin—a successful application of mutasynthesis. In: Hutchinson CR, McAlpine J (eds) Developments in industrial microbiology-BMP 97. Society for Industrial Microbiology, Fairfax, pp 43–48
193. Vezina C, Kudelski A, Sehgal SN (1975) Rapamycin (AY 22,989), a new antifungal antibiotic. I. Taxonomy of the producing streptomycete and isolation of the active principle. J Antibiot 28:721–726
194. Kino T, Hatanaka H, Hashimoto M, Nishiyama M, Goto T, Okuhara M, Kohsaka M, Aoki H, Imanaka H (1987) FK-506, a novel immunosuppressant isolated from *Streptomyces.* I. Fermentation, isolation and physico-chemical and biological characteristics. J Antibiot 40:1249–1255
195. Jung S, Moon S, Lee K, Park Y-J, Yoon S et al (2009) Strain development of *Streptomyces* sp. for tacrolimus production using sequential adaptation. J Ind Microbiol Biotechnol 36:1467–1471
196. Chen X, Wei P, Fan L, Yang D, Zhu X, Shen W et al (2009) Generation of high-yield rapamycin-producing strains through protoplast-related techniques. Appl Microbiol Biotechnol 83:507–512
197. Kuscer E, Coates N, Challis I, Gregory M, Wilkinson B, Sheridan R, Petkovic H (2007) Roles of *rapH* and *rapG* in positive regulation of rapamycin biosynthesis in *Streptomyces hygroscopicus.* J Bacteriol 189:4756–4763
198. Jin ZH, Xu B, Lin SZ, Jin QC, Cen PL (2009) Enhanced production of spinosad in *Saccharopolyspora spinosa* by genome shuffling. Appl Biochem Biotechnol 159:655–663
199. Leader B, Baca QJ, Golan DE (2008) Protein therapeutics: a summary and pharmacological classification. Nat Rev Drug Discov 7:21–39
200. Demain AL, Vaishnav P (2009) Production of recombinant proteins by microbes and higher organisms. Biotechnol Adv 27:297–306
201. Rayder RA (2008) Expression systems for process and product improvement. Bioprocess Int 6:4–9
202. Choi BK, Bobrowicz P, Davidson RC, Hamilton SR, Kung DH, Li H et al (2003) Use of combinatorial genetic libraries to humanize N-linked glycosylation in the yeast *Pichia pastoris.* Proc Natl Acad Sci USA 100:5022–5027
203. Yuan L, Kurek I, English J, Keenan R (2005) Laboratory-directed protein evolution. Microbiol Mol Biol Rev 69:373–392
204. Terpe K (1996) Overview of bacterial expression systems for heterologous protein production: from molecular and biochemical fundamentals to commercial systems. Appl Microbiol Biotechnol 72:211–223
205. Swartz J (1996) *Escherichia coli* recombinant DNA technology. In: Neidhardt FC (ed) *Esherichia coli* and *Salmonella*: cellular and molecular biology, 2nd edn. American Society of Microbiology, Washington, DC, pp 1693–1771
206. Wong MS, Wu S, Causey TB, Bennett GN, San K-Y (2008) Reduction of acetate accumulation in *Escherichia coli* cultures for increased recombinant protein production. Metab Eng 10:97–108
207. Morrow KJ (2009) Grappling with biologic manufacturing concerns. Genet Eng Biotechnol News 29(5):54–55
208. Choi JH, Lee SJ, Lee SJ, Lee SY (2003) Enhanced production of insulin-like growth factor I fusion protein in *Escherichia coli* by coexpression of the down-regulated genes identified by transcriptome profiling. Appl Environ Microbiol 69:4737–4742
209. Barnard GC, Henderson GE, Srinivasan S, Gerngross TU (2004) High level recombinant

protein expression in *Ralstonia eutropha* using T7 RNA polymerase based amplification. Protein Expr Purif 38:264–271
210. Squires CH, Lucy P (2008) Vendor voice: a new paradigm for bacterial strain engineering. Bioprocess Int 6:22–27
211. Sreekrishana K, Nelles L, Potenz R, Cruze J, Mazzaferro P et al (1989) High level expression, purification, and characterization of recombinant human tumor necrosis factor synthesized in the methylotrophic yeast *Pichia pastoris*. Biochemistry 28:4117–4125
212. Werten MWT, van den Bosch TJ, Wind RD, Mooibroek H, De Wolf FA (1999) High-yield secretion of recombinant gelatins by *Pichia pastoris*. Yeast 15:1087–1096
213. Clare JJ, Rayment FB, Ballantine SP, Sreekrishna K, Romanos MA (1991) High level expression of tetanus toxin fragment C in *Pichia pastoris* strains containing multiple tandem integrations of the gene. Nat Biotechnol 9:455–460
214. Nevalainen KMH, Te'o VSJ, Bergquist PL (2005) Heterologous protein expression in filamentous fungi. Trends Biotechnol 23: 468–474
215. Morrow KJ (2007) Strategic protein production. Genet Eng Biotechnol News 27:50–54
216. Gellison G, Janowicz ZA, Weydemann U, Melber K, Strasser AWM, Hollenberg CP (1992) High-level expression of foreign genes in *Hansenula polymorpha*. Biotechnol Adv 10:179–189
217. Meyer V (2008) Genetic engineering of filamentous fubgi—progress, obstacles and future trends. Biotechnol Adv 26:177–185
218. Lubertozzi D, Keasling JD (2009) Developing *Aspergillus* as a host for heterologous expression. Biotechnol Adv 27:53–75
219. Van Hartinsveldt W, van Zeijl CM, Harteeld GM, Gouka RJ, Suykerbuyk M, Luiten RG et al (1993) Cloning, characterization and overexpression of the phytase-encoding gene (phyA) of *Aspergillus niger*. Gene 127:87–94
220. Ward PP, Piddlington CS, Cunningham GA, Zhou X, Wyatt RD, Conneely OM (1995) A system for production of commercial quantities of human lactoferrin: a broad spectrum natural antibiotic. Nat Biotechnol 13:498–503
221. Christensen T, Woeldike H, Boel E, Mortensen SB, Hjortshoej K, Thim L et al (1988) High level expression of recombinant genes in *Aspergillus oryzae*. Nat Biotechnol 6:1419–1422
222. Headon DR, Wyatt RD (1995) Human lactoferrin from *Aspergillus* spp. SIM News 45:113–117
223. Verdoes JC, Punt PJ, Burlingame R, Bartels J, van Dijk R, Slump E, Meens M, Joosten R, Emalfarb M (2007) A dedicated vector for efficient library construction and high throughput screening in the hyphal fungus *Chrysosporium lucknowense*. Ind Biotechnol 3:48–57
224. Andersen DC, Krummen L (2002) Recombinant protein expression for therapeutic applications. Curr Opin Biotechnol 13:117–123
225. Wrotnowski C (1998) Animal cell culture; novel systems for research and production. Genet Eng News 18(3):13–37
226. Griffin TJ, Seth G, Xie H, Bandhakavi S, Hu W-S (2007) Advancing mammalian cell culture engineering using genome-scale technologies. Trends Biotechnol 25:401–408
227. Decaria P, Smith A, Whitford W (2009) Many considerations in selecting bioproduction culture media. Bioprocess Int 7:44–51
228. Scott C, Montgomery SA, Rosin LJ (2007) Genetic engineering leads to microbial, animal cell, and transgenic expression systems. BIO Internat Convention, pp. 27–34
229. Morrow KJ (2007) Improving protein production processes. Genet Eng Biotechnol News 27:44–47
230. Ryll T (2008) Antibody production using mammalian cell culture—how high can we push productivity? SIM Annual Meeting Program & Abstract, San Diego, S146, p.101
231. Meyer HP, Biass J, Jungo C, Klein J, Wenger J, Mommers R (2008) An emerging star for therapeutic and catalytic protein production. Bioprocess Int 6:10–21
232. CocoMartin JM, Harmsen MM (2008) A review of therapeutic protein expression by mammalian cells. Bioprocess Int 6:28–33
233. Jarvis LM (2008) A technology bet. DSM's pharma product unit leverages its biotech strength to survive in a tough environment. Chem Eng News 86:30–31
234. Agathos SN (1991) Production scale insect cell culture. Biotechnol Adv 9:51–68
235. Luckow VA, Summers MD (1988) Trends in the development of baculovirus expression vectors. Nat Biotechnol 6:47–55
236. Miller LK (1988) Baculoviruses as gene expression vectors. Annu Rev Microbiol 42:177–199
237. Wilkinson BE, Cox M (1998) Baculovirus expression system: the production of proteins for diagnostic, human therapeutic or vaccine use. Genet Eng News 18, 35(Nov)
238. Maiorella B, Harano D (1988) Large scale insect cell culture for recombinant protein production. Nat Biotechnol 6:1406–1409

239. Morrow KJ Jr (2007) Improving protein production processes. Genet Eng News 27(5):50–54
240. Knight P (1991) Baculovirus vectors for making proteins in insect cells. ASM News 57:567–570
241. Falch E (1991) Industrial enzymes—developments in production and application. Biotechnol Adv 9:643–658
242. Cowan D (1996) Industrial enzyme technology. Trends Biotechnol 14:177–178
243. Vohra A, Satyanarayana T (2003) Phytases: microbial sources, production, purification, and potential biotechnological applications. Crit Rev Biotechnol 23:29–60
244. Vaishnav P, Demain AL (2009) Industrial biotechnology overview. In: Schaechter M, Lederberg J (eds) Encyclopedia of microbiology, 3rd edn. Elsevier, Oxford, p 335
245. Wackett LP (1997) Bacterial biocatalysis: stealing a page from nature's book. Nat Biotechnol 15:415–416
246. Palva I (1982) Molecular cloning of α-amylase gene from *Bacillus amyloliquefaciens* and its expression in *Bacillus subtilis.* Gene 19: 81–87
247. O'Neill GP, Kilburn DG, Warren RAJ, Miller RC (1986) Overproduction from a cellulase gene with a high guanosine-plus-cytosine content in *Escherichia coli.* Appl Environ Microbiol 52:737–743
248. Shoemaker S, Schweickart V, Ladner M, Gelfand D, Kwok S, Myambo K et al (1983) Molecular cloning of exo-cellobiohydrolase I derived from *Trichoderma reesei* strain L27. Nat Biotechnol 1:691–696
249. Van Brunt J (1986) Fungi: the perfect hosts? Biotechnology 4:1057–1062
250. Mondou F, Shareck F, Morosoli R, Kleupfel D (1986) Cloning of the xylanase gene of *Streptomyces lividans.* Gene 49:323–329
251. Van den Burg B, Vriend G, Veltman O, Venema G, Eijsink VGH (1998) Engineering an enzyme to resist boiling. Proc Natl Acad Sci USA 95:2056–2060
252. Ness JE, Welch M, Giver L, Bueno M, Cherry JR, Borchert TV et al (1999) DNA shuffling of subgenomic sequences of subtilisin. Nat Biotechnol 17:893–896
253. Jaeger KE, Reetz MT (2000) Directed evolution of enantioselective enzymes for organic chemistry. Curr Opin Chem Biol 4:68–73
254. Suenaga H, Mitsokua M, Ura Y, Watanabe T, Furukawa K (2001) Directed evolution of biphenyl dioxygenase: emergence of enhanced degradation capacity for benzene, toluene, and alkylbenzenes. J Bacteriol 183: 5441–5444
255. Song JK, Rhee JS (2001) Enhancement of stability and activity of phospholipase A(1) in organic solvents by directed evolution. Biochim Biophys Acta 1547:370–378
256. Raillard S, Krebber A, Chen Y, Ness JE, Bermudez E, Trinidad R et al (2001) Novel enzyme activities and functional plasticity revealed by recombining highly homologous enzymes. Chem Biol 8:891–898
257. Kurtzman AL, Govindarajan S, Vahle K, Jones JT, Heinrichs V, Patten PA (2001) Advances in directed protein evolution by recursive genetic recombination: applications to therapeutic proteins. Curr Opin Biotechnol 12:361–370
258. Marshall SH (2002) DNA shuffling: induced molecular breeding to produce new generation long-lasting vaccines. Biotechnol Adv 20:229–238
259. Locher CP, Soong NW, Whalen RG, Punnonen J (2004) Development of novel vaccines using DNA shuffling and screening strategies. Curr Opin Mol Ther 6:34–39
260. Tobin MB, Gustafsson C, Huisman GW (2000) Directed evolution: the 'rational' basis for 'irrational' design. Curr Opin Struct Biol 10:421–427
261. Crameri A, Whitehorn A, Stemmer WPC (1996) Improved green fluorescent protein by molecular evolution using DNA shuffling. Nat Biotechnol 14:315–319

Chapter 2

Microbial Carotenoids

Preejith Vachali, Prakash Bhosale, and Paul S. Bernstein

Abstract

Carotenoids are among the most widely distributed pigments in nature, and they are exclusively synthesized by plants and microorganisms. These compounds may serve a protective role against many chronic diseases such as cancers, age-related macular degeneration, and cardiovascular diseases and also act as an excellent antioxidant system within cells. Recent advances in the microbial genome sequences and increased understanding about the genes involved in the carotenoid biosynthetic pathways will assist industrial microbiologists in their exploration of novel microbial carotenoid production strategies. Here we present an overview of microbial carotenogenesis from biochemical, proteomic, and biotechnological points of view.

Key words: Carotenoids, Carotene, Xanthophylls, Biosynthesis, Carotenogenesis, *Crt* genes

1. Introduction

Carotenoids are naturally occurring terpenoid pigments, consisting of isoprene residues and a polyene chain of conjugated double bonds. These pigments are responsible for the wide variety of orange-red colors seen in nature that absorb light in the wavelength range of 300–600 nm. The absorbance is directly related to the number of conjugated double bonds and functional groups present in the structure. Structurally and functionally, carotenoids can be broadly classified into hydrocarbons (HC, carotenes) and oxygenated derivatives (xanthophylls). These molecules are formed by the head to head condensation of two geranylgeranyl diphosphate molecules (GGDP) (C20 HC), which results in a basic symmetrical acyclic C40 HC structure called phytoene ($C_{40}H_{56}$) (1). The remainder of the natural carotenoids is derived from this basic molecule by a variety of biochemical reactions, mainly mediated by enzymes in plants and microorganisms. Animals are usually the dietary recipients at the other end of the food chain and

José-Luis Barredo (ed.), *Microbial Carotenoids From Fungi: Methods and Protocols*, Methods in Molecular Biology, vol. 898,
DOI 10.1007/978-1-61779-918-1_2,

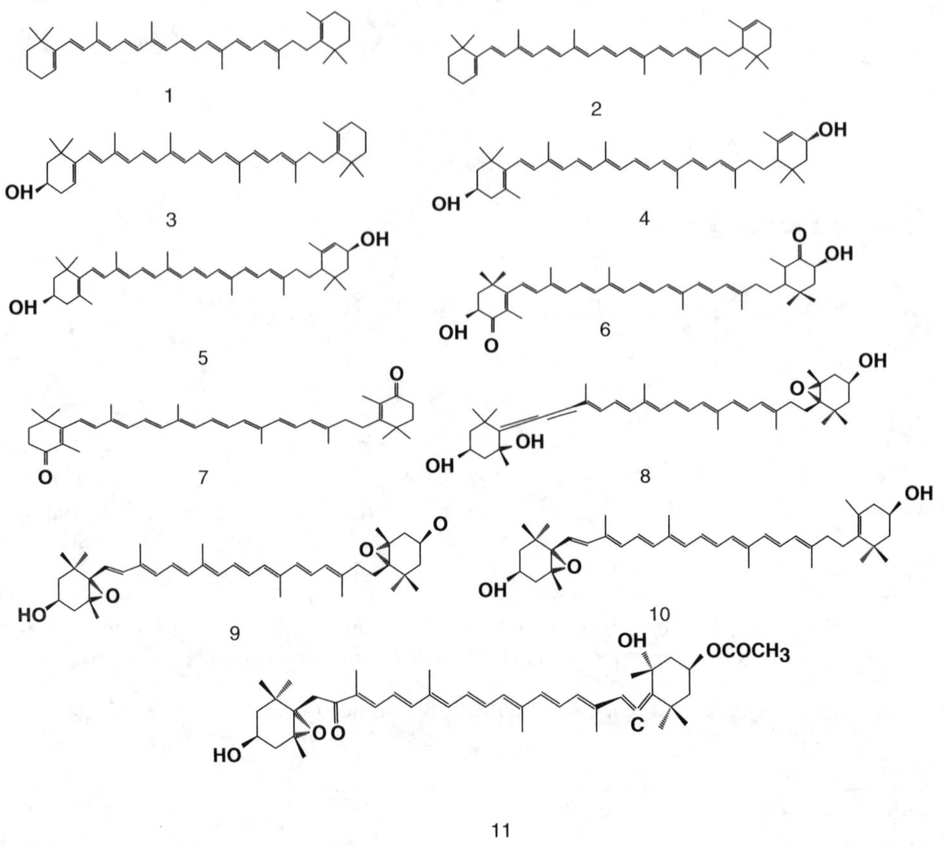

Fig. 1. Structures of important carotenoids produced by microorganisms. (1) β-Carotene, (2) α-Carotene, (3) β-Cryptoxanthin, (4) Lutein, (5) Zeaxanthin, (6) Astaxanthin, (7) Canthaxanthin, (8) Neoxanthin, (9) Violaxanthin, (10) Antheraxanthin, and (11) Fucoxanthin.

have limited ability to metabolically transform carotenoids (2, 3). Figure 1 shows the chemical structures of important carotenoids.

Carotenoid biosynthesis likely originated in ancient anoxygenic photosynthetic microorganisms. As these microbes evolved, carotenoid biosynthetic pathways have also branched out, resulting in structurally different carotenoids. The earliest record of oxygenic microbial carotenoid biosynthesis is attributed to cyanobacteria, which can be traced back 3.5 billion years based on fossil and molecular evidences (4). Prochlorophytes, another group of oxygenic photosynthetic bacteria, are reported to have carotenoid pigments similar to algae and eukaryotic plants, suggesting evolutionary relations with these groups (3). Single or multiple endosymbiotic associations with cyanobacteria or purple photosynthetic bacteria might be responsible for the chloroplast and pigment diversity in higher organisms (5). The change in the earth's atmosphere from anaerobic to aerobic condition is a major environmental factor responsible for the biosynthesis of structurally diverse carotenoids by oxygen-dependent enzymes (3).

Carotenoids are molecules of great interest in many scientific disciplines because of their unique properties, wide distribution and diverse functions (6). In photosynthetic organisms, carotenoids serve as light harvesting pigments. In many organisms, their major role is to act as an antioxidant by neutralizing free radicals and thereby preventing potential oxidative damage to the cells (3, 4, 7–15). Liu et al. demonstrated that a *Staphylococcus aureus* mutant with disrupted carotenoid biosynthesis is more susceptible to oxidant killing, indicating that carotenoids could act as a virulence factor (16). Carotenoids could reduce the penetration of singlet oxygen by decreasing membrane fluidity (17). Recent reports by Kamila et al. showed that polar carotenoids such as zeaxanthin could mediate transmembrane proton transfer in vivo (18).

The health benefits of carotenoids are becoming increasingly evident. Carotenoids play an important role as pro-vitamin A compounds. Of the more than 600 carotenoids that have been identified, approximately 30–50 are believed to have vitamin A activity. The most well-known compounds of this group are β-carotene and α-carotene. Beyond their pro-vitamin A role, epidemiological evidence and experimental results suggest antioxidant functions of dietary carotenoids that can prevent onset of many diseases such as arteriosclerosis, cataracts, age-related macular degeneration, multiple sclerosis, bone abnormalities, and most importantly, cancers, each of which may be initiated by free radical damage (19–24). Currently, a majority of the carotenoids available on the global market is produced by multi-step chemical synthesis or by solvent-based chemical extraction from their nonmicrobial natural sources (25–27).

There are very few carotenoid-based products on the market that originate from microbial sources. The two most prominent microalgal sources of carotenoids which have sustained competition from synthetic manufactures are β-carotene from *Dunaliella salina* and astaxanthin from *Haematococcus pluvialis* (28–32). There has been escalating interest in the microbial sources of other carotenoids with significant health benefits, which could be attributed to consumer preferences for natural additives over synthetic sources and also to the potential cost effectiveness of production by industrial microbial biotechnology (2, 6, 14, 33–35).

2. Microbial Carotenogenesis

Microbial carotenoid biosynthesis is a well-regulated mechanism which is dependent on the biochemical makeup of the microorganism's environmental conditions and cultural stress incurred during growth (6, 36). Several microbial biosynthetic pathways have been proposed and experimentally confirmed in the last five decades by well-known carotenoid researchers (36–46).

The evolution of carotenoid biosynthetic pathways is a continuous process and extends beyond designated carotenoid biosynthetic pathways (47). Generally, carotenoid biosynthesis starts with the bioprecursor called isopentenyl pyrophosphate (IPP) (48–50). The biosynthetic pathway to IPP known as the mevalonic acid (MVA) pathway starts from the key precursor acetyl-CoA. The first step in the MVA pathway involves the conversion of acetyl-CoA to 3-hydroxy-3-methyl glutaryl CoA (HMG-CoA) catalyzed by HMG-CoA synthase. HMG-CoA gets converted into a C6 compound MVA. MVA is then converted into a C5 IPP by a succession of reactions involving phosphorylation by MVA kinase followed by decarboxylation (51, 52).

The condensation of one molecule of dimethylallyl diphosphate (DMADP) and three molecules of IPP by pyrenyltransferase produces a C20 diterpene GGDP compound. Two molecules of GGDP condense head to head to form the first colorless carotenoid, phytoene (15-*cis*-7, 8, 11, 12, 7′, 8′, 11′, 12′-octahydro-ψ, ψ-carotene). Subsequent desaturation of phytoene results in the formation of C40 acyclic carotenoids such as neurosporene or lycopene. The desaturation process is a multistep process which varies between microbes. The most common of all is the four step desaturation process which leads to the formation of lycopene via intermediate steps of phytofluene and 3,4-didehydrolycopene. However, in some purple photosynthetic bacteria such as the *Rhodobacter* species, neurosporene, spheroidene, and hydroxy-spheroidene are produced as final products of the desaturation process (53).

All *trans*-lycopene acts as the precursor for many commercially important acyclic and cyclic carotenoids and oxygenated carotenoids, which are also called xanthophylls. The formation of carotene(s) such as β-carotene is considered to be the most common step in microbial carotenogenesis. Desaturated lycopene gets cyclized at both ends and forms a β-carotene or α-carotene molecule. These reactions are catalyzed by a β- or ε-cyclase in some of the green alga (54). The formation of xanthophylls involves sequential oxidations of post-carotene molecules yielding -hydroxy, -epoxy, and -oxo groups. The introduction of hydroxyl (–OH) groups at the positions of C3 and C3′ of the ionone rings leads to the formation of zeaxanthin and lutein, which are C3, C3′-dihydroxy derivatives of β,β-carotene and β,ε-carotene, respectively (55). A monohydroxy carotenoid, β-cryptoxanthin, acts as an intermediate in the biosynthesis of the dihydroxy carotenoids such as zeaxanthin. The carotenoids with keto (C=O) functional groups such as canthaxanthin and astaxanthin are formed by the introduction of keto groups at C4 and C4′ with or without hydroxylation at C3 and C3′. The formation of keto carotenoids from β-carotene has been well studied in algae, yeast, and nonphotosynthetic bacteria (55, 56). Violaxanthin, neoxanthin, and fucoxanthin were formed by

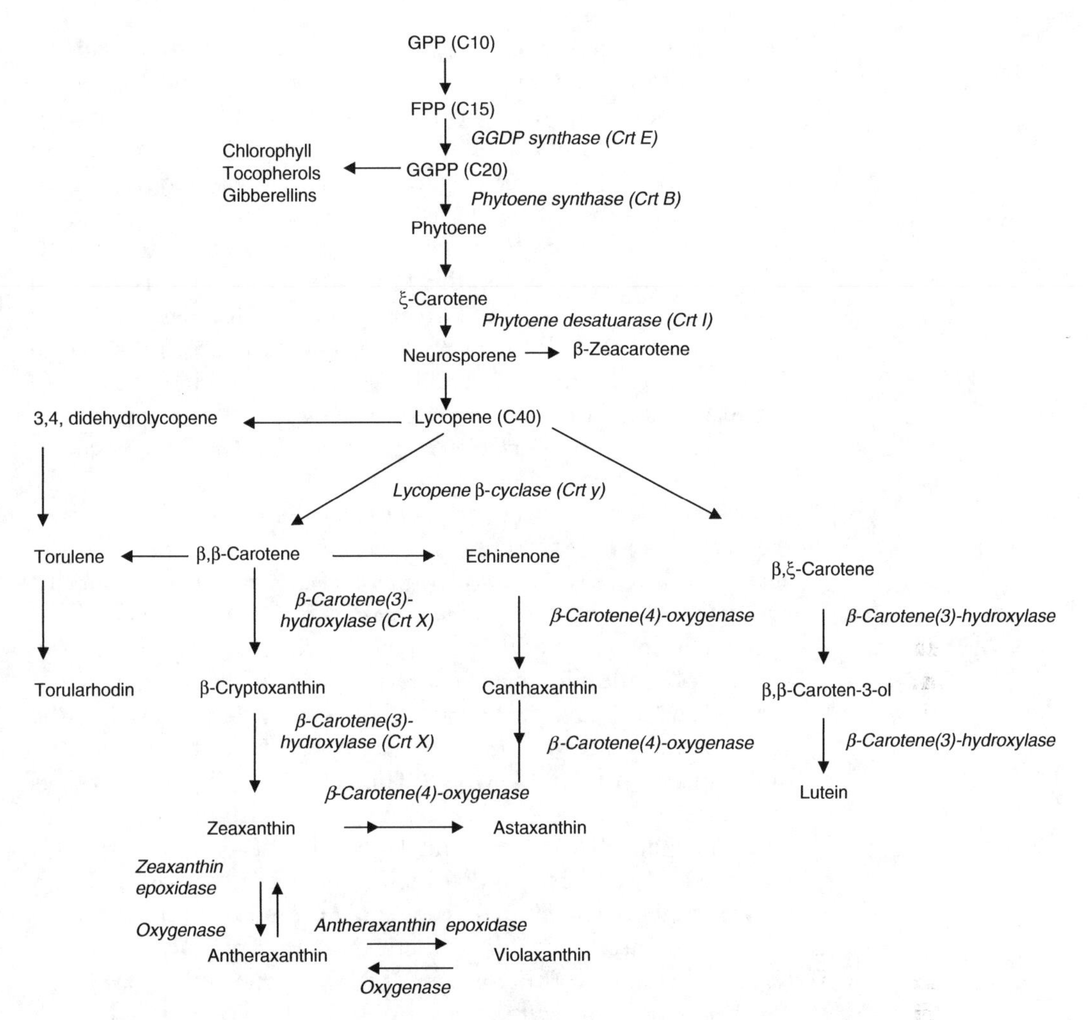

Fig. 2. Carotenoid biochemical pathway compiled from several reports on microbial production (1, 2, 6, 36, 38, 39, 50, 61, 65, 101, 104).

epoxidation at the 5, 6 and 5′, 6′ positions (55). Antia et al. and Yamamoto et al. studied the interconversion of violaxanthin, antheraxanthin, and zeaxanthin in green algae through epoxidation (2, 57, 58). Figure 2 represents the schematics of microbial carotenogenesis.

3. Proteomics of Microbial Carotenogenesis

Most of the enzymes involved in the MVA pathway are soluble proteins, whereas the enzymes involved in the later steps of the carotenogenic pathway are mostly membrane-bound proteins. The lipophilic products of carotenoid pathways partition to cytoplasmic or organelle membranes (59). Bacteria and yeast contain many isoprenoid compounds such as dolichols, quinones, and ergosterols (60, 61). However, farnesyl pyrophosphate (FPP) was identified as the first substrate encoded by the *crt* genes cluster, which is a common precursor for many carotenoid biosynthetic pathways (62, 63). Misawa et al. suggested that by incorporating carotenogenic genes into these systems, it might be possible to partially direct the carbon flux for the biosynthesis of these isoprenoid compounds to the pathway for carotenoid production (63). They also proposed that elucidation of *crt* gene structure and function would help carotenoid researchers to explore the biosynthesis of the main carotenoids. This could be achieved by the appropriate combinations of the *Erwinia crt* genes using FPP as the precursor in microbial systems (63).

Much later, Markus et al. 2002 (64) cloned and characterized the genes coding for all the enzymes involved in the conversion of acetyl-CoA to farnesyl diphosphate (FPP) in the zeaxanthin-producing bacterium *Paracoccus zeaxanthinifaciens.* They identified two genes encoding enzymes catalyzing the condensation of two acetyl-CoA molecules to acetoacetyl-CoA. The gene cluster named the mevanolate operon regulates the six enzymes involved in the conversion of acetyl-CoA and acetoacetyl-CoA to isopentenyl diphosphate (IPP) and DMAPP. The genes encoding the enzymes catalyzing two consecutive condensations, IPP and DMAPP to geranyl diphosphate (GPP) and IPP and GPP to FPP, were also identified. It was reported that these genes were not clustered with any other genes encoding an enzyme of the isoprenoid pathway.

In the carotenoid biosynthetic pathway, phytoene synthase (*crtB*), phytoene desaturase (*crtI*), and lycopene cyclase (*crtY*) are the three essential enzymes responsible for the biosynthesis of both acyclic and cyclic carotenoids. In the early 1990s, Sandmann's group identified and mapped the major *Erwinia herbicola* genes responsible for the cyclization and hydroxylation of carotenoids (36). Much later, Phadwal reviewed the molecular phylogenies of the *crt* genes involved early in the carotenoid pathway (65). The work summarized the phylogenetic evolution of phytoene synthase (*crtB*), phytoene desaturase (*crtI*), and lycopene cyclase (*crtY*) among bacteria and their functional significance in microbial carotenogenesis. Acyclic hydroxyl carotenoids from the carotene hydratase (*crtC*) and carotene 3,4-desaturase (*crtD*) genes of *Rhodobacter* were also summarized by Steiger et al. (66). Several enzymes, such as carotene hydroxylases (*crtZ*), carotene oxygenase (*crtW*), carotene isomerase (*crtH*), and so on, have been linked with bacterial carotenogenesis. However, recent work shows that carotene hydroxylases (*crtZ*) and its cDNA sequence were also isolated from green alga such as *H. pluvialis.* The carotene hydroxylase isolated from flavobacterim was pursued for the biosynthesis of unique hydroxyl carotenoids (67). The hydroxylation reaction among carotenoid producing microbes is considered common phenomena, but ketolation is quite limited, and is characteristic of selected microbes. β-Carotene ketolase enzymes (*crtW* and *crtO*) have been identified from methylotrophic bacterium such as *Methylomans*. The incorporation of the genes encoding these enzymes has resulted in a higher percentage of canthaxanthin in these strains (68). Recombinant DNA technology has enabled biotechnologists to make new carotenoids for a variety of uses (69).

4. Biotechnology of Microbial Carotenoids

Several studies have been conducted highlighting the importance of carotenoids in healthcare as well as the nutraceutical and food industries (2, 8, 34, 70–72). The market demand for carotenoids continue to rise as more and more clinical research studies surface, revealing the various health benefits of carotenoids (33, 34, 73–76). Of the 600 naturally occurring carotenoids, only a few have proven useful in human- and animal-based industries, and these have primarily received focus on their abilities to act as antioxidants and light-screening ingredients.

The carotenoid market is expected to increase to $919 million by 2015 with an annual growth rate of 2.3% (77). Most of these carotenoids are available from synthetic sources; however, synthetic pigments have been perceived to cause hazardous effects to human health at high dose ranges and have been subsequently warned by the Food and Drug Administration (78). This resulted in the hunt for a process of pigment production by alternative natural sources (79). There is a growing demand for microbial sources of pigments as an alternative.

Currently, β-carotene and astaxanthin are industrially produced from microbial sources and are widely used in food and feed industries (2, 80, 81). The major limitation on the use of microbial systems for commercial production is the low yield, slow growth, and high production cost compared with chemical synthesis. However, strain improvement strategies such as optimization of growth conditions and preparation of mutants in conjunction with metabolic engineering techniques could improve carotenoid productivity. In the following section, we discuss the major utilities and available microbial sources of some commercially important carotenoids.

4.1. Astaxanthin (3S,3S′-Dihydroxy-β,β-carotene-4,4′-dione)

Astaxanthin is the most commonly occurring red carotenoid in marine and aquatic animals (82). It is responsible for the pink color of salmon flesh and also gives coloration to crustacean shells. It is known to scavenge free radicals and quench singlet oxygen (83, 84). It can enhance the immune system (85) protect skin from radiation injury, cancer (86, 87) and block reactions induced by other chemicals and toxins. It is widely used in the pharmaceutical, cosmetic, and nutraceutical industries (88). Two major microorganisms, which have been commercially exploited for astaxanthin production, are microalga *H. pluvialis*, and heterobasidiomycetous yeast *Xanthophyllomyces dendrorhous* (2, 89–91).

Recently, de la Fuente et al. reported an improved semi-industrial process for astaxanthin production by the fermentation of *X. dendrorhous.* A volumetric yield of 350 mg/L astaxanthin was reported with 800-L scale (92). Although *X. dendrorhous* has been studied by various researchers for the past three decades, it still attracts interest in various biotechnological industries (89, 92, 93).

Alternately, the freshwater unicellular alga, *H. pluvialis*, accumulates astaxanthin in its aplanospores under stressful conductions and is considered to be one of the richest sources of this carotenoid. Ranjbar et al. reported photoautotrophic conditions in a bubble column with the fed-batch addition of nutrients for the production of astaxanthin by *H. pluvialis.* A combination of the fed-batch addition of nutrients and dilution of broth for nutrient deficiency was proposed as the most promising method for attainment of high cell and astaxanthin concentrations in a bubble column photo-bioreactor. The final concentration of astaxanthin was reported to be 390 mg/L which was several times higher than anything ever previously reported (94). Sandesh Kamath et al. reported a 23–59% increase in the total carotenoid and astaxanthin contents by implementing a strain improvement strategy for *H. pluvialis* with chemical and UV mutation (35).

Apart from these two major carotenoid producing microorganisms, *Brevibacterium linens* (95) and *Agrobacterium aurantiacum* (96), the marine bacterium *Paracoccus haeundaensis* (97) and *Mycobacterium lacticola* (98) are also reported to produce astaxanthin, but are not considered commercially significant sources.

4.2. β-Carotene (β, β-Carotene)

β-Carotene is an important compound because of its role as an antioxidant, and as precursor of vitamin A in food and feed products (99, 100). In 2004, the worldwide market value of β-carotene was US $242 million and was proposed to reach US $253 million by 2009 (77, 101). Currently, more than 90% of commercialized β-carotene is produced through chemical synthesis (102).

β-Carotene is produced primarily by microalgae, fungi and yeasts, as well as some species of bacteria and lichens (103, 104). Commercially available β-carotene is produced mainly from the genus *Dunaliella* (101, 105). Since 1980, *Dunaliella* powder and extracts (yielding dried biomass and natural β-carotene) have been available in Israel, China, the USA, Australia, and Mexico (3, 101, 106).

Beside *Dunaliella*, the greatest yields have been obtained by the mating of (+) and (−) strains of *Blakeslea trispora* (103) resulting in yields that are comparable to those of chemical processing (107).

The US Department of Agriculture's process reported a yield of 17 mg of β-carotene per gram of mycelium and recently improved to about 30 mg of β-carotene per gram of mycelium and about 3 g/L (107). *Phycomyces* is used as a model system by many researchers to study the regulation of the biosynthesis of the pigment β-carotene in fungi as it accumulates β-carotene, at a lower extent than *B. trispora*, in the lipid globules of their mycelia (108). Sang-Hwal Yoon et al. reported a novel approach by the combinatorial expression of the whole bacterial mevalonate pathway for the production of β-carotene in *Escherichia coli*. The recombinant *E. coli* DH5α harboring the whole MVA pathway and β-carotene synthesis genes produced a β-carotene yield of 465 mg/L at a glycerol concentration of 2% (w/v) (102). However, the regulatory concern surrounding the use of recombinant strains is still a major roadblock for the success of microbial biotechnology.

4.3. β-Cryptoxanthin (3R-β, β-Caroten-3-ol)

β-Cryptoxanthin is a xanthophyll carotenoid with the potential to act as provitamin A and has been reported to improve bone health. A study conducted in experimental rats demonstrated that β-cryptoxanthin also stimulates unique anabolic bone calcification (109). Yamaguchi et al. (75) claimed that β-cryptoxanthin, which may promote osteogenesis, could be of value as an active ingredient in the treatment of bone diseases. Recent studies also indicate that β-cryptoxanthin is highly preventative against prostate cancer (76, 110), lung cancer (111–113), colon cancer (114), and rheumatoid arthritis (115).

While there is currently little to no commercial demand for microbial β-cryptoxanthin, it holds great potential for future investigation, production, and use. Information on microbial production of β-cryptoxanthin is very limited, mainly due to the lack of proper microbial sources and feasible culture conditions (2, 116). *B. linens*, which is traditionally known for its cheese ripening process, accumulates β-cryptoxanthin in low amounts (117). *Flavobacterium lutescens* and *Flavobacterium multivorum* are reported to produce β-cryptoxanthin under optimized media conditions (116, 118). However, these microbial sources are not yet competitive in comparison to the naturally occurring citrus and capsicum based sources of β-cryptoxanthin (119).

4.4. Canthaxanthin (4,4′ Diketo-β-carotene)

Canthaxanthin is a diketo-carotenoid, which was first isolated from edible mushrooms (3). It is widely used as a colorant in food, feed additives for egg yolk, fish, and crustacean farms and also in the cosmetic industry as a tanning agent for human skin (2, 120). Canthaxanthin is reported to prevent UV-induced immune suppression in mice, protect against skin cancer in experimental animals and also to be useful in the treatment of skin diseases such as photodermatosis (2). In vitro studies demonstrated that canthaxanthin has greater antioxidant activity than its non-oxygenated analog such as β-carotene due to the presence of keto groups at the 4 and 4′ positions in the β-ionone ring (121). Only a few microbial sources of canthaxanthin have been reported. Askar et al. (122) identified an extremely halophilic bacteria, *Haloferax alexandrinus*, with a 0.69 mg/g cellular accumulation of canthaxanthin. Other bacterial producers of canthaxanthin include *Corynebacterium michiganense*, *Micrococcus roseus*, *Brevibacterium* sp. strain KY 4313, *Gordonia jacobaea*, and *Dietzia natronolimnaea* HS-1 (123). It was also discovered that various green microalgae such as *Chlorella pyrenoidosa*, *Chlorella zofingiensis*, *Chlorella emersonii*, and *Dictyococcus cinnabarinus* produce canthaxanthin under various growth conditions (123). Recently, it was reported that under submerged fermentation, a mutant strain of *Aspergillus carbonarius* produces canthaxanthin with a yield of 32 mg/g (124). The current market demand for canthaxanthin is fulfilled by chemical synthesis (125). However, much attention is being devoted by researchers to discovering a microbial source for canthaxanthin and developing novel production strategies for promising microbial strains (123, 126, 127).

4.5. Fucoxanthin (3′-(Acetyloxy)-6′,7′-didehydro-5,6-epoxy-5,5′,6,6′,7,8-hexahydro-3,5′-dihydroxy-8-oxo-β,β-carotene)

Fucoxanthin is a naturally occurring xanthophyll carotenoid found in brown algae and edible brown seaweeds that has fascinated the nutraceutical and food industries recently due to its unique health benefits such as antiobesity, antidiabetes, etc., which have not been reported with other carotenoids (74). Fucoxanthin intake has been shown to promote fat metabolism, particularly around the abdominal area (33, 128, 129). An antidiabetic effect of fucoxanthin was also reported; a 0.2% fucoxanthin supplementation decreased the blood glucose and plasma insulin concentrations in experimental mice (130). A study conducted in KKAy mice, a model for obese/type II diabetes, showed that fucoxanthin could enhance the amount of DHA in the liver of mice fed with soybean oil without direct fish oil supplementation (131). Fucoxanthin is shown to have anticancer properties (132, 133). Fucoxanthin isolated from brown algae has also been found to act as an anticoagulant (71), anti-inflammatory (34), antioxidant (73), and antimicrobial (134).

Undaria pinnatifida and *Lamaria* sp. are the most popular edible seaweeds in Japan and many other Southeast Asian countries, and fucoxanthin accounts for >10% of the total carotenoids in these seaweeds (74). Other brown sea weeds that have been reported to produce fucoxanthin include *Hijikia fusiformis*, *Ecklonia stolonifera*, and *Sargassum siliquastrum* (135–137).

4.6. Lutein [(3R,3′R,6′R)-β,ε-Carotene-3,3′-diol]

Lutein is one of the fastest growing carotenoids on the market (77). Currently, the natural commercial source of lutein is the solvent extract of marigold (*Tagetes erecta*) petals (125, 138). However, the lutein content of marigold petals is low, 0.03% dry wt., and it contains several esters with similar polarity, making them difficult to separate from each other (139).

In recent years, several microalgae have been studied as potential lutein sources, such as *Chlamydomonas reinhardtii* (140), *Muriellopsis* sp. (141), *Chlorella protothecoides* (142), and *Scenedesmus almeriensis* (139); however, microbial sources still lack commercial potential mainly due to lack of studies involving strain improvement and high-volume bioreactors.

4.7. Zeaxanthin [(3R,3′R)-β, β-Carotene-3,3′-diol]

Zeaxanthin is an isomer of lutein, and its commercial demand exists in parallel to lutein mainly in the ocular health market. Zeaxanthin coexists with its several optical isomers in natural sources such as corn, alfalfa, yellow peppers, egg yolks, and marigold flowers (143, 144). Among the microbial sources, marine bacterium *Flavobacterium* species are well documented for their zeaxanthin production (107, 145–148). Unlike lutein, which is typically present in photosynthetic microorganisms, zeaxanthin occurs in cyanobacteria (149) and also in some non-photosynthetic bacteria (143). In the non-photosynthetic bacteria, zeaxanthin sometimes presents in the form of glycoside esters (143, 150). Other microbial sources include *Dunaliella* sp., which produces zeaxanthin under various stress and gene manipulation conditions (151, 152) and *Microcystis aeruginosa* (153).

Lutein and zeaxanthin together have many potential uses in the pharmaceutical and nutraceutical industries. Various studies suggest that they play an important role in the prevention of cancer (154), age-related macular degeneration (AMD) (70, 155), and enhancement of immunity. They are also used as colorants in food and in the cosmetic industry (156).

Table 1 summarizes the major microbial producers of carotenoids vs. natural sources. In general, the microbial yield is quite low compared with other nonmicrobial sources such as plants or animals. Microbial biosynthesis has an economic niche for those carotenoids which have complex structures that make them difficult to synthesize chemically. With proper strain improvement strategies and fermentation technologies, it is possible to produce higher levels of pure and isomer-free carotenoids from microbial sources.

5. Conclusions

Microbial carotenogenesis is a well-studied phenomenon that has been, and will continue to be, researched for years for its regulation and functionality using several biochemical approaches. Over the years, numerous groups of microorganisms have been characterized by their specific compositions of carotenoids. The characterization of genes and proteins involved in the biochemical pathways coupled with the use of genetic and metabolic engineering tools to improve the selective pathways has directed the biotechnologist toward the

Table 1
Natural and microbial sources of carotenoids

Carotenoids	Natural sources[a]	Content[a]	Microbial sources	Yield[b]	References
Astaxanthin	Krill	120 mg/kg	*Xanthophyllomyces dendrorhous*	1,080 μg/g; 4.7 mg/g dry cell matter (420 mg/L)	(6, 92, 94)
	Arctic Shrimp	1.2 g/kg	*Haematococcus pluvialis*	350 mg/L	
	Adonis annua	300–500 mg/kg		390 mg/L	
				22.7 mg/g	
β-Carotene	Carrots	183 μg/kg	*Blakeslea trispora*	420 μg/g	(6)
	Mango	131 μg/kg	*Dunaliella salina*	10.35 mg/L	(143)
	Sweet potato	95 μg/kg	*Streptomyces chrestomcyeticus subsp. Rubescens*	NG	(104)
	Pumpkin	69 μg/kg	*Rhodotorula glutinis mutant 32*	250 mg/L	
	Apricots	26 μg/kg			
	Cantaloupe	16 μg/kg			
β-Cryptoxanthin	Red bell peppers	22,050 μg/kg	*Brevibacterium linens*	0.3 μg/mL	(2)
	Mango	15,500 μg/kg	*Flavobacterium lutescens*	770 mg/kg	(119)
	Papaya	2,250 μg/kg			(116)
	Cilantro	4,040 μg/kg			
	Oranges	3,240 μg/kg			
	Corn	1,190 μg/kg			
	Watermelon	1,030 μg/kg			

Fucoxanthin	Edible Brown Sea Vegetable (Kelp)	3.5 g/kg	*Undaria pinnatifida*	30 g/kg	(157)
			Sargassum fusiforme	79.5 g/kg	(157)
			Laminaria japonica	122.1 μg/g	(158)
Lutein	Corn	12,720 μg/kg	*Chlorella zofingiensis*	21 μg/mL	(2)
	Marigold flower	0.3 g/kg	*Chlorella protothecoides CS-41*	225 μg/mL	(27)
	Collard	78,250 μg/kg	*Muriellopsis sp.*	35 μg/mL	(138)
	Kale	156,250 μg/kg	*Scenedesmus almeriensis*	0.7 g/L	(139)
	Spinach	11,607 μg/kg			
Zeaxanthin	Corn	5,280 μg/kg	*Dunaliella salina*	6 mg/g	(2)
	Collard	2,660 μg/kg	*Phormidium laminosum*	5.9 mg/mg	(144)
	Persimmon Japanese	4,880 μg/kg	*Flavobacterium multivorum*	10.65 μg/mL	(138)
	Spinach raw	3,310 μg/kg	*Microcystis aeruginosa*	0.962 mg/g	(153)

NG not given
[a]Adapted from Database,1998 (159)
[b]Cellular accumulation and volumetric data are added as reported

hyperproduction of carotenoids from microbial processes. Although the developed processes for microbial production of selected carotenoids appears to be very promising, gaining a deeper understanding and further development of the fermentation process in future years will be necessary before microbial sources become a realistic alternative to synthetic carotenoids.

Acknowledgments

This work was supported by National Institute of Health Grant EY-11600. We thank Kelly Nelson for critical reading of the manuscript.

References

1. Sandmann G (1994) Carotenoid biosynthesis in microorganisms and plants. Eur J Biochem 223:7–24
2. Bhosale P, Bernstein PS (2005) Microbial xanthophylls. Appl Microbiol Biotechnol 68:445–455
3. Johnson E, Schroeder W (1996) Microbial carotenoids. In: Fiechter A (ed) Advances in biochemical engineering/biotechnology. Springer, Berlin, pp 119–178
4. Liang C et al (2006) Carotenoid biosynthesis in cyanobacteria: structural and evolutionary scenarios based on comparative genomics. Int J Biol Sci 2:197–207
5. Gray MW (1989) The evolutionary origins of organelles. Trends Genet 5:294–299
6. Bhosale P (2004) Environmental and cultural stimulants in the production of carotenoids from microorganisms. Appl Microbiol Biotechnol 63:351–361
7. Cantrell A et al (2003) Singlet oxygen quenching by dietary carotenoids in a model membrane environment. Arch Biochem Biophys 412:47–54
8. Tinkler JH et al (1994) Dietary carotenoids protect human cells from damage. J Photochem Photobiol B 26:283–285
9. Truscott TG et al (1995) The interaction of carotenoids with reactive oxy-species. Biochem Soc Trans 23:252S
10. Krinsky NI et al (2003) Biologic mechanisms of the protective role of lutein and zeaxanthin in the eye. Annu Rev Nutr 23:171–201
11. Bohm F et al (1998) Enhanced protection of human cells against ultraviolet light by antioxidant combinations involving dietary carotenoids. J Photochem Photobiol B 44:211–215
12. Altermann W, Kazmierczak J (2003) Archean microfossils: a reappraisal of early life on Earth. Res Microbiol 154:611–617
13. Schopf JW (1993) Microfossils of the Early Archean Apex chert: new evidence of the antiquity of life. Science 260:640–646
14. Edge R et al (1997) The carotenoids as antioxidants—a review. J Photochem Photobiol B 41:189–200
15. El-Agamey A et al (2004) Are dietary carotenoids beneficial? Reactions of carotenoids with oxy-radicals and singlet oxygen. Photochem Photobiol Sci 3:802–811
16. Liu GY et al (2005) Staphylococcus aureus golden pigment impairs neutrophil killing and promotes virulence through its antioxidant activity. J Exp Med 202:209–215
17. Subczynski WK et al (1991) Effect of polar carotenoids on the oxygen diffusion-concentration product in lipid bilayers. An EPR spin label study. Biochim Biophys Acta 1068:68–72
18. Kamila K et al (2008) Can membrane-bound carotenoid pigment zeaxanthin carry out a transmembrane proton transfer? Biochim Biophys Acta 1778:2334–2340
19. Peto R et al (1981) Can dietary β-carotene materially reduce human cancer rates? Nature 290:201–208
20. Greenberg ER et al (1990) A clinical trial of beta-carotene to prevent basal-cell and squamous-cell cancers of the skin. N Engl J Med 323:789–795

21. Stahl W, Sies H (1996) Lycopene: a biologically important carotenoid for humans? Arch Biochem Biophys 336:1–9
22. Hennekens CH (1997) Beta-carotene supplementation and cancer prevention. Nutrition 13:697–699
23. Moeller SM et al (2000) The potential role of dietary xanthophylls in cataract and age-related macular degeneration. J Am Coll Nutr 19:522S–527S
24. Bone RA et al (2000) Lutein and zeaxanthin in the eyes, serum and diet of human subjects. Exp Eye Res 71:239–245
25. Haigh GW (1994) High purity beta-carotene. US Patent 5,310,554
26. Khachik F (2009) Process for isolation, purification, and recrystallization of lutein from saponified marigold oleoresin and uses thereof. US Patent RE409,31
27. Ausich RL, Sanders DJ (1997) Process for the formation, isolation and purification of comestible xanthophyll crystals from plants. Patent 5648564
28. Boussiba S, Vonshak A (2000) Procedure for large-scale production of astaxanthin from *Haematococcus*. US Patent 602,270,1
29. Venkatesh NS et al (2005) Medium for the production of betacarotene and other carotenoids from *Dunaliella salina* (ARL 5) and a strain of *Dunaliella salina* for production of carotenes using the novel media. US patent 693,645,9
30. Park EK et al (2001) Effects of medium compositions for the growth and the astaxanthin production of *Haematococcus pluvialis*. Sanop Misaengmul Hakhoechi 29:227–233
31. Orosa M et al (2001) Comparison of the accumulation of astaxanthin in *Haematococcus pluvialis* and other green microalgae under N-starvation and high light conditions. Biotechnol Lett 23:1079–1085
32. Orset S, Young AJ (1999) Low-temperature-induced synthesis of α-carotene in the microalga *Dunaliella salina* (chlorophyta). J Phycol 35:520–527
33. Nakazawa Y et al (2009) Comparative evaluation of growth inhibitory effect of stereoisomers of fucoxanthin in human cancer cell lines. J Funct Foods 1:88–97
34. Heo SJ et al (2010) Evaluation of anti-inflammatory effect of fucoxanthin isolated from brown algae in lipopolysaccharide-stimulated RAW 264.7 macrophages. Food Chem Toxicol 48:2045–2051
35. Sandesh KB et al (2008) Enhancement of carotenoids by mutation and stress induced carotenogenic genes in *Haematococcus pluvialis* mutants. Bioresour Technol 99:8667–8673
36. Sandmann G (1991) Biosynthesis of cyclic carotenoids: biochemistry and molecular genetics of the reaction sequence. Physiol Plant 83:186–193
37. Sandmann G (2001) Genetic manipulation of carotenoid biosynthesis: strategies, problems and achievements. Trends Plant Sci 6:14–17
38. Armstrong GA (1997) Genetics of eubacterial carotenoid biosynthesis: a colorful tale. Annu Rev Microbiol 51:629–659
39. Schmidt DC (2000) Engineering novel carotenoids in microorganisms. Curr Opin Biotechnol 11:255–261
40. Ducrey Sanpietro LM, Kula MR (1998) Studies of astaxanthin biosynthesis in *Xanthophyllomyces dendrorhous* (*Phaffia rhodozyma*). Effect of inhibitors and low temperature. Yeast 14:1007–1016
41. Lee PC, Schmidt DC (2002) Metabolic engineering towards biotechnological production of carotenoids in microorganisms. Appl Microbiol Biotechnol 60:1–11
42. Chemler J et al (2006) Biosynthesis of isoprenoids, polyunsaturated fatty acids and flavonoids in *Saccharomyces cerevisiae*. Microb Cell Fact 5:20
43. Armstrong GA (1995) Genetic analysis and regulation of carotenoid biosynthesis: structure and function of the *crt* genes and gene products. In: Blankenship RE et al (eds) Advances in photosynthesis. Kluwer Academic, Dordrecht, pp 1135–1157
44. Chamovitz D et al (1993) Molecular and biochemical characterization of herbicide-resistant mutants of cyanobacteria reveals that phytoene desaturation is a rate-limiting step in carotenoid biosynthesis. J Biol Chem 268: 17348–17353
45. Hirschberg J, Chamovitz D (1994) Carotenoids in cyanobacteria. In: Bryant DA (ed) The molecular biology of cyanobacteria. Kluwer Academic, Dordrecht, pp 559–579
46. Hodgson DA, Murillo FJ (1993) Genetics of regulation and pathway of synthesis of carotenoids. In: Dworkin M, Kaiser D (eds) Myxobacteria II. American Society for Microbiology, Washington, DC, pp 157–181
47. Umeno D et al (2005) Diversifying carotenoid biosynthetic pathways by directed evolution. Microbiol Mol Biol Rev 69:51–78
48. Simpson KL et al (1964) Biosynthesis of yeast carotenoids. J Bacteriol 88:1688–1694
49. Goodwin TW (1980) Biosynthesis of carotenoids. In: Goodwin TW (ed) The biochemistry of the carotenoids. Chapman and Hall, London, pp 33–76
50. Goodwin TW (1993) Biosynthesis of carotenoids: an overview. In: Packer L (ed) Methods

in enzymology. Carotenoids, part B: metabolism, genetics and biosynthesis. Academic, San Diego, pp 330–340

51. Lynen F (1967) Biosynthetic pathways from acetate to natural products. Pure Appl Chem 14:137–167
52. Bloch KE (1983) Sterol structure and membrane function. CRC Crit Rev Biochem 14:47–92
53. Takaichi S (1999) Carotenoids and carotenogenesis in anoxygenic photosynthetic bacteria. In: Frank HA et al (eds) The photochemistry of carotenoids. Kluwer Academic, Dordrecht, pp 39–69
54. Shaish A et al (1992) Biosynthesis of β-carotene in Dunaliella. In: Lester P (ed) Methods in enzymology. Academic, New York, pp 439–444
55. Britton G (1998) Overview of carotenoid biosynthesis. In: Britton G et al (eds) Carotenoids: biosynthesis and metabolism. Birkhauser Verlag, Basel, pp 13–140
56. Fraser PD et al (1997) In vitro characterization of astaxanthin biosynthetic enzymes. J Biol Chem 272:6128–6135
57. Antia N, Cheng JY (1983) Evidence for anomalous xanthophyll composition in a clone of *Dunaliella tertiolecta* (Chlorophyceae). Phycology 22:235–242
58. Yamamoto HY et al (1999) Biochemistry and molecular biology of the xanthophyll cycle. In: Frank HA et al (eds) Advances in photosynthesis: the photochemistry of carotenoids. Kluwer Academic, Dordrecht, pp 293–303
59. Bramely PM (1985) The in vitro biosynthesis of carotenoids. Adv Lipid Res 21:243–219
60. Sherman MM et al (1989) Isolation and characterization of isoprene mutants of *Escherichia coli.* J Bacteriol 171:3619–3628
61. Misawa N, Shimada H (1997) Metabolic engineering for the production of carotenoids in non-carotenogenic bacteria and yeasts. J Biotechnol 59:169–181
62. Sandmann G, Misawa N (1992) New functional assignment of the carotenogenic genes *crtB* and *crtE* with constructs of these genes from *Erwinia species.* FEMS Microbiol Lett 69:253–257
63. Shimada H et al (1998) Increased carotenoid production by the food yeast *Candida utilis* through metabolic engineering of the isoprenoid pathway. Appl Environ Microbiol 64:2676–2680
64. Humbelin M et al (2002) Genetics of isoprenoid biosynthesis in *Paracoccus zeaxanthinifaciens.* Gene 297:129–139
65. Phadwal K (2005) Carotenoid biosynthetic pathway: molecular phylogenies and evolutionary behavior of *crt* genes in eubacteria. Gene 345:35–43
66. Steiger S et al (2002) Heterologous production of two unusual acyclic carotenoids, 1,1′-dihydroxy-3,4-didehydrolycopene and 1-hydroxy-3,4,3′,4′-tetradehydrolycopene by combination of the *crtC* and *crtD* genes from *Rhodobacter and Rubrivivax.* J Biotechnol 97:51–58
67. Rählert N et al (2009) A *crtA*-related gene from *Flavobacterium* P99-3 encodes a novel carotenoid 2-hydroxylase involved in myxol biosynthesis. FEBS Lett 10:1605–1610
68. Tang X et al (2007) Improvement of a *CrtO*-type of β-carotene ketolase for canthaxanthin production in *Methylomonas sp.* Metab Eng 9:348–354
69. Misawa N et al (1991) Production of beta-carotene in *Zymomonas mobilis* and *Agrobacterium tumefaciens* by introduction of the biosynthesis genes from *Erwinia uredovora.* Appl Environ Microbiol 57:1847–1849
70. Snodderly DM (1995) Evidence for protection against age-related macular degeneration by carotenoids and antioxidant vitamins. Am J Clin Nutr 62:1448S–1461S
71. Athukorala Y et al (2007) Anticoagulant activity of marine green and brown algae collected from Jeju island in Korea. Bioresour Technol 98:1711–1716
72. Meléndez-Martínez AJ et al (2004) Nutritional importance of carotenoid pigments. Arch Latinoam Nutr 54:149–154
73. Je JY et al (2009) Antioxidant activity of enzymatic extracts from the brown seaweed *Undaria pinnatifida* by electron spin resonance spectroscopy. LWT-Food Sci Technol 42:874–878
74. Maeda H et al (2008) Seaweed carotenoid, fucoxanthin, as a multi-functional nutrient. Asia Pac J Clin Nutr 17(Suppl 1):196–199
75. Yamaguchi M, Shizuoka JP (2006) Osteogenesis promoter containing beta-cryptoxanthin as the active ingredient. US Patent 200,601,061,15
76. Binns CW et al (2004) The relationship between dietary carotenoids and prostate cancer risk in Southeast Chinese men. Asia Pac J Clin Nutr 13:S117
77. Ulrich M (2008) The Global Market for Carotenoids, BBC Market Research
78. Klaui H, Bauerfeind CJ (1981) Carotenoids as food colors. In: Bauerfeind JC (ed) Carotenoids as colorants. Academic, New York, pp 48–292

79. Francis FJ (2000) Carotenoids as food colorants. Cereal Food World 45:198–203
80. Jacobson GK et al (2000) Astaxanthin overproducing strains of *Phaffia rhodozyma*. Method for their cultivation and their use in animal feeds. Patent 6,015,684
81. Lorenz RT, Cysewski GR (2000) Commercial potential for *Haematococcus* microalgae as a natural source of astaxanthin. Trends Biotechnol 18:160–167
82. Miao F et al (2006) Characterization of astaxanthin esters in *Haematococcus pluvialis* by liquid chromatography-atmospheric pressure chemical ionization mass spectrometry. Anal Biochem 352:176–181
83. Mortensen A et al (1997) Comparative mechanisms and rates of free radical scavenging by carotenoid antioxidants. FEBS Lett 418:91–97
84. Miki W (1991) Biological functions and activities of animal carotenoids. Pure Appl Chem 63:141–146
85. Jyonouchi H et al (1993) Studies of immunomodulating actions of carotenoids. II. Astaxanthin enhances in vitro antibody production to T-dependent antigens without facilitating polyclonal B-cell activation. Nutr Cancer 19:269–280
86. Mayne ST (1996) Beta-carotene, carotenoids, and disease prevention in humans. FASEB J 10:690–701
87. Chew BP et al (1999) A comparison of the anticancer activities of dietary beta-carotene, canthaxanthin and astaxanthin in mice in vivo. Anticancer Res 19:1849–1853
88. Guerin M et al (2003) *Haematococcus* astaxanthin: applications for human health and nutrition. Trends Biotechnol 21:210–216
89. Johnson EA (2003) *Phaffia rhodozyma*: colorful odyssey. Int Microbiol 6:169–174
90. Bubrick P (1991) Production of astaxanthin from *Haematococcus*. Bioresour Technol 38:237–239
91. Zheng YG et al (2006) Large-scale production of astaxanthin by *Xanthophyllomyces dendrorhous*. Food Bioproducts Process 84:164–166
92. de la Fuente JL et al (2010) High-titer production of astaxanthin by the semi-industrial fermentation of *Xanthophyllomyces dendrorhous*. J Biotechnol 148:144–146
93. Lee JH et al (2008) Fermentation kinetic studies for production of carotenoids by *Xanthophyllomyces dendrorhous*. J Biotechnol 136:S732–S732
94. Ranjbar R et al (2008) High efficiency production of astaxanthin by autotrophic cultivation of *Haematococcus pluvialis* in a bubble column photobioreactor. Biochem Eng J 39:575–580
95. Krubasik P, Sandmann G (2000) A carotenogenic gene cluster from *Brevibacterium linens* with novel lycopene cyclase genes involved in the synthesis of aromatic carotenoids. Mol Gen Genet 263:423–432
96. Misawa N et al (1990) Elucidation of the *Erwinia uredovora* carotenoid biosynthetic pathway by functional analysis of gene products expressed in *Escherichia coli*. J Bacteriol 172:6704–6712
97. Lee JH et al (2003) Isolation of cDNAs for gonadotropin-II of flounder (*Paralichthys olivaceus*) and its expressions in adult tissues. J Microbiol Biotechnol 13:710–716
98. Fang TJ, Chiou TY (1996) Batch cultivation and astaxanthin production by a mutant of the red yeast *Phaffia rhodozyma* NCHU-FS501. J Ind Microbiol Biotechnol 16:175–181
99. Borowitzka LJ (1992) Beta-Carotene production using algal biotechnology. J Nutr Sci Vitaminol (Tokyo) 38:248–250
100. Paiva Sergio AR, Russell RM (1999) Beta-Carotene and other carotenoids as antioxidants. J Am Coll Nutr 18:426–433
101. Ye Z et al (2008) Biosynthesis and regulation of carotenoids in *Dunaliella*: progresses and prospects. Biotechnol Adv 26:352–360
102. Yoon SH et al (2009) Combinatorial expression of bacterial whole mevalonate pathway for the production of beta-carotene in *E. coli*. J Biotechnol 140:218–226
103. Ciegler A (1965) Microbial carotenogenesis. In: Wayne WU (ed) Advances in applied microbiology. Academic, New York, pp 1–34
104. Bhosale P, Gadre RV (2001) Production of β-carotene by a *Rhodotorula glutinis* mutant in sea water medium. Bioresour Technol 76:53–55
105. Raja R et al (2007) Exploitation of *Dunaliella* for β-carotene production. Appl Microbiol Biotechnol 74:517–523
106. Borowitzka MA (1999) Commercial production of microalgae: ponds, tanks, tubes and fermenters. J Biotechnol 70:313–321
107. Ninet L, Renault J (1979) Carotenoids. In: Peppier HJ, Perlman D (eds) Microbiol technology, 2nd edn. Academic, New York, pp 529–253
108. Reyes P et al (1964) The mechanism of beta-ionone stimulation of carotenoid and ergosterol biosynthesis in *Phycomyces blakesleeanus*. Biochim Biophys Acta 90:578–592
109. Uchiyama S et al (2004) Anabolic effect of beta-cryptoxanthin on bone components in

the femoral tissues of aged rats in vivo and in vitro. J Health Sci 50:491–496

110. Giovannucci E et al (1995) Intake of carotenoids and retinol in relation to risk of prostate cancer. J Natl Cancer Inst 87:1767–1776

111. Yuan JM et al (2001) Prediagnostic levels of serum β-cryptoxanthin and retinol predict smoking-related lung cancer risk in Shanghai, China. Cancer Epidemiol Biomarkers Prev 10:767–773

112. Yuan JM et al (2003) Dietary cryptoxanthin and reduced risk of lung cancer: the Singapore Chinese Health Study. Cancer Epidemiol Biomarkers prev 12:890–898

113. Kohno H et al (2001) Inhibitory effect of mandarin juice rich in beta-cryptoxanthin and hesperidin on 4-(methylnitrosamino)-1-(3-pyridyl)-1-butanone-induced pulmonary tumorigenesis in mice. Cancer Lett 174: 141–150

114. Tanaka T et al (2000) Suppression of azoxymethane-induced colon carcinogenesis in male F344 rats by mandarin juices rich in beta-cryptoxanthin and hesperidin. Int J Cancer 88:146–150

115. Pattison DJ et al (2004) The role of diet in susceptibility to rheumatoid arthritis: a systematic review. J Rheumatol 31:1310–1319

116. Serrato JO et al (2006) Production of β-cryptoxanthin, a provitamin-A precursor, by *Flavobacterium lutescens.* J Food Sci 71:E314–E319

117. Guyomarc'h F et al (2000) Production of carotenoids by *Brevibacterium linens*: variation among strains, kinetic aspects and HPLC profiles. J Ind Microbiol Biotechnol 24:64–70

118. Bhosale P, Bernstein PS (2004) Beta-carotene production by *Flavobacterium multivorum* in the presence of inorganic salts and urea. J Ind Microbiol Biotechnol 31:565–571

119. Khachik F et al (2007) Partial synthesis of (3R,6′R)-alpha-cryptoxanthin and (3R)-beta-cryptoxanthin from (3R,3′R,6′R)-lutein. J Nat Prod 70:220–226

120. Baker Rémi TM (2001) Canthaxanthin in aquafeed applications: is there any risk? Trends Food Sci Technol 12:240–243

121. Palozza P, Krinsky NI (1992) Astaxanthin and canthaxanthin are potent antioxidants in a membrane model. Arch Biochem Biophys 297:291–295

122. Asker O (2002) Production of canthaxanthin by *Haloferax alexandrinus* under non-aseptic conditions and a simple, rapid method for its extraction. Appl Microbiol Biotechnol 58: 743–750

123. Nasri N et al (2010) Use of response surface methodology in a fed-batch process for optimization of tricarboxylic acid cycle intermediates to achieve high levels of canthaxanthin from *Dietzia natronolimnaea* HS-1. J Biosci Bioeng 109:361–368

124. Krupa D et al (2010) Extraction, purification and concentration of partially saturated canthaxanthin from *Aspergillus carbonarius.* Bioresour Technol 101:7598–7604

125. Ausich RL (1997) Commercial opportunities for carotenoid production by biotechnology. Pure Appl Chem 69:2169–2173

126. Khodaiyan F et al (2008) Optimization of canthaxanthin production by *Dietzia natronolimnaea* HS-1 from cheese whey using statistical experimental methods. Biochem Eng J 40:415–422

127. Lotan T, Hirschberg J (1995) Cloning and expression in *Escherichia coli* of the gene encoding *beta-C-4-oxygenase*, that converts beta-carotene to the ketocarotenoid canthaxanthin in *Haematococcus pluvialis.* FEBS Lett 364:125–128

128. Maeda H et al (2005) Fucoxanthin from edible seaweed, *Undaria pinnatifida*, shows antiobesity effect through UCP1 expression in white adipose tissues. Biochem Biophys Res Commun 332:392–397

129. Maeda H et al (2007) Effect of medium-chain triacylglycerols on anti-obesity effect of fucoxanthin. J Oleo Sci 56:615–621

130. Maeda H et al (2007) Dietary combination of fucoxanthin and fish oil attenuates the weight gain of white adipose tissue and decreases blood glucose in obese/diabetic KK-Ay mice. J Agric Food Chem 55:7701–7706

131. Tsukui T et al (2007) Fucoxanthin and fucoxanthinol enhance the amount of docosahexaenoic acid in the liver of KKAy obese/diabetic mice. J Agric Food Chem 55: 5025–5029

132. Hosokawa M et al (2004) Fucoxanthin induces apoptosis and enhances the antiproliferative effect of the PPARgamma ligand, troglitazone, on colon cancer cells. Biochim Biophys Acta 1675:113–119

133. Das SK et al (2005) Fucoxanthin induces cell cycle arrest at G0/G1 phase in human colon carcinoma cells through up-regulation of p21WAF1/Cip1. Biochim Biophys Acta 1726:328–335

134. Yun SM et al (2007) Isolation and identification of an antibacterial substance from sea mustard, *Undaria pinnatifida*, for *Streptococcus mutans.* Korean Soc Food Sci Nutr 36:149–154

135. Cahyana AH et al (1992) Pyropheophytin A as an antioxidative substance from the marine alga, Arame (*Eicenia bicyclis*). Biosci Biotechnol Agrochem 18:1533–1535
136. Yan X et al (1999) Fucoxanthin as the major antioxidant in *Hijikia fusiformis*, a common edible seaweed. Biosci Biotechnol Biochem 63:605–607
137. Kang HS et al (2003) A new phlorotannins from the brown alga *Ecklonia stolonifera*. Chem Pharm Bull 51:1012–1014
138. Khachik F (2005) Process for extraction and purification of lutein, zeaxanthin and rare carotenoids from marigold flowers and plants. US Patent 6,262,284
139. Sánchez JF et al (2008) Influence of culture conditions on the productivity and lutein content of the new strain *Scenedesmus almeriensis*. Process Biochem 43:398–405
140. Francis GW et al (1975) Variations in the carotenoid content of *Chlamydomonas reinhardii* throughout the cell cycle. Arch Microbiol 104:249–254
141. Del Campo JA et al (2001) Lutein production by *Muriellopsis sp.* in an outdoor tubular photobioreactor. J Biotechnol 85:289–295
142. Shi XM et al (2000) Heterotrophic production of biomass and lutein by *Chlorella protothecoides* on various nitrogen sources. Enzyme Microb Technol 27:312–318
143. Nelis HJ, De Leenheer AP (1991) Microbial sources of carotenoid pigments used in foods and feeds. J Appl Microbiol 70:181–191
144. Khachik F et al (1989) Separation, identification, and quantification of the major carotenoids in extracts of apricots, peaches, cantaloupe, and pink grapefruit by liquid chromatography. J Agric Food Chem 37:1465–1473
145. McDERMOTT CB et al (1973) Effect of inhibitors on zeaxanthin synthesis in a *Flavobacterium*. J Gen Microbiol 77: 161–171
146. Alcantara S, Sanchez S (1999) Influence of carbon and nitrogen sources on *Flavobacterium* growth and zeaxanthin biosynthesis. J Ind Microbiol Biotechnol 23:697–700
147. Masetto A et al (2001) Application of a complete factorial design for the production of zeaxanthin by *Flavobacterium sp.* J Biosci Bioeng 92:55–58
148. Sajilata MG et al (2010) Development of efficient supercritical carbon dioxide extraction methodology for zeaxanthin from dried biomass of *Paracoccus zeaxanthinifaciens*. Sep Purif Technol 71:173–177
149. Fresnedo O et al (1991) Carotenoid composition in the cyanobacterium *Phormidium laminosum*. Effect of nitrogen starvation. FEBS Lett 282:300–304
150. Yokoyama T et al (1995) Thermozeaxanthins, new carotenoid-glycoside-esters from thermophilic eubacterium *Thermus thermophilus*. Tetrahedron Lett 36:4901–4904
151. Salguero A et al (2005) UV-A mediated induction of carotenoid accumulation in *Dunaliella bardawil* with retention of cell viability. Appl Microbiol Biotechnol 66:506–511
152. Jin ES et al (2003) A mutant of the green alga *Dunaliella salina* constitutively accumulates zeaxanthin under all growth conditions. Biotechnol Bioeng 81:115–124
153. Chen F et al (2005) Isolation and purification of the bioactive carotenoid zeaxanthin from the microalga *Microcystis aeruginosa* by high-speed counter-current chromatography. J Chromatogr A 1064:183–186
154. Ziegler RG et al (1996) Importance of α-carotene, β-carotene, and other phytochemicals in the etiology of lung cancer. J Natl Cancer Inst 88:612–615
155. Beatty S et al (2004) Macular pigment optical density and its relationship with serum and dietary levels of lutein and zeaxanthin. Arch Biochem Biophys 430:70–76
156. Hadden WL et al (1999) Carotenoid composition of marigold (*Tagetes erecta*) flower extract used as nutritional supplement. J Agric Food Chem 47:4189–4194
157. Li Y, Li L (2010) Method for producing fucoxanthin. US Patent 201,001,522,86
158. Wang WJ et al (2005) Isolation of fucoxanthin from the Rhizoid of *Laminaria japonica*. J Integr Plant Biol 47:1009–1015
159. Holden JM et al (1999) Carotenoid content of US foods: an update of the database. J Food Compost Anal 12:169–196

Chapter 3

Biosynthesis, Extraction, Purification, and Analysis of Trisporoid Sexual Communication Compounds from Mated Cultures of *Blakeslea trispora*

C. Schimek and J. Wöstemeyer

Abstract

The zygomycete *Blakeslea trispora* produces high amounts of the general zygomycete β-carotene-derived sexual signal compounds, the trisporoids. These can be isolated from the culture medium and purified by extraction with organic solvents followed by thin layer chromatography. Concentration is determined spectrophotometrically using specific extinction coefficients established for some members of this compound family. The effect of the extraction and activity of the isolated compounds is best tested physiologically, exploiting the ability of trisporoids to induce the formation of sexually committed hyphae, the zygophores, in other zygomycete species. Methods for *B. trispora* culture, trisporoid extraction, and further analyses of trisporoids are described in this chapter.

Key words: *Blakeslea trispora*, Trisporic acid, Liquid extraction, Phase separation, Absorbance, Thin layer chromatography, Zygophore induction assay

1. Introduction

Trisporoids are C_{18} or C_{19} β-carotene-derived compounds synthesized by fungi of the class *Zygomycetes* as sexual signals involved in partner finding and in regulation of the early steps of the mating reaction. The first identified major compound of this class was trisporic acid (1) named for the source organism *Blakeslea trispora*. Diverse trisporoids have been isolated from *Mucor mucedo* and *Phycomyces blakesleeanus* as well, and the trisporoid biosynthesis enzymes and genes coding for such enzymes were identified in even more species (2–5). Besides their function in mating regulation, trisporoids also positively affect the production of β-carotene in *B. trispora*, *M. mucedo*, and *P. blakesleeanus* (6–9). Trisporoid biosynthesis is

José-Luis Barredo (ed.), *Microbial Carotenoids From Fungi: Methods and Protocols*, Methods in Molecular Biology, vol. 898,
DOI 10.1007/978-1-61779-918-1_3,

strictly dependent on mating situations. While individually growing strains of either mating type (+) or (–) only produce minor amounts of some of the biosynthesis intermediates, high amounts of trisporoids, and especially the trisporic acids formed as the end products of the biosynthesis pathway, are only found when the two compatible mating types are grown together for several days. Until now, no evidence for negative feedback, e.g. product inhibition of trisporoids synthesis, has been found. On the other hand, the mycelia seem to produce trisporoids only after competence for mating has been reached (2). This is apparently correlated to (1) a certain age of the mycelium and (2) the production of sufficient biomass to support the further steps of sexual development. Each species was found to produce trisporoids in several derivative series marked by the letters A to E added to the compound name. Trisporic acid B exhibits a keto function at C-13 of the structural backbone, while Trisporic acid C contains a hydroxyl group at the same position. The D and E derivatives contain additional hydroxyl groups at positions 2 and 3 of the ionone ring, and A stands for the completely unmodified structures (3). The resulting polarity differences can be exploited in the separation of these derivatives.

The highest production of trisporic acid has been observed in mated cultures of *B. trispora*. Only in this species, too, trisporoids seem to be freely released into the culture medium in submerged culture. In *M. mucedo*, the trisporoids, produced at several order of magnitude lower amounts, are apparently predominantly retained within the hyphae, and extraction from the mycelium would be necessary.

Trisporoids are extractable into organic solvents. For the routine procedures, solvents like trichloromethane or ethyl acetate are preferred because of their comparatively low boiling point which allows easy removal of the solvent from the extracted trisporoids by vacuum evaporation.

Commercial standards for identification of trisporoids by HPLC or GC-MS are not available. The best routine technique for identification is therefore a combination of thin layer chromatography (TLC) and physiological testing for biological activity (5, 10). *M. mucedo*, especially the (–) mating type, reacts highly sensitive on externally applied trisporoids. Substance amounts down to 0.2 ng elicit the formation of committed sexual hyphae, the zygophores, in the vicinity of the application site.

2. Materials

Prepare all media solutions using demineralised/deionised water at a conductivity of <25 μS/cm. Inorganic salts should be of analytical grade, and the sugars of biochemical grade. If further work is intended with the extracted trisporoids, use solvents of HPLC/

spectroscopy analytical grade for extraction and TLC. All prepared media components must be sterilized immediately to prevent undesired microbial growth; the sterile media can be stored in closed vessels at room temperature for several weeks at least. Take care to use sterile techniques in handling of cultures and sterile media. Never wear gloves while working near an open flame. For all procedures involving organic solvents, use a fume hood and wear protective goggles. Follow the relevant waste disposal regulations for disposing culture and solvent waste material.

2.1. Fungal Strains

The following strains are used in this work (see Notes 1 and 2):

1. *B. trispora* (+) CBS 130.59 (Centraalbureau for Schimmelcultures, CBS-KNAW Fungal Biodiversity Centre, Utrecht, The Netherlands).
2. *B. trispora* (+) ATCC 14271 (American Type Culture Collection, Manassas, VA, USA).
3. *B. trispora* (+) NRRL 2456 (Agricultural Research Service Culture Collection, Peoria, IL, USA).
4. *B. trispora* (–) CBS 131.59 (Centraalbureau for Schimmelcultures, CBS-KNAW Fungal Biodiversity Centre, Utrecht, The Netherlands).
5. *B. trispora* (–) ATCC 14272 (American Type Culture Collection, Manassas, VA, USA).
6. *B. trispora* (–) NRRL 2457 (Agricultural Research Service Culture Collection, Peoria, IL, USA).
7. *M. mucedo* (+) CBS 144.24 (Centraalbureau for Schimmelcultures, CBS-KNAW Fungal Biodiversity Centre, Utrecht, The Netherlands).
8. *M. mucedo* (+) ATCC 38693 (American Type Culture Collection, Manassas, VA, USA).
9. *M. mucedo* (+) NRRL 3635 (Agricultural Research Service Culture Collection, Peoria, IL, USA).
10. *M. mucedo* (–) CBS 145.24 (Centraalbureau for Schimmelcultures, CBS-KNAW Fungal Biodiversity Centre, Utrecht, The Netherlands).
11. *M. mucedo* (–) CBS 109.16 (Centraalbureau for Schimmelcultures, CBS-KNAW Fungal Biodiversity Centre, Utrecht, The Netherlands).
12. *M. mucedo* (–) IMI 078408 (CAB International Mycological Institute Culture Collection, Wallingford, Oxfordshire, UK).
13. *M. mucedo* (–) IMI 133298 (CAB International Mycological Institute Culture Collection, Wallingford, Oxfordshire, UK).
14. *M. mucedo* (–) NRRL 3634 (Agricultural Research Service Culture Collection, Peoria, IL, USA).

15. *M. mucedo* (−) VKM F-1355 (All-Russian Collection of Microorganisms, G.K. Skryabin Institute of Biochemistry and Physiology of Microorganisms, Pushchino Biological Research Centre, Pushchino, Russia).

2.2. Culture Media and Vessels

1. Solution SUP-A: Weigh 10 g glucose and 12 g powdered agar, and transfer to an autoclavable 1-L glass bottle (see Note 3). Add 800 mL water and mix thoroughly to prevent the formation of agar lumps.
2. Solution SUP-B: Weigh 1 g ammonium chloride, 4 g potassium dihydrogen phosphate (KH_2PO_4), 0.9 g dipotassium hydrogen phosphate (K_2HPO_4), 0.25 g magnesium sulfate heptahydrate ($MgSO_4{\cdot}7H_2O$), and 5 g yeast extract, and transfer to an autoclavable 250-mL glass bottle. Add 200 mL water and mix until all solids have dissolved.
3. 0.1 M Calcium chloride: Weigh 14.7 g calcium chloride dihydrate and transfer to a beaker. Add 80 mL water and mix until the calcium chloride has been dissolved. Fill into a measuring cylinder and adjust to 100 mL with water. Filter-sterilize using a 0.45-μm pore membrane filter (see Note 4). Fill the filtrate into a sterilized bottle. Store at room temperature.
4. Solidified supplemented minimal medium (SSUP) (11) (see Note 5): Sterilize solutions SUP-A and SUP-B by autoclaving for 20 min at 121°C (see Note 6). Remove from the autoclave after cooling down (see Note 7). Add 1 mL of sterile 0.1 M calcium chloride solution to solution SUP-B and mix. Add solution SUP-B to solution SUP-A and mix again. Pour 25 mL portions into 90-mm plastic Petri dishes with aeration vents. 1 L of medium will yield about 40 plates. Leave on the table until they solidify, then pack into a plastic bag to prevent drying out and store at room temperature.
5. Liquid supplemented minimal medium (LSUP) (11) (see Note 5). Prepare similar to SSUP, but omit the agar powder of solution A. Store at room temperature.
6. Maltose solution (see Note 8). Weigh 20 g maltose and 0.1 g ammonium dihydrogen phosphate [$(NH_4)H_2PO_4$]; transfer to a 1-L glass bottle and add 1 L water. Mix and autoclave for 20 min at 121°C. Store at room temperature.
7. Sterile 0.5-L Erlenmeyer flasks.
8. Material for sterile transfer of mycelia: Sterile fine-meshed steel sieve and beaker or alternatively sterile paper coffee filters, funnel, and flask. Put the sieve inside the beaker; cover with aluminum foil and heat-sterilize at 160°C for 4 h. Alternatively, put some coffee filters inside a large, autoclavable lidded glass box; wrap tightly in aluminum foil and autoclave for 20 min at 121°C (see Note 6). Ethanol, technical grade, and steel tweezers.

9. Solution IM-A: Weigh 20 g maltose and 15 g agar. Add 800 mL water and mix thoroughly to prevent the formation of agar lumps.
10. Solution IM-B: Weigh 10 g potassium nitrate, 5 g potassium dihydrogen phosphate, 2.5 g magnesium sulfate heptahydrate, and 1 g yeast extract. Add 200 mL water and mix until all solids have dissolved.
11. Induction medium (IM): Proceed as for solution 4, but without addition of calcium chloride. Store the plates at room temperature.

2.3. Solvent Extraction of Trisporoids

1. Trichloromethane–2-propanol mixture: Mix 1 L trichloromethane with 50 mL 2-propanol. Store at room temperature (see Note 9).
2. Separating funnel.
3. Brown glass storage bottles.
4. Solid water-free sodium sulfate.
5. Rotary vacuum evaporator and evaporation flasks.

2.4. UV Spectrophotometric Determination of Trisporoid Concentration

1. UV/Vis spectrophotometer.
2. 1-mL Quartz glass cuvettes.

2.5. Enrichment of Trisporic Acids (12)

1. 4 g/100 mL Sodium hydrogen carbonate. Weigh 40 g sodium hydrogen carbonate, transfer into a beaker and dissolve in 800 mL water. Adjust to 1 L in a graduated cylinder, mix well, and fill the solution into a bottle for storage. Store at room temperature.

2.6. Components for Thin Layer Chromatography

1. Mobile phase: trichloromethane: ethyl acetate: *tert*-butyl methyl ether: acetic acid 50:30:15:5. Measure 50 mL trichloromethane with a graduated 100-mL cylinder. Add 30 mL ethyl acetate, 15 mL *tert*-butyl methyl ether, and 5 mL acetic acid. Fill into a bottle, close, and mix by shaking until homogeneous. Prepare freshly and keep in a tightly closed bottle until used (see Note 10).
2. Solid phase (sorbent): Silica 60 F_{254} analytical TLC plates, 20×20 cm, glass base (see Note 11).
3. Lidded chromatography tank to fit the TLC plates.

2.7. Test for Physiological Activity: Zygophore Induction Assay

1. Disks of strong filtration paper (5 mm diameter).
2. Dissecting microscope.

3. Methods

Trisporoids are sensitive to light. UV light, either from the sun or from fluorescent tubes, will lead to rapid degradation. Cover the mated cultures on the way from the culture room to the lab for extraction. Take care to protect all samples from light: work with the lamps in the fume hood switched off and keep the room light as low as possible. Work under dim light or red protective light. Protect flasks and bottles from light whenever possible. We use brown glass bottles for drying and storing the extracts. Carry out all procedures at room temperature.

3.1. Working Cultures

1. Cut a 0.5 cm^2 agar block with adhering mycelial mat out of a tested stock culture (see Note 12) and place in the middle of a Petri dish containing maintenance medium SSUP.
2. Cultivate at approximately 20°C and ambient day–night conditions (see Note 13), until the surface of the medium is fully covered with mycelium and asexual spores have formed (see Note 14).
3. Use the working cultures to start liquid cultures. Use separate Petri dishes for each mating type.

3.2. Liquid Cultures: Starting Cultures

1. Transfer 100 mL each of liquid culture growth medium (LSUP) into an even number of culture vessels.
2. Inoculate half of the vessels with *B. trispora* (+) mating type, and the other half with *B. trispora* (–) mating type.
3. Cut 0.5 cm^2 pieces of mycelial mat from mature working cultures and transfer three pieces to each vessel. The supporting agar blocks may be transferred as well. Make sure that the inoculum is submerged in the culture medium.
4. Cultivate on a shaker for 3 days at approximately 20°C and ambient day–night conditions (see Note 13). The growing mycelium will form rounded pebbles of different size, sometimes also long ribbons are observed.

3.3. Liquid Cultures: Production Cultures

1. Transfer 100 mL each of liquid culture maintenance medium (maltose solution) to fresh culture vessels. Prepare half the number of culture vessels as were used for growing the starting cultures.
2. Prepare the workplace for transferring the mycelia: Set up a large sterile beaker and put the steel sieve on top.
3. Pour one (+) and one (–) starting culture into the sieve to drain off the culture medium.
4. Rinse with 50 mL liquid culture maintenance medium.

5. Gently press the mycelium with tweezers to remove excess liquid and transfer the mycelial mass into a culture vessel with liquid culture maintenance medium.
6. Continue until all cultures have been combined (see Note 15).
7. Cultivate the production cultures for 7 days in the dark on a shaker at approximately 20°C (see Note 16).

3.4. Trisporoid Extraction

1. To harvest the cultures, separate the mycelium from the culture liquid using a steel sieve. Sterile techniques are now not longer required.
2. Squeeze the mycelial mass to recover most of the liquid. Discard the mycelium.
3. Adjust the culture liquid to pH 8.0 using 1 M potassium hydroxide. Control the process using indicator paper (see Note 17).
4. Fill the culture liquid into a separating funnel.
5. Add half a volume of trichloromethane–2-propanol.
6. Invert several times to extract the culture liquid and allow the two phases to separate completely.
7. Collect the organic phase now containing the trisporoids (lower phase with trichloromethane–2-propanol and upper phase with ethyl acetate) into a brown glass bottle.
8. Recover the aqueous phase and adjust to pH 2.0 with 1 M hydrochloric acid. Control the process using indicator paper (see Note 18).
9. Repeat the extraction as above.
10. Discard the extracted culture liquid (aqueous phase) (see Note 19).
11. To dry the extracts, add water-free sodium sulfate to the two culture extracts at a ratio of 10 g/100 mL to bind residual water.
12. Leave the extracts to dry overnight at room temperature.
13. Concentrate the extracts by rotary vacuum evaporation. Filtrate each extract into an evaporation flask with ground socket using a funnel and filtration paper to remove the sodium sulfate. Adjust the evaporation conditions according to the solvent used. Evaporate until all solvent has been removed.
14. After evaporation, swill the evaporation flask with an appropriate volume (between 1 and 4 mL) of pure ethanol to dissolve the trisporoids. Allow the ethanol to settle at the bottom of the flask for several minutes before transferring the solution into 1.5 or 4 mL brown glass screw top vials with PTFE seals. Store at –20°C.

3.5. UV Spectrophotometric Determination of Trisporoid Concentration

1. Warm up the spectrophotometer according to the manufacturer's instructions.
2. Prepare serial dilutions (1:10, 1:100, 1:500, 1:1,000, ...) of the basic and the acidic trisporoid extract using pure ethanol as diluent.
3. Measure the absorbance of the 1:100 dilution at 285, 300, and 325 nm, using 1-mL quartz glass cuvettes.
4. Proceed to the next dilution if the absorbance is higher than 0.8.
5. Based on these measurements, the concentration of trisporoids contained in the extracts can be determined as the specific extinction coefficients of several trisporoid compounds are known (13): (1) trisporic acid: $E_{325\,nm}^{10\,\mu g/mL\,cm} = 572$. (2) 4-dihydromethyl trisporate: $E_{285\,nm}^{10\,\mu g/mL\,cm} = 547$ (see Note 20).

3.6. Enrichment of Trisporic Acids

1. Extract the trichloromethane fraction of the second acid extraction again with an equal volume of sodium hydrogen carbonate solution. Afterwards, the organic phase will contain mainly neutral trisporoids while trisporic acids have separated into the sodium hydrogen carbonate.
2. Adjust the sodium hydrogen carbonate fraction to pH 2 and extract again as in steps 4–7 of Subheading 3.4. The resulting organic phase contains the enriched trisporic acids.
3. Filtrate through a filter bag containing 10 g water-free sodium sulfate into an evaporation flask and concentrate as in steps 13 and 14 of Subheading 3.4.

3.7. Thin Layer Chromatography

Take care to minimize opening times of a filled chromatography tank. Use only close-fitting lids. Never move the tank during development of the plate.

1. Fill the solvent mixture into the chamber and keep standing in the closed chamber for 1 h to allow saturation of the chamber atmosphere (see Note 21).
2. Take a TLC plate out of its package by touching it only at the sides with finger tips. Never touch the sorbent surface with bare fingers (see Note 22).
3. Mark the spots for sample application with a soft pencil. Be careful not to damage the surface of the sorbent layer. Leave 1.5 cm free at the sides and the bottom of the plate. Multiple lanes should be 1–1.5 cm apart. Do not write or sign anything below the sample application spots; necessary information can be put at the upper end of the plate.
4. Apply 20 μg of a trisporoid mixture or 2–5 mg of purified trisporoids to each lane. Ideally, only 1–2 μL should be applied per spot. If necessary, higher volumes (10–20 μL) can be used,

but these should be applied as a series of small volumes (up to 2 μL). Let the spot always dry completely before adding the next portion (see Note 23).

5. Let the plate dry for 15 min after all samples have been applied.
6. Put the plate into the chromatography tank and develop for 1 h.
7. Take the plate out of the chamber and mark the solvent front immediately with a soft pencil.
8. Leave the plate for 15 min under the fume hood for evaporating the solvent mixture.
9. Wrap the dry plate to protect the trisporoids from light until documentation.
10. Visualize the TLC result by putting the TLC plate face downwards onto a standard 254 or 312 nm UV transilluminator and take a photograph. The bands may also be marked on the glass backside with a waterproof felt-tip pen (see Note 24).

3.8. Isolation of Individual Trisporoid Compounds

1. Individual trisporoids can be isolated by preparative TLC. Apply 1–4 mg of total trisporoids in a thin continuous line or a series of dots only 4 mm apart, and develop the plate as described previously.
2. For detection, cover the central part of the plate so that no UV light reaches the silica gel, and mark the position of the relevant bands only at both outside lanes on the glass backside.
3. Extrapolate the position of the bands between the marks.
4. Remove the sorbent layer in this inner region with a spatula or a razor blade.
5. Elute the sorbent three times with a triple volume of ethanol, pool the eluates, filtrate to remove silica particles, and concentrate the extract if necessary as in steps 13 and 14 of Subheading 3.4. Store at −20°C.

3.9. Test for Physiological Activity: Zygophore Induction Assay

Zygophore induction is most reliably tested in *M. mucedo*. Zygophores in this species are easily recognizable as they are quite different from other aerial hyphae (see Note 25). Both mating types of *M. mucedo* are able to produce zygophores, but from our strain pair, the (−) mating type reacts much better and makes much more zygophores per amount of trisporoids than the (+) mating type.

1. Inoculate IM plates with mycelial blocks cut from working cultures of *M. mucedo* (+) and (−) strains. Put the inoculum near the rim of the Petri dish. Incubate at 20°C for 4–5 days in the dark until half the medium surface is covered with mycelium.

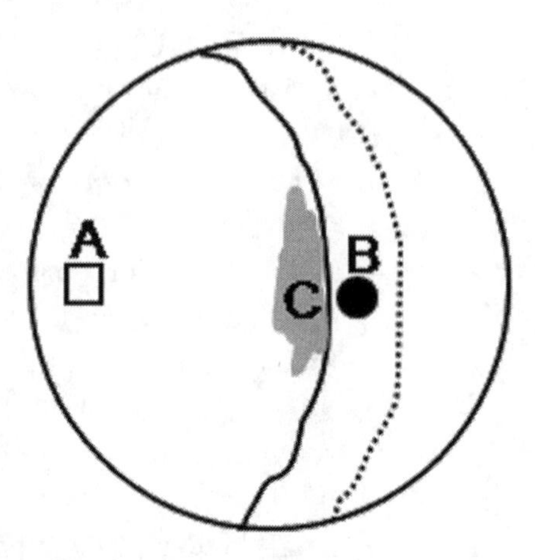

Fig. 1. Zygophore induction assay. Zygophores develop within several hours after stimulation with trisporoids applied in front of a growing mycelium. (**A**) The place where the inoculum has been put. (**B**) The position of the sample disk. (**C**) The area where zygophores will develop and carotene production is increased. *Solid line*: mycelial growth front at the beginning of the assay. *Dotted line*: mycelial growth front 18 h later, at the time of evaluation.

Prepare four cultures of each mating type for each sample to be tested.

2. Spread the required number of 5 mm diameter disks of strong filtration paper on a clean surface (see Note 26).
3. Soak each filter with 10 μL of ethanolic trisporoid dilution. Probably, you will have to test several dilutions. *M. mucedo* reacts to nanogram amounts (0.2–100 ng) of trisporoids.
4. Prepare similarly treated controls using pure ethanol.
5. Let the filters dry completely to make sure that you are measuring indeed the effect of the trisporoid and not that of the solvent.
6. Put each filter about 2 mm before the growth front of the *M. mucedo* mycelium onto the blank medium surface (see Fig. 1).
7. On the next day, the mycelium has grown around the filter disk. In the area of the previous mycelial front, zygophores have formed. In this area, you will also find the mycelium more yellow than elsewhere, as trisporoids also induce carotene production. Count the zygophores directly behind the filter (in the direction towards the place of inoculation) using the largest setting of your dissecting microscope (see Note 27).

4. Notes

1. Strains for trisporoid extraction: *B. trispora* (+) (CBS 130.59, ATCC 14271, NRRL 2456) and *B. trispora* (−) (CBS 131.59, ATCC 14272, NRRL 2457). Strains for testing the physiological activity of trisporoid-containing extracts: *M. mucedo* (+) (CBS 144.24, ATCC 38693, NRRL 3635) and *M. mucedo* (−) (CBS 145.24, IMI 078408, VKM F-1355), or *M. mucedo* (−) (CBS 109.16, ATCC 38694, IMI 133298, NRRL 3634).
2. First, you need to establish that the strain pair intended for this experiment is truly compatible for mating reactions. Observe cocultures of the two strains: if the carotene production increases, i.e., if the mycelium, or part thereof becomes more yellow, they will be a compatible mating pair even if no sexual development can be observed, which is sometimes hard to see. Other mating pairs of zygomycetes might also be used for trisporoid extraction, but to our experience relevant amounts are only produced in strains also visibly producing beta-carotene.
3. For more solid medium, 15 g of agar may be used.
4. Different devices for filter sterilization are commercially available. Follow the instructions of the supplier.
5. This medium supports prolonged mycelial growth.
6. The total time for autoclave sterilization is much longer. It also includes the time for heating up, equilibration, and cooling down. Depending on the type of apparatus, an autoclave time of 20 min may translate into a total time of 2–3 h for the complete procedure. For autoclave sterilization, never use tightly closed vessels. This will trap air inside the vessel, and as air is a very poor heat conductor, effective sterilization conditions will not be reached throughout the vessel. Moreover, the expanding air may lead to breakage of the bottle. An autoclave uses steam for sterilization, and the steam must be able to penetrate everywhere. Therefore, with screw top bottles, screw down the lid only loosely. Tighten it only after sterilization has been completed. Alternatively, the bottles may be covered with strong, folded alumina foil or commercially available loose slip lids. Follow all safety regulations for using an autoclave. Open the autoclave only, after the contents is cooled down at least below 80°C.
7. If the medium is not to be poured immediately, it may be stored in a heated water bath at 50–55°C up to several hours. Agar solidifies at about 40°C.

8. This liquid culture maintenance medium does not allow prolonged mycelial growth but will support basic metabolism for several days. It is useful for the induction of developmental switches.
9. Alternatively, ethyl acetate may be used for extraction. Handle organic solvents inside a fume hood. Keep in tightly closed bottles to prevent evaporation. Use a specialized aerated solvent cupboard for prolonged storage.
10. Prepare daily fresh and avoid repeated opening of the bottle. Otherwise, the ratio between the components will change, as each solvent exhibits a different equilibrium vapor pressure.
11. For a tank admitting 20×20 cm TLC plates, 100 mL solvent mixture is needed. The plates should only be submerged to a height of about 1 cm at the beginning of chromatography, and the solvent must not be completely used up after development.
12. Reduce subcultivation to the absolute minimum. Wherever possible, go back to a certified stock culture and start your working cultures from these. Sometimes, it helps to start working cultures from spore stocks. To prepare a spore stock, pipet 10 mL of 20% glycerol onto a Petri dish containing a mature sporulated culture. Carefully distribute the glycerol using the pipette or a glass rod. Mature sporangia will break and release the spores into the glycerol. Slightly tilt the Petri dish and remove as much of the liquid as possible. Fill aliquots of the spore suspension into Eppendorf type reaction tubes and store at –20°C. Take care not to transfer mycelium or medium particles. Using a spore stock, inoculate the working culture by plating 0.2 mL of the diluted spore stock per Petri dish.
13. *B. trispora* requires some light for germination.
14. *B. trispora* cultures for inoculating liquid cultures should be between 2 and 4 weeks old.
15. Only take as much cultures from the shaker for transfer as can be handled in a few minutes. Prolonged interruption of agitation will cause oxygen limitation and may lead to undesired and incalculable side effects.
16. Trisporoid production commences soon after combination of the two mating types and continues for many days (about 14). The trisporoids will accumulate in the culture liquid. During the first days, predominantly intermediates are formed, with increasing cultivation, the amount of trisporic acids increases (14). For trisporic acid extraction, cultivation should last at least 5 days.
17. Trisporic acids will be deprotonated in basic pH conditions and will therefore not separate into the neutral solvent. This

extraction step removes most of the neutral trisporoid biosynthesis intermediates from the culture liquid.

18. At acidic pH, trisporic acids will be fully protonated and can now enter into the neutral solvent. With this step, the trisporic acids are extracted from the culture liquid.
19. Take care that no residual solvent reaches the sewerage.
20. Read as follows: The absorbance of a 10 μg/mL trisporic acid/4-dihydromethyl trisporate solution measured at 325/285 nm, respectively, using a cuvette with 1 cm path length, equals 572/547. Using the measured absorbance values of the diluted trisporoids, the concentration in your extracts can be calculated. This procedure gives no exact values, because (1) trisporoids extracts contain a mixture of trisporoid compounds, some of them without known specific extinction coefficient, and (2) several coextracted other substances, e.g., ergosterol or other UV-absorbing metabolites, will also add to the absorbance measurements. Nevertheless, the values are helpful to assess the quality of an extraction and provide a means to adjust different samples to similar concentrations, e.g., for TLC or physiological tests. With a recording UV/Vis spectrophotometer, the absorbance spectra in the range between 200 and 400 nm provide more information on the quality of the extract (5).
21. Chamber saturation can be enhanced by lining the chamber inside with strong filtration paper.
22. Packages containing more than one plate should only be opened far enough to allow removal of single plates. Never unwrap a parcel of plates completely, as the sorbent will take up and accumulate contaminations from the air.
23. Best results are obtained with special fixed volume glass capillaries, but a micropipette may also be used. Take care not to damage the sorbent surface with the application tool.
24. Trisporoids absorb UV irradiation and therefore appear as dark spots on the green-fluorescing background provided by the fluorescent indicator mixed into the sorbent layer. Document the result fast, as the trisporoids will degrade under UV irradiation and the bands will fade rapidly.
25. *M. mucedo* zygophores are small, smooth-walled aerial hyphae with rounded tips. They do not form any bulges and do not develop sporangia at the top. The total length is about 0.5 mm, somewhat longer for our (+) mating type. Usually, they do not branch.
26. We prepare the filter disks using an office punch and spread them on freshly uncovered laboratory film.

27. In an area of a few square millimeters, several hundred zygophores may develop. Best for counting are numbers between 5 and $25/mm^2$. Too densely grown zygophores are impossible to count; with too low numbers, it is hard to reliably recognize a given hyphae as a zygophore.

References

1. Caglioti L, Cainelli G, Camerino B, Mondelli R, Prieto A, Quilico A, Salvatori T, Selva A (1966) The structure of trisporic-C acid. Tetrahedron 22(Suppl 7):175–187
2. Wöstemeyer J, Schimek C (2007) Trisporic acid and mating in zygomycetes. In: Heitman J, Kronstad JW, Taylor JW, Casselton LA (eds) Sex in fungi. Molecular determination and evolutionary implications. ASM Press, Washington, pp 431–443
3. Schimek C, Wöstemeyer J (2006) Pheromone action in the fungal groups chytridiomycota, and zygomycota, and in the oomycota. In: Kües U, Fischer R (eds) The mycota I: growth, differentiation and sexuality. Springer, Heidelberg, pp 213–229
4. Schultze K, Schimek C, Wöstemeyer J, Burmester A (2005) Sexuality and parasitism share common regulatory pathways in the fungus *Parasitella parasitica*. Gene 348:33–44
5. Schimek C, Kleppe K, Saleem A-R, Voigt K, Burmester A, Wöstemeyer J (2003) Sexual reactions in *Mortierellales* are mediated by the trisporic acid system. Mycol Res 107:736–747
6. Gooday GW (1978) Functions of trisporic acid. Philos Trans R Soc Lond B 284:509–520
7. Kuzina V, Cerdá-Olmedo E (2006) Modification of sexual development and carotene production by acetate and other small carboxylic acids in *Blakeslea trispora* and *Phycomyces blakesleeanus*. Appl Environ Microbiol 72:4917–4922
8. Schmidt AD, Heinekamp T, Matuschek M, Liebmann B, Bollschweiler C, Brakhage AA (2005) Analysis of mating-dependent transcription of *Blakeslea trispora* carotenoid biosynthesis genes *carB* and *carRA* by quantitative real-time PCR. Appl Microbiol Biotechnol 67:549–555
9. Thomas DM, Goodwin TW (1967) Studies on carotenogenesis in *Blakeslea trispora*. I. General observations on synthesis in mated and unmated strains. Phytochemistry 6:355–360
10. Schimek C, Petzold A, Schultze K, Wetzel J, Wolschendorf F, Burmester A, Wöstemeyer J (2005) 4-Dihydromethyltrisporate dehydrogenase, an enzyme of the sex hormone pathway in *Mucor mucedo*, is constitutively transcribed but its activity is differently regulated in (+) and (−) mating types. Fungal Genet Biol 42:804–812
11. Wöstemeyer J (1985) Strain-dependent variation in ribosomal DNA arrangement in *Absidia glauca*. Eur J Biochem 146:443–448
12. Sutter RP, Whitaker JP (1981) Zygophore-stimulating precursors (pheromones) of trisporic acids active in (−) *Phycomyces blakesleeanus*. J Biol Chem 256:2334–2341
13. Nieuwenhuis M, van den Ende H (1975) Sex specificity of hormone synthesis in *Mucor mucedo*. Arch Microbiol 102:167–169
14. Schachtschabel D, Menzel KD, Krauter G, David A, Roth M, Horn U, Boland W, Wöstemeyer J, Schimek C (2010) Production and derivate composition of trisporoids in extended fermentation of *Blakeslea trispora*. Appl Microbiol Biotechnol 88:241–249

Chapter 4

Isolation of Mutants and Construction of Intersexual Heterokaryons of *Blakeslea trispora*

Enrique Cerdá-Olmedo and Bina J. Mehta

Abstract

The Mucoral fungus *Blakeslea trispora* is used for the industrial production of β-carotene and lycopene. Two genetic techniques have been used to increase carotene accumulation: the isolation of mutants and the formation and segregation of heterokaryons. Because all life stages are multinucleated, recessive mutants are isolated after exposure to *N*-methyl-*N′*-nitro-*N*-nitrosoguanidine, a strong mutagen and inactivator of nuclei. Intersexual heterokaryons are obtained easily, because they are formed spontaneously during sexual interaction. Here are the pertaining methods, based on those previously developed for *Phycomyces blakesleeanus*, a related and better-known fungus.

Key words: *Blakeslea trispora*, Multinucleated spores, Mutants, Heterothallism, Intersexual heterokaryons

1. Introduction

Carotenes are used by the food, feed, cosmetics, and pharmaceutical industries because of their attractive colors and their beneficial effects on human and animal health. Natural strains of the filamentous fungi *Blakeslea trispora* and *Phycomyces blakesleeanus* (Order Mucorales, Subphylum Mucoromycotina, Zygomycota) accumulate β-carotene in their vegetative mycelia and can be mutated to produce its biosynthetic precursors, particularly lycopene. They are attractive to industry because of their frugal and flexible growth conditions and because large increases in carotene content can be brought about by modifying their genotype and the culture conditions. Industry prefers *Blakeslea* (1) to the better-known *Phycomyces*, and the methods developed with the latter (2, 3) are the basis for those used with the former (4) that are described here.

The β-carotene biosynthetic pathway is similar in all Mucorales, but its regulation varies so much that our knowledge of *Phycomyces*

José-Luis Barredo (ed.), *Microbial Carotenoids From Fungi: Methods and Protocols*, Methods in Molecular Biology, vol. 898, DOI 10.1007/978-1-61779-918-1_4, © Springer Science+Business Media New York 2012

does not often apply to *Blakeslea*. This has been shown by mutational analyses of structural and regulatory genes of *P. blakesleeanus*, *B. trispora*, and *Mucor circinelloides* and by the effects of physical and chemical agents on the carotene content of the mycelia (4–7 and references therein).

B. trispora has neither been crossed nor transformed with exogenous DNA. *P. blakesleeanus* has been subjected to classical genetic analyses via complementation and recombination (2, 8). *M. circinelloides* allows the interruption, insertion, and replacement of genes and the inhibition of gene expression by small RNAs (9, 10). Any biological problem can be investigated in parallel with these fungi, but results obtained with one of them should not be extrapolated to the others without verification.

The Mucorales are coenocytes. All stages of the *Blakeslea* life cycle are multinucleated and can be heterokaryotic, that is, nuclei with different genetic information may coexist in a cell. The coenocytic nature of *Blakeslea* is a hindrance for the isolation of recessive mutants and for the purification of dominant mutations to homokaryosis. The distribution of the number of nuclei in the vegetative spores (sporangiospores, or spores for short) depends on the strain (4, 11) and varies very little with the culture conditions, so that strains with low values are preferable in the laboratory.

Heterokaryons arise by mutation and may be formed artificially. Thus, spontaneous mutants detected by their phenotype are heterokaryons for dominant mutations. Because spores contain random samples of the nuclei in the mycelium, heterokaryons segregate progeny with different proportions of the constituent nuclei, and thus homokaryons can be obtained by subculturing and screening (4).

Many mutagens damage DNA and inactivate nuclei to the extent that they cannot reproduce further (12). Some surviving spores become functionally uninucleated, having lost all of their nuclei except one, and can express recessive mutations (13). The method below is appropriate for the screening of common color mutants (4, 11) and can be adapted to other purposes. When applied to strain F921 (average 6.4 nuclei per spore), about 12% of the surviving spore population should be functionally uninucleated and allow the isolation of all kinds of viable mutants, including those with recessive mutations (4). The multinucleated survivors constitute an unproductive workload, which can be avoided by applying the mutagen to the spores of a heterokaryon in which a minority of the nuclei carry a selectable recessive marker; the survivors are first selected for the marker and then screened for the desired recessive mutations. The marker that was used with *Phycomyces*, resistance to 5-deazariboflavin, cannot be used with *Blakeslea*, which is naturally resistant.

Heterokaryons permit analyses of the degree of dominance and recessivity of mutations, functional complementation, and descriptions of the effect of allele dosage on phenotype (2, 14). They have been used to identify genes from mutations and to obtain information on their function. *Phycomyces* lends itself particularly well for these methods, which can be extended to *Blakeslea*.

Many Mucorales, including *Blakeslea*, are heterothallic, i.e., the mycelia belong to either of two sexes, called (+) and (−). When mycelia of opposite sex grow near each other, they exchange chemical signals that initiate sexual development and, in the case of many strains, stimulate carotene production (sexual carotenogenesis), leading to large increases in carotene content. This was the first known case of chemical signaling between organisms (15) and is mediated by apocarotenoids derived from a double fragmentation of β-carotene (16). *Blakeslea* has been used often in this field of research (1, 4, 17, 18). Industry exploits sexual carotenogenesis to extract carotenes from "mated cultures" of two strains of opposite sex. However, stimulation in the industrial process may be hampered by suboptimal distribution of the mycelia of both strains.

The concept and the name "heterokaryon" originated from early research with *Phycomyces* (19). Artificial heterokaryons of this fungus have been prepared with various methods (2), which often take advantage from the large size of the sporangiophores and may be difficult to adapt to *Blakeslea*. Intersexual heterokaryons contain nuclei of opposite sex in the same cytoplasm and allow sexual carotenogenesis to occur in a single strain. Instability results from variations in the relative proportion of the constituent nuclei and from the vegetative segregation of homokaryons. Segregation could be impeded by introducing recessive lethal mutations in the constituent genomes (20) and mycelia with the best nuclear proportions could be obtained by subculturing small mycelial fragments (4). Intersexual heterokaryons are easy to obtain because they are formed spontaneously during sexual interaction (21). This is the fundamental method used for *Blakeslea* (4).

2. Materials

2.1. Strains

1. *B. trispora* Thaxter F921 (−) wild-type strain (22, 23) (Vsiesoyuznaya Kollektsiya Mikroorganizmov (VKM), Moscow, Russia) (see Note 1).
2. *B. trispora* Thaxter F986 (+) wild-type strain (22, 23) (VKM, Moscow, Russia) (see Note 1).

2.2. Culture Media

All solutions and media are prepared with distilled water. The culture media described here were designed originally for strains of *P. blakesleeanus* and work well for *B. trispora*.

1. Minimal agar: Prepare two separate solutions. Solution A: 20 g D-glucose and 15 g agar in 500 mL of distilled water. Solution B: 2 g L-asparagine H_2O, 5 g KH_2PO_4, 0.5 mg $MgSO_4 \cdot 7H_2O$, and 20 mL micronutrient stock solution in 500-mL water. Autoclave separately at 100 kPa (about 1 atm) overpressure for 20 min, mix well and use.
2. Minimal agar, acid: Minimal agar acidified to pH 3.4 with HCl after autoclaving and mixing Solution A and Solution B (see Note 2).
3. Micronutrient stock solution (50×): 2.8 g/L $CaCl_2$, 0.05 g/L-thiamine HCl, 0.1 g/L citric acid H_2O, 75 mg/L $Fe(NO_3)_3 \cdot 9H_2O$, 50 mg/L $ZnSO_4 \cdot 7H_2O$, 15 mg/L $MnSO_4 \cdot H_2O$, 2.5 mg/L $CuSO_4 \cdot 5H_2O$, and 2.5 mg/L $Na_2MoO_4 \cdot 2H_2O$ in water. For long-term storage at room temperature, add a few drops of chloroform.
4. Rich agar: Minimal agar with an additional 1 g yeast extract in Solution A.
5. Rich agar, acid: Rich agar acidified to pH 3.4 with HCl after autoclaving and mixing Solution A and Solution B (see Note 2).
6. Potato-dextrose agar: Boil 200 g fresh, peeled and diced potatoes in 1.5 L water for 1 h. Filter and add to the extract 20 g D-glucose, 1 mg thiamine.HCl and 15 g agar. Bring volume to 1 L with distilled water and autoclave. Alternatively, commercial, dehydrated potato dextrose agar may be used (sometimes with poor results).

2.3. Miscellaneous

1. *N*-methyl-*N*′-nitro-*N*-nitrosoguanidine (MNNG) stock solution: It is prepared under a safety hood (see Note 3). Pour a small amount of MNNG (a few milligrams) into a pre-weighed tube. Weigh the tube again and bring the concentration to 1 mg/mL with water. Close the tube tightly and shake to accelerate solution. Place aliquots of about 0.2 mL in sterile Eppendorf tubes (1.5 mL capacity) and keep frozen in the dark until use. The solution is yellow and bleaches if the compound is inactivated; the absorbance of a 1 mg/mL solution at 400 nm through a 1-cm light path is 1.1. Unfrozen solution aliquots should not be reused (see Note 4).
2. Sodium thiosulfate solution: 20 g/L $Na_2S_2O_3 \cdot 10H_2O$.
3. Tween-80 solution: 1 mL/L polyoxyethylene sorbitan monooleate in water, sterilized in the autoclave.

3. Methods

3.1. Spore Harvest

1. Harvest the spores from sporulated cultures (4–5 days after incubation on Minimal agar at 30°C) by gently washing the mycelial surface with sterile Tween-80 solution (see Note 5).
2. Centrifuge the resulting spore suspension at 2,000 × *g* (or more) for 1 min.
3. Wash the spore pellet twice with sterile Tween-80 solution and resuspend in sterile Tween-80 solution.

3.2. Strain Conservation

1. Spore suspensions in Tween-80 solution lose viability at 4°C in a few days. For long-term conservation at −20°C, sterile glycerol is added to a final concentration of 200 mL/L.
2. Strain collections are kept as lyophils. Sterilize lyophil tubes (see Note 6) after placing in them a small piece of paper with the penciled strain name and a cotton plug. Suspend the spores in a sterile protein solution (e.g., blood serum, serum albumin solution, or low-fat milk reconstituted from powder), put 0.5 mL in each tube, and follow the instructions of a lyophilizer to freeze them (at about −50°C), dry them (at a few Pa), and close them with a blowtorch. Write the strain name in outside tags and keep the collection in a cold room or at room temperature.

3.3. Strain Cultivation

1. Spread 10^4 spores onto minimal agar plates and incubate in the dark for 4 days (see Note 7). The recommended temperature for growth is 30°C and for sporulation 23°C; genetic procedures can be carried out satisfactorily at the usual room temperatures. Medium, inoculum size, temperature, and incubation time may have to be changed in some experiments.

3.4. Mutagenesis

1. Keep ready a beaker (500 mL) with 200 mL of sodium thiosulfate solution.
2. Centrifuge for 1 min at 2,000 × *g* (or more) two aliquots of a spore suspension, each containing 5×10^6 recently collected, viable spores/mL. Resuspend the pellets in 0.95 mL Tween-80 solution in sterile Eppendorf tubes (2 mL). Remove 50 μL from each tube to dilution tubes with Tween-80 solution for viable spore counts on minimal agar.
3. Add 0.1 mL Tween-80 solution to one of the tubes and 0.1 mL MNNG solution to the other and incubate them for 30 min at 30°C in the dark with very gentle shaking. The final concentration of MNNG would be 0.1 g/L. The contaminated micropipette tip is disposed in the thiosulfate solution.

4. Once the incubation is completed, remove 50 μL from each tube for viable spore counts and centrifuge the rest for 1 min. Discard the control tube.
5. Decant the supernatant in the sodium thiosulfate solution (see Note 8).
6. Resuspend the spore pellet in 1 mL Tween-80 solution, shake well and centrifuge again.
7. Resuspend the spore pellet in 1 mL Tween-80 solution, shake well and spread 50 μL aliquots on Rich acid agar plates.
8. Incubate the plates in the dark at 30°C for 4–8 days (see Note 9).

3.5. Screening and Purification of Mutants

1. Identify putative mutant colonies and subculture them individually on acid agar at 30°C in the dark. The inoculum can be a mycelial fragment cut with the tips of fine tweezers (see Note 10) or spores transferred with sterile toothpicks. Do the same with some wild-type colonies as controls.
2. Subculture a colony from each putative mutant by spreading spores on acid agar to obtain separate colonies.
3. Check phenotype and subculture at least once more. If the colonies are not uniform, subculture until all colonies are similar in morphology, growth, and color.
4. Harvest spores and keep in the collection under a permanent identification number and a genotype (see Note 11).

3.6. Construction of Intersexual Heterokaryons

1. Inoculate mycelial fragments (see Note 10) of two strains of opposite sex on opposite sides of Potato-dextrose agar plates (see Note 12).
2. Incubate the plates in the dark until bright yellow bands appear along the line where the mycelia meet.
3. Transfer mycelial fragments from the bright yellow bands to Minimal agar (see Note 13).
4. After incubation in the dark, check for patches of intersexual mycelia (see Note 14).
5. Subculture mycelial fragments of these intersexual mycelia on minimal agar plates and look for relatively uniform bright color and velvety appearance.
6. Keep the intersexual heterokaryotic mycelia in the collection (see Note 15).

3.7. Segregation and Purification of Intersexual Heterokaryons

1. Place small fragments of the intersexual mycelia in a sterile Eppendorf tube with 0.2 mL of sterile 0.55 M sorbitol solution (see Note 16) and about 20–25 mg sterile sea sand.
2. Shake the tube vigorously in a vortex mixer. Shake thrice for 20 s and keep the tube in ice for 10 s in between. Check for

hyphal breakage under the microscope and repeat if necessary.

3. Spread about 40 μL of the shaken suspension on minimal acid agar plates and incubate them for about 4 days in dark.
4. The colonies that result are heterogeneous, from homokaryons to heterokaryons with various proportions of the constituent strains. Choose and keep the individual colonies of interest.
5. If necessary, subculture mycelial fragments of intersexual mycelia until they are relatively uniform in color and keep them in the collection (see Note 17).

4. Notes

1. Strains of *B. trispora*, most of them coming from tropical environments, are present in many public collections. These methods were standardized for the strains indicated and may have to be modified for strains with different environmental preferences (e.g., optimal temperature) or genetic background (e.g., number of nuclei per spore).
2. The mycelia of *B. trispora* grow indefinitely over solid substrates. The acidified medium restricts growth to distinct colonies and facilitates titration of viable spores (defined as colony forming units), isolation of mutants, and other genetical procedures. Determine the needed amount of HCl solution on an aliquot of the molten agar preparation cooled to just above gelification.
3. MNNG is a potent carcinogen to be used with extreme care. It is instantly inactivated by thiosulfate and more slowly by organic matter and light. Its low volatility and high chemical reactivity diminish the risks of accidental exposure and environmental contamination. Contaminated material should be immersed in the thiosulfate solution before washing or disposal.
4. To control accidental thawing and refreezing, put 1 mL water in a similar tube, let it freeze, place a tack on the ice surface, and keep with the other tubes.
5. The spores of *B. trispora* are ellipsoidal with tufts of fine flagella at both ends. They stick to each other and to many surfaces, making it difficult to obtain suspensions of free spores. They separate and float freely in a mild detergent solution.
6. Use lyophil tubes (110 × 7 mm) of good quality (e.g., Pyrex), because common glass tends to develop fissures.
7. Spores require no special activation to germinate. Mycelial development varies with the strains, and some require 1 or 2 more days for sporulation.

8. If the spore pellet is disturbed or not properly settled, it may decant along with the supernatant. In this case, aspire the supernatant slowly with a pipette with a sterile disposable tip (1 mL), taking care not to touch the pellet. The aspirated supernatant and the used tip are disposed in the thiosulfate solution.
9. This method is appropriate for mutants to be screened, for example, for changes in mycelial color or biochemical properties. For rarer mutants that can be selected, the volumes should be scaled up in adequate containers, such as 10-mL centrifuge tubes with a conical bottom end. The spore concentration should not be higher than 10^7 spores/mL to avoid the inactivation of MNNG during exposure.
10. Adequate inocula are mycelial mat fragments (about 3 × 3 mm, about 0.5 mg dry mass) cut with the tips of fine tweezers. Sterilize the tweezers tips by dipping them in ethanol and passing them through a low flame.
11. Strains are named with one or more capital letters that identify the original collection followed by a number. Thus, SB39 is a strain in the Sevilla *Blakeslea* collection. Each strain is assumed to have its own genotype. Genotypes are described by the sex, either (+), (−), or (0), and the names of the mutations, all in italics. Mutation names consist of a low-case three-letter code, followed by a hyphen or a capital letter, and an isolation number. The three-letter code is usually chosen to remind of a group of related phenotypes. Thus, the genotype of SB40 is *car-7* (+); *car* designates all the genes whose mutations cause quantitative or qualitative changes in carotene content. The hyphen indicates an undefined gene. If the gene becomes known, the hyphen will be replaced by a capital letter to specify that gene. Three letter codes, gene letters and mutation numbers should be kept unique by contacting other researchers working in the field. A new round of mutagenesis would change the strain designation and add one or more mutation names to the ones already present in the genotype.
12. Sexual reactions occur on Minimal agar, but are stimulated by acetate and by potato extract (18).
13. Intersexual heterokaryons are produced spontaneously in the "sexual tissue" at the contact zone of mycelia of opposite sex. Vegetative sporulation is inhibited at and near the contact zone. Some strains, both isolated from nature and obtained in the laboratory, are neutral, (0); they do not react sexually with other strains in any known conditions.
14. Intersexual heterokaryons are recognized easily by their bright yellow color and peculiar velvety appearance. The bright color is due to sexual carotenogenesis and the velvety appearance is

due to the development of tiny aerial, contorted hyphae called pseudophores.

15. Heterokaryons are designated with the names of the constituent strains separated by a low asterisk, for example, SB39*SB32. For intersexual heterokaryons, the (+) strain is written first.
16. The osmotic balance improves viability when mycelia are broken during vigorous shaking.
17. The mycelia of intersexual heterokaryons of *Phycomyces* are less uniform in color than regular heterokaryons, unless the constituent strains share most of their genetic backgrounds. There are no isogenic strains of *Blakeslea* whose genomes differ in sex only. Intersexual heterokaryons may have to be repurified to improve sexual stimulation and the concomitant increase in carotene accumulation.

Acknowledgment

This work was supported by Junta de Andalucía (CVI-03901) and the Spanish Government (BIO2009-12486 and 11131).

References

1. Ciegler A (1965) Microbial carotenogenesis. Adv Appl Microbiol 7:1–34
2. Cerdá-Olmedo E, Lipson ED (eds) (1987) *Phycomyces*. Cold Spring Harbor Laboratory, New York
3. Cerdá-Olmedo E (2001) *Phycomyces* and the biology of light and color. FEMS Microbiol Rev 25:503–512
4. Mehta BJ, Obraztsova IN, Cerdá-Olmedo E (2003) Mutants and intersexual heterokaryons of *Blakeslea trispora* for production of β-carotene and lycopene. Appl Environ Microbiol 69:4043–4048
5. Avalos J, Cerdá-Olmedo E (2004) Fungal carotenoid production. In: Arora DK (ed) Handbook of fungal biotechnology, vol 20. Marcel Dekker, New York, pp 367–378
6. Rodríguez-Sáiz M, Paz B, De la Fuente JL, López-Nieto MJ, Cabri W, Barredo JL (2004) *Blakeslea trispora* genes for carotene biosynthesis. Appl Environ Microbiol 70:5589–5594
7. Silva F, Navarro E, Peñaranda A, Murcia-Flores L, Torres-Martínez S, Garre V (2008) A RING-finger protein regulates carotenogenesis via proteolysis-independent ubiquitylation of a *white collar-1*-like activator. Mol Microbiol 70:1026–1036
8. Cerdá-Olmedo E (1975) The genetics of *Phycomyces blakesleeanus*. Genet Res 25:285–296
9. Silva F, Torres-Martínez S, Garre V (2006) Distinct *white collar-1* genes control specific light responses in *Mucor circinelloides*. Mol Microbiol 61:1023–1037
10. Haro JP, Calo S, Cervantes M, Nicolás FE, Torres-Martínez S, Ruiz-Vázquez RM (2009) A single *dicer* gene is required for efficient gene silencing associated with two classes of small antisense RNAs in *Mucor circinelloides*. Eukaryot Cell 8:1486–1497
11. Mehta BJ, Cerdá-Olmedo E (1995) Mutants of carotene production in *Blakeslea trispora*. Appl Microbiol Biotechnol 42:836–838
12. Cerdá-Olmedo E, Reau P (1970) Genetic classification of the lethal effects of various agents on heterokaryotic spores of *Phycomyces*. Mutat Res 9:369–384
13. Roncero MIG, Zabala C, Cerdá-Olmedo E (1984) Mutagenesis in multinucleate cells: the effects of N-methyl-N′-nitro-N-nitrosoguanidine on *Phycomyces* spores. Mutat Res 125:195–204
14. De la Guardia MD, Aragón CMG, Murillo FJ, Cerdá-Olmedo E (1971) A carotenogenic

enzyme aggregate in *Phycomyces*: evidence from quantitative complementation. Proc Natl Acad Sci USA 68:2012–2015

15. Burgeff H (1924) Untersuchungen über Sexualität und Parasitismus bei Mucorineen 1. Botanische Abhandlungen 4:1–135
16. Polaino S, Herrador MM, Cerdá-Olmedo E, Barrero AF (2010) Splitting of β-carotene in the sexual interaction of *Phycomyces*. Org Biomol Chem 8:4229–4231
17. Sutter RP (1970) Trisporic acid synthesis in *Blakeslea trispora*. Science 168:1590–1592
18. Kuzina V, Cerdá-Olmedo E (2006) Modification of sexual development and carotene production by acetate and other small carboxylic acids in *Blakeslea trispora* and *Phycomyces blakesleeanus*. Appl Environ Microbiol 72:4917–4922
19. Burgeff H (1914) Untersuchungen über Variabilität, Sexualität und Erblichkeit bei *Phycomyces nitens* Kunze, I. Flora 107:259–316
20. Murillo FJ, Calderón IL, López-Díaz I, Cerdá-Olmedo E (1978) Carotene-superproducing strains of *Phycomyces*. Appl Environ Microbiol 36:639–642
21. Gauger WL, Peláez MI, Álvarez MI, Eslava AP (1980) Mating type heterokaryons in *Phycomyces blakesleeanus*. Exp Mycol 4:56–64
22. Thaxter R (1914) New or peculiar zygomycetes. III. *Blakeslea*, *Dissophora*, *Haplosporangium*, nova genera. Bot Gaz (Crawfordsville) 58:353–366
23. Zycha H, Siepmann R, Linnemann G (1969) Mucorales, Eine Beschreibung aller Gattungen und Arten dieser Pilzegruppe. Cramer, Lehre

Chapter 5

Molecular Tools for Carotenogenesis Analysis in the Zygomycete *Mucor circinelloides*

Santiago Torres-Martínez, Rosa M. Ruiz-Vázquez, Victoriano Garre, Sergio López-García, Eusebio Navarro, and Ana Vila

Abstract

The carotene producer fungus *Mucor circinelloides* is the zygomycete more amenable to genetic manipulations by using molecular tools. Since the initial development of an effective procedure of genetic transformation, more than two decades ago, the availability of new molecular approaches such as gene replacement techniques and gene expression inactivation by RNA silencing, in addition to the sequencing of its genome, has made *Mucor* a valuable organism for the study of a number of processes. Here we describe in detail the main techniques and methods currently used to manipulate *M. circinelloides*, including transformation, gene replacement, gene silencing, RNAi, and immunoprecipitation.

Key words: Mucor, Transformation, Gene replacement, Gene silencing, RNAi, Immunoprecipitation, Ubiquitylation

1. Introduction

Zygomycetes are a basal class of filamentous fungi that are receiving a growing attention due to their evolutionary distance to other fungi. In recent years, several molecular processes have been dissected, such as mechanisms involved in regulation of light responses (1, 2), determination and evolutionary implications of genes involved in sex determination (3, 4), methylation of their genomes (5), the mechanism of gene silencing (6–9), and the implication of endogenous small RNAs in gene regulation (10). They are also the main source of fungal carotenoids in industry, and could be considered as the most worthwhile alternative to applying molecular approaches to obtain carotene-overproducing strains of industrial interest.

José-Luis Barredo (ed.), *Microbial Carotenoids From Fungi: Methods and Protocols*, Methods in Molecular Biology, vol. 898, DOI 10.1007/978-1-61779-918-1_5,

The main advantages of using filamentous fungi over yeast or bacteria are the high levels of produced metabolite and final biomass. Structural genes involved in the β-carotene biosynthesis in the zygomycetes *Blakeslea trispora* (11), *Phycomyces blakesleeanus* (12, 13), and *Mucor circinelloides* (14, 15) have been cloned and characterized. Among these, *M. circinelloides* has proved to be a good organism for a molecular approach to the study of carotenogenesis regulation. Several molecular tools, including genetic transformation (16, 17), gene replacement (18), and RNA-mediated gene silencing (19), are available for this fungus, allowing the manipulation of its genome. Furthermore, its genome sequence is already available (http://genome.jgi-psf.org/Mucci2/Mucci2.home.htmL). The availability of these molecular tools has opened new ways to study carotene biosynthesis regulation in *M. circinelloides* and is also contributing to the design of strategies for the study of many different processes by a growing number of research groups. Here we describe the procedures to obtain protoplasts and to transform *Mucor*, including a gene replacement method and the analysis of transformants, the construction of a specific vector to silence genes, procedures to isolate low molecular weight RNAs and, finally, the methods currently used to prepare cell lysate and to analyze specific proteins by immunoprecipitation.

2. Materials

2.1. Transformation, Gene Replacement, and Transformant Analysis

1. *M. circinelloides* CBS277.49 (CBS-KNAW Fungal Biodiversity Center, Utrecht, The Netherlands).
2. *M. circinelloides* R7B (20) (see Note 1).
3. SPB (0.1 M sodium phosphate buffer pH 6.5): 1.42 g of Na_2HPO_4 in a final volume of 100 mL double-distilled water (solution 1); 1.38 g of NaH_2PO_4 monohydrate in a final volume of 100 mL double-distilled water (solution 2). Take 100 mL of solution 2 and add solution 1 slowly until pH 6.5 is reached (about 53 mL).
4. PS buffer: Mix 18.22 g of sorbitol and 20 mL of SPB. Make up to 200 mL with double-distilled water (final sorbitol concentration: 0.5 M).
5. Lysing enzymes (Sigma-Aldrich, St. Louis, MO, USA).
6. Chitosanase RD (US Biological, Swampscott, MA, USA).
7. 0.5 M Sorbitol: Dissolve 22.77 g of sorbitol in 200 mL double-distilled water. Make up to 250 mL with double-distilled water.
8. Gene Pulser Cuvette (0.2-cm electrode gap) (Bio-Rad, Hercules, CA, USA).
9. Bio-Rad Gene Pulser Xcell (Bio-Rad, Hercules, CA, USA).

10. YPG: 3 g/L of yeast extract, 10 g/L of peptone, and 20 g/L of glucose (see Note 2).
11. YPG-agar: YPG and 20 g/L agar (see Note 2).
12. YPGS: Add 91.1 g of sorbitol to the YPG medium before to adjust the volume to 1,000 mL with distilled water (final sorbitol concentration: 0.5 M).
13. YNB: 1.5 g/L of ammonium sulfate, 1.5 g/L of glutamic acid, 0.5 g/L of yeast nitrogen base (w/o ammonium sulfate and amino acids), and 10 g/L of glucose. After autoclaving add thiamine and niacine at a final concentration of 1 μg/mL (see Note 2).
14. YNB-agar: YNB and 20 g/L agar (see Note 2).
15. YNBS: Add 91.1 g/L of sorbitol to the YNB medium before to adjust the volume to 1,000 mL with distilled water (final sorbitol concentration: 0.5 M).
16. MMC: 10 g/L of casaminoacids, 0.5 g/L of yeast nitrogen base (w/o ammonium sulfate and amino acids), and 20 g/L of glucose. After autoclaving add thiamine and niacine at a final concentration of 1 μg/mL (see Note 2).
17. MMC-agar: MMC and 15 g/L agar (see Note 2).
18. MMCS: Add 91.1 g/L of sorbitol to the MMC medium before to adjust the volume to 1,000 mL with distilled water (final sorbitol concentration: 0.5 M).
19. DMSO (dimethyl sulfoxide).
20. Biotaq DNA Polymerase (Bioline Ltd, London, UK).
21. Mixture for colony PCR: 0.5 μL of Biotaq enzyme (5 U/μL), 2.5 μL of 10× Biotaq specific PCR buffer, 2.5 μL of 2 mM dNTP's, 1 μL of 50 mM $MgCl_2$, 1 μL of a 10 pmol/μL solution of each primer, and 15.25 μL of double-distilled water (see Note 3).

2.2. Gene Function Analysis by RNA Silencing

1. pr5: 5′-CCGCGGTCGACGGATCCTACGATGCGCCTG CTG-3′. Restriction site for *Bam*HI is underlined.
2. pr3a: 5′-CGGGGGTCGACTGAACGCGGAATACTTCAGG-3′. Restriction site for *Sal*I is underlined.
3. pr3b: 5′-GCGGGCTCGAGTTGCACCCACAAAGAATAG-3′. Restriction site for *Xho*I is underlined.
4. pBlueScript SK+ (Stratagene, La Jolla, CA, USA).
5. pEUKA4 (21).
6. Trizol (Invitrogen, Carlsbad, CA, USA).
7. Diethyl pyrocarbonate (DEPC)-treated water: Add 1 mL of DEPC to 1,000 mL of double-distilled water (0.1% DEPC v/v) and stir at room temperature for at least 3 h to bring the DEPC into the solution. Let the solution incubate for 12 h at 37°C and autoclave for 15 min to remove any trace of DEPC (see Note 4).

8. 4 M NaCl: 234 g of NaCl and double-distilled water to 1,000 mL.
9. PEG 8000: Poly(ethylene glycol) 8000 (Sigma-Aldrich, St Louis, MO, USA).
10. 50% PEG 8000 (w/v): 50 g of PEG 8000; bring the final volume to 100 mL with double-distilled water and autoclave.
11. 0.5 M EDTA pH 8.0: Add 186.1 g of disodium EDTA·$2H_2O$ to 800 mL of H_2O. Stir vigorously on a magnetic stirrer. Adjust the pH to 8.0 with NaOH (~20 g of NaOH pellets). Dispense into aliquots and sterilize by autoclaving. The disodium salt of EDTA will not go into solution until the pH of the solution is adjusted to ~8.0 by the addition of NaOH.
12. 1× TAE: Make a concentrated (50×) stock solution of TAE by weighing out 242 g of Trizma base and dissolving in approximately 750 mL distilled water. Carefully add 57.1 mL glacial acetic acid and 100 mL of 0.5 M EDTA (pH 8.0), and adjust the solution to a final volume of 1 L. This stock solution can be stored at room temperature. The pH of this buffer is not adjusted and should be about 8.5. The working solution of 1× TAE buffer is made by diluting the stock solution by 50× in double-distilled water (final concentrations: 40 mM Tris acetate and 1 mM EDTA).
13. 1.5% Agarose gel: Add 1.5 g of RNase-free agarose to 100 mL 1× TAE. Heat the solution to boiling in the microwave to dissolve the agarose.
14. 3 M Sodium acetate pH 5: 24.61 g of sodium acetate (anhydrous) and 80 mL double-distilled water. Adjust to pH 5.0 with glacial acetic acid. Adjust volume to 100 mL with double-distilled water and autoclave.
15. 2× Formamide loading buffer: To prepare 20 mL, mix 19 mL of formamide (final concentration: 95% v/v), 720 μL of 0.5 M EDTA pH 8.0 (final concentration: 18 mM), 50 μL of 10% SDS (final concentration: 0.025% w/v), 5 mg of xylene cyanol (0.025% w/v), 5 mg of bromophenol blue (0.025% w/v), and 230 μL double-distilled water. Store at −20°C.
16. 10× TBE: 108 g of Trizma base, 55 g of boric acid, 9.5 g of EDTA, and double-distilled water to 1,000 mL. Adjust pH to 8.35.
17. 0.5× TBE: To prepare 1,000 mL, add 950 mL double-distilled water to 50 mL of 10× TBE.
18. AccuGel™ 19:1–40%: A 19:1 acrylamide to bisacrylamide stabilized solution (National Diagnostics, Atlanta, GA, USA).
19. 0.5× TBE/7 M urea/15% acrylamide gel: Mix in a 50-mL Falcon tube 16.8 g of Urea, 10 mL DEPC-treated water, 2 mL of 10× TBE, and 15.53 mL of AccuGel™ 19:1–40%. Stir at

room temperature to dissolve urea and add DEPC-treated double-distilled water to 40 mL. Add (in this order) 320 μL of 10% (w/v) ammonium persulphate (APS) and 42.6 μL TEMED. Mix well and pour immediately, using a pipette, into the assembled glass plate sandwich, following indications of the Protean II xi Cell Instruction Manual. Keep at 4°C. Polymerization time is about 30 min. After that, remove the comb and fill the wells with 0.5× TBE. With a syringe, remove the bubbles inside the wells. Then, the gel is ready to be assembled in the electrophoresis chamber.

20. Protean® II xi Gel System (Bio-Rad, Hercules, CA, USA).
21. 3 MM Whatman paper (Whatman Ltd, Maidstone, Kent, UK).
22. Hybond™-N+ (GE Healthcare, Waukesha, WI, USA).
23. Semidry electroblotter unit (Sigma-Aldrich, St Louis, MO, USA).
24. UV-Crosslinker (Hoefer, Inc., Holliston, MA, USA).
25. MAXIscript® T7 (Applied Biosystems/Ambion, Austin, TX, USA).
26. TE: 10 mM Tris–HCl, pH 8.0, and 1 mM EDTA.
27. Sephadex G-50 (GE Healthcare, Waukesha, WI, USA).
28. Alkaline buffer: 0.317 g of Na_2CO_3 and 0.168 g of $NaHCO_3$, and double-distilled water to 25 mL (final concentrations: 120 mM Na_2CO_3 and 80 mM $NaHCO_3$). Autoclave, make aliquots, and store at −20°C.
29. Prehybridization/hybridization buffer: To prepare 20 mL of buffer, dissolve 1.4 g of SDS in 9.1 mL double-distilled water, by incubating at 65°C. Once dissolved, add 1.5 mL of 4 M NaCl, 1 mL of 1 M Na_2HPO_4–NaH_2PO_4, pH 7.1, 400 μL of 50× Denhardt's solution, and 8 mL of formamide (final concentrations: 7% SDS, 0.3 M NaCl, 0.05 M Na_2HPO_4–NaH_2PO_4, pH 7, 1× Denhardt's solution).
30. 50× Denhardt's solution: 1% (w/v) bovine serum albumin (BSA), 1% (w/v) of Ficoll 400, 1% (w/v) of polyvinylpyrollidone (PVP) in double-distilled water.
31. 20× SSC solution: 175.32 g of NaCl and 88.23 g of trisodium citrate; adjust to pH 7 and bring to a final volume of 1,000 mL with double-distilled water.
32. 10% SDS solution: 20 g of SDS in a final volume of 200 mL double-distilled water. Resuspend at 50°C (do not autoclave).
33. SDS/SSC solution: Mix 100 mL of 20× SSC, 20 mL of 10% SDS, and 880 mL of double-distilled water (see Note 5).
34. Salmon sperm DNA (deoxyribonucleic acid sodium salt from salmon testes) (Sigma-Aldrich, St Louis, MO, USA).
35. Salmon sperm DNA 5 mg/mL: Weigh 100 mg of salmon sperm DNA and add DEPC-treated double-distilled water to

20 mL (final concentration: 5 mg/mL). If necessary, stir the solution on a magnetic stirrer for 2–4 h at room temperature to help the DNA to dissolve. Autoclave and dispense into aliquots in microfuge tubes. Store at –20°C.

36. Kodak BioMax™ MS film (Sigma-Aldrich, St Louis, MO, USA).
37. 1 M Tris–HCl pH 7.5 (100 mL): Mix 12.1 g of Trizma base with 80 mL ddH_2O. Adjust pH with HCl (pH is temperature dependent; make sure it is at room temperature before making final pH adjustments). Add dH_2O to 100 mL and autoclave.
38. RNase buffer: 400 μL 1 M Tris–HCl pH 7.5, 200 μL 0.5 M EDTA, 300 μL 4 M NaCl, and double-distilled water to 20 mL.
39. RNase A (Sigma-Aldrich, St Louis, MO, USA).

2.3. Cell Lysate Preparation, Immunoprecipitation, and Ubiquitilation Analysis

1. Semidry electroblotter unit (Sigma-Aldrich, St Louis, MO, USA).
2. Mini-Protean 3 Cell (Bio-Rad Laboratories, Inc., Hercules, CA, USA).
3. Protease Inhibitor Cocktail (Sigma-Aldrich, St Louis, MO, USA).
4. Benzamidine (Sigma-Aldrich, St Louis, MO, USA).
5. Sodium deoxycholate (Sigma-Aldrich, St Louis, MO, USA).
6. Igepal CA-630 (Sigma-Aldrich, St Louis, MO, USA).
7. Lysis buffer (washing buffer 1): 2.5 mL of 1 M Tris–HCl pH 7.5 (121.14 g/L), 1.875 mL of 4 M NaCl (233 g/L), 0.5 mL of Igepal CA-630, and 0.25 g of sodium deoxycholate. Add double-distilled water to 50 mL (final concentrations: 50 mM Tris–HCl, pH 7.5; 150 mM sodium chloride; 1% Igepal CA-630; 0.5% sodium deoxycholate).
8. Protein A-Agarose Fast Flow (Sigma-Aldrich, St Louis, MO, USA).
9. Washing buffer 2: 2.5 mL of 1 M Tris–HCl pH 7.5 (121.14 g/L), 6.25 mL of 4 M NaCl (233 g/L), 0.05 mL of Igepal CA-630, and 0.025 g of sodium deoxycholate. Add double-distilled water to 50 mL (final concentrations: 50 mM Tris–HCl, pH 7.5; 500 mM sodium chloride; 0.1% Igepal CA-630; 0.05% sodium deoxycholate).
10. Washing buffer 3: 2.5 mL of 1 M Tris–HCl pH 7.5 (121.14 g/L), 0.05 mL of Igepal CA-630, and 0.025 g of sodium deoxycholate. Add double-distilled water to 50 mL (final concentrations: 50 mM Tris–HCl, pH 7.5; 0.1% Igepal CA-630; 0.05% sodium deoxycholate).
11. 5× Protein loading buffer: 375 microL 3M Tris–HCl pH 6.8, 2.5 mL of glycerol, 0.25 g of SDS, 0.0025 g of bromophenol blue, and 0.0125 g of dithiothreitol (DTT), add double distilled water to 5 mL (final concentrations: 0.225 M Tris–HCl

pH 6.8; 50% glycerol; 5% SDS; 0.05% bromophenol blue; 0.25% DTT).

12. ProtoGel™ 30%: A 37.5:1 acrylamide to bisacrylamide stabilized solution (National Diagnostics, Atlanta, GA, USA).
13. 7% Resolving gel: 1 mL of 3 M Tris–HCl pH 8.85, 1.04 mL of ProtoGel™ 30%, 1.96 mL of double-distilled water, and 40 μL of 10% SDS. Add (in this order) 32 μL of 25% (w/v) APS and 4 μL of TEMED. Mix well and pour immediately following indications of Subheading 3.
14. Stacking gel: 0.5 mL of 0.5 M Tris–HCl pH 6.8, 0.4 mL of ProtoGel™ 30%, 1.1 mL of double-distilled water, and 20 μL of 10% SDS. Add (in this order) 16 μL of 25% (w/v) APS and 4 μL of TEMED. Mix well and pour immediately following indications of Subheading 3.
15. 10× Running buffer: 60.56 g of Tris, 144.14 g of glycine, and 10 g of SDS. Add distilled water to 1,000 mL and adjust to pH 8.3 (final concentrations: 0.5 M Tris–HCl, 1.92 M glycine, 1% SDS).
16. BenchMark prestained protein ladder (Invitrogen, San Diego, CA, USA).
17. Amersham ECL Western Blotting System (GE Healthcare, Waukesha, WI, USA).
18. Protran nitrocellulose membrane (Whatman Ltd, Maidstone, Kent, UK).
19. Plastic wrap.
20. Western blotting buffer: 5.8 g of Trizma base, 2.9 g of glycine, 200 mL methanol, 0.37 g of SDS, and distilled water to 1,000 mL. Adjust to pH 8.8 (final concentrations: 48 mM Trizma base, 29 mM glycine, 20% methanol, 0.037% SDS).
21. 10× PBS buffer: 81.82 g of NaCl, 2.46 g of KCl, 14.2 g of Na_2HPO_4, and 2.45 g of KH_2PO_4. Add double-distilled water to 1,000 mL (final concentrations: 1.4 M NaCl, 33 mM KCl, 100 mM Na_2HPO_4, 18 mM KH_2PO_4).
22. PBST buffer: 100 mL of 10× PBS, 900 mL double-distilled water, and 1 mL of Tween 20.
23. Mouse anti-ubiquitin monoclonal IgG_1 antibody (Santa Cruz Biotechnology Inc., Santa Cruz, CA, USA).
24. Blocking buffer: Dissolve 0.5 g of nonfat dry milk in 10 mL of PBST buffer (final concentration: 5%).
25. ECL™ Western Blotting System (GE Healthcare, Waukesha, WI, USA).
26. Hyperfilm ECL (GE Healthcare, Waukesha, WI, USA).
27. Multigrade paper developer.
28. Rapid fixer.

3. Methods

3.1. Protoplasts Preparation and Transformation

1. Collect fresh spores (no more than 1 week old) and resuspend them in YPG medium pH 4.5. Adjust the final spore concentration to 10^7 spores/mL.
2. Incubate overnight at 4°C, without shaking.
3. Incubate the spores at 26°C with shaking (300 rpm), until germ tube length becomes about four times the swollen spore diameter. This usually takes 3–4 h.
4. Wash the cells twice by centrifugation in PS buffer pH 6.5 at $340 \times g$ for 5 min.
5. Resuspend the pellet in 4 mL of PS buffer. Transfer the germinated spore solution to a 50-mL Erlenmeyer flask.
6. Add 5 mg of lysing enzymes dissolved in 1 mL PS buffer, and 100 μL (0.15 U) of Chitosanase RD (dissolved in PS buffer). Incubate at 30°C with gentle shaking (60 rpm) for about 90 min (see Note 6).
7. Transfer the 5 mL solution to a screw cap centrifuge tube and fill the tube with cold 0.5 M sorbitol. Wash twice by centrifugation in cold 0.5 M sorbitol at $91 \times g$ for 5 min.
8. Resuspend the pellet gently in 800 μL of cold 0.5 M sorbitol. This 800 μL solution allows for eight different transformation experiments.
9. Each tube of transformation mixture must contain 100 μL protoplast solution and 10 μL DNA sample (1 μg total DNA for circular plasmid or 3 μg total DNA for lineal fragments; DNA must be dissolved in double-distilled water). Use as a control 10 μL of double-distilled water instead of DNA.
10. Mix and transfer to the electroporation cuvette.
11. Apply an electrical pulse using the following conditions: field strength of 0.8 kV, capacitance of 25 μF, and constant resistance of 400 Ω.
12. Immediately after the pulse, remove the cuvette and add 1 mL cold YPGS 4.5. Keep on ice until all cuvettes have been pulsed.
13. Transfer the liquid of each cuvette to 1.5-mL microcentrifuge tubes.
14. Incubate for 1 h at 26°C and 100 rpm.
15. Centrifuge at $91 \times g$ for 5′ and gently resuspend the pellet in a final volume of 400–600 μL YNBS 4.5.
16. Inoculate plates of the adequate medium containing 0.5 M sorbitol with 200 μL solution (see Notes 7 and 8).
17. Incubate in the dark at 26°C for 3–4 days.

3.2. Gene Replacement

To obtain null mutants by gene replacement, it is necessary to design replacement fragments (RF) to disrupt the target gene by homologous recombination. These RF must contain a selectable marker gene flanked by DNA sequences (about 1 kb each) upstream and downstream of the target gene (Fig. 1a). In *M. circinelloides* two main marker genes, *leuA1* and *pyrG*, are normally used. The 5′ and 3′ flanking regions and the marker gene with its promoter sequence can be obtained by PCR amplification, using the appropriated primers, and then connected by an overlapping PCR (see Note 9). The RF thus obtained can be cloned in a vector to obtain enough DNA for transformation, although the recombinant plasmid must be digested to release a linear RF to be used for transformation. It is recommended to apply the following rapid PCR amplification procedure (colony PCR) to detect if gene replacement has been successful (Fig. 1b).

1. After one or two rounds of spore recycling of transformants, pick up a small amount (about 1 cm^2) of transformant mycelium (without agar) with a tweezer, freeze on liquid nitrogen and resuspend in 50 μL DMSO (see Note 10).
2. Set up amplification reactions with 1.25 μL of the DMSO solution and PCR mixture for a final volume of 25 μL. Depending on primer length and sequence, the annealing temperature could vary between 53 and 55°C. A standard program is 30 cycles, each one consisting of 95°C for 30 s, 53°C for 30 s, and 72°C for 1 min/kb. Finally, analyze the PCR results by gel electrophoresis.
3. Continue the spore recycling only with those transformants showing a successful gene replacement.

3.3. Silencing Vector Construction

RNA silencing, or RNA interference (RNAi), is being increasingly considered as a worthwhile alternative for manipulating fungal gene expression, since it does not imply sequence DNA modification and, probably most important, it allows to study genes whose lack of expression could be lethal for the organism. To silence a gene, it is necessary designing the adequate "silencing vector", and also demonstrate that silencing is being produced. Below are explained the methods used in *M. circinelloides* to construct a "silencing vector" and to detect the small RNA molecules (siRNAs) that are the hallmarks of RNA silencing, which in the case of *M. circinelloides* are 21- and 25-nt long (Fig. 2).

The "silencing vector" contains two main features. One of them is a selectable marker gene in *Mucor*, for instance the gene *leuA1*, which complements the *leu*$^-$ auxotrophy of R7B strain and its derivatives (see Note 8). The other one is a construction (transgene) containing inverted repeated sequences of the gene to be silenced cloned under a strong promoter, as the *gpd1* promoter. The plasmid pEUKA4 (21) contains the *leuA1* gene and the *gpd1*

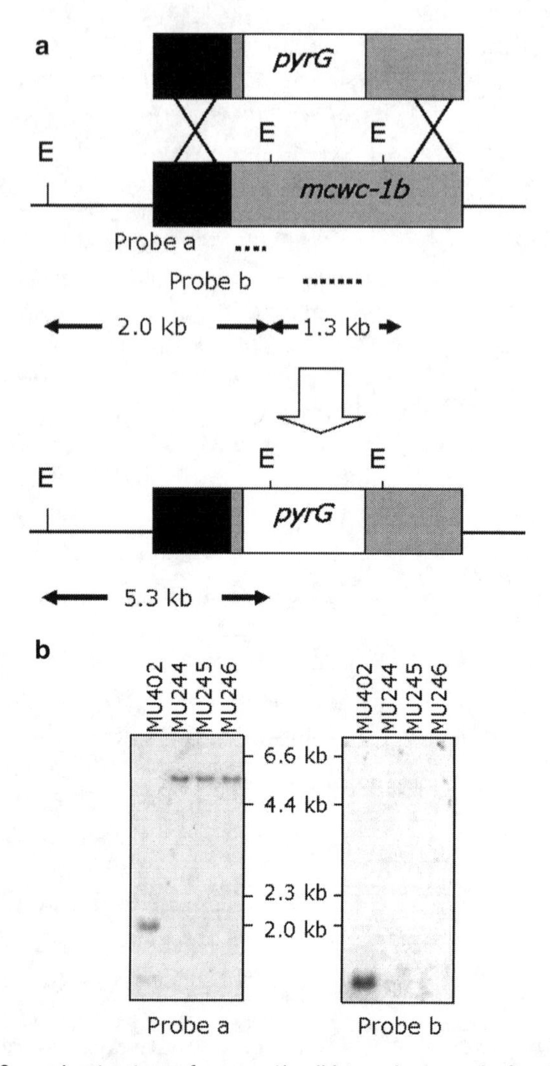

Fig. 1. (**a**) Genomic structure of *mcwc-1b* wild-type locus and after homologous recombination of the replacement fragment. Positions of the probes used and expected sizes of *Eco*RI fragments detected by the probes are indicated. E, *Eco*RI. (**b**) Genomic DNA (0.5 μg) from recipient strain in transformation (MU402) and *mcwc-1b* knockout mutants (MU244, MU245, and MU246) was digested with *Eco*RI and hybridized with probe *a* (0.7 kb *Bam*HI fragment of the *mcwc-1b* gene), and subsequently with probe *b* (1 kb *Eco*RV–*Apa*I fragment of the *mcwc-1b* gene (2)). The positions and sizes of the fragments of the DNA molecular weight marker (λ *Hin*dIII) are indicated in the middle of both Southern blots.

promoter. Transgene sequences must be displayed as an inverted repeat (see Note 11). Figure 2 outlines an example of "silencing vector" construction. In this case, the transgene contains sequences of the *M. circinelloides carB* gene (accession number AJ238028), a gene involved in the biosynthesis of β-carotene. Thus, silencing of this gene produces albino transformants, due to the lack of *carB* (phytoene dehydrogenase) gene activity.

The construction of the silencing vector is as follows:

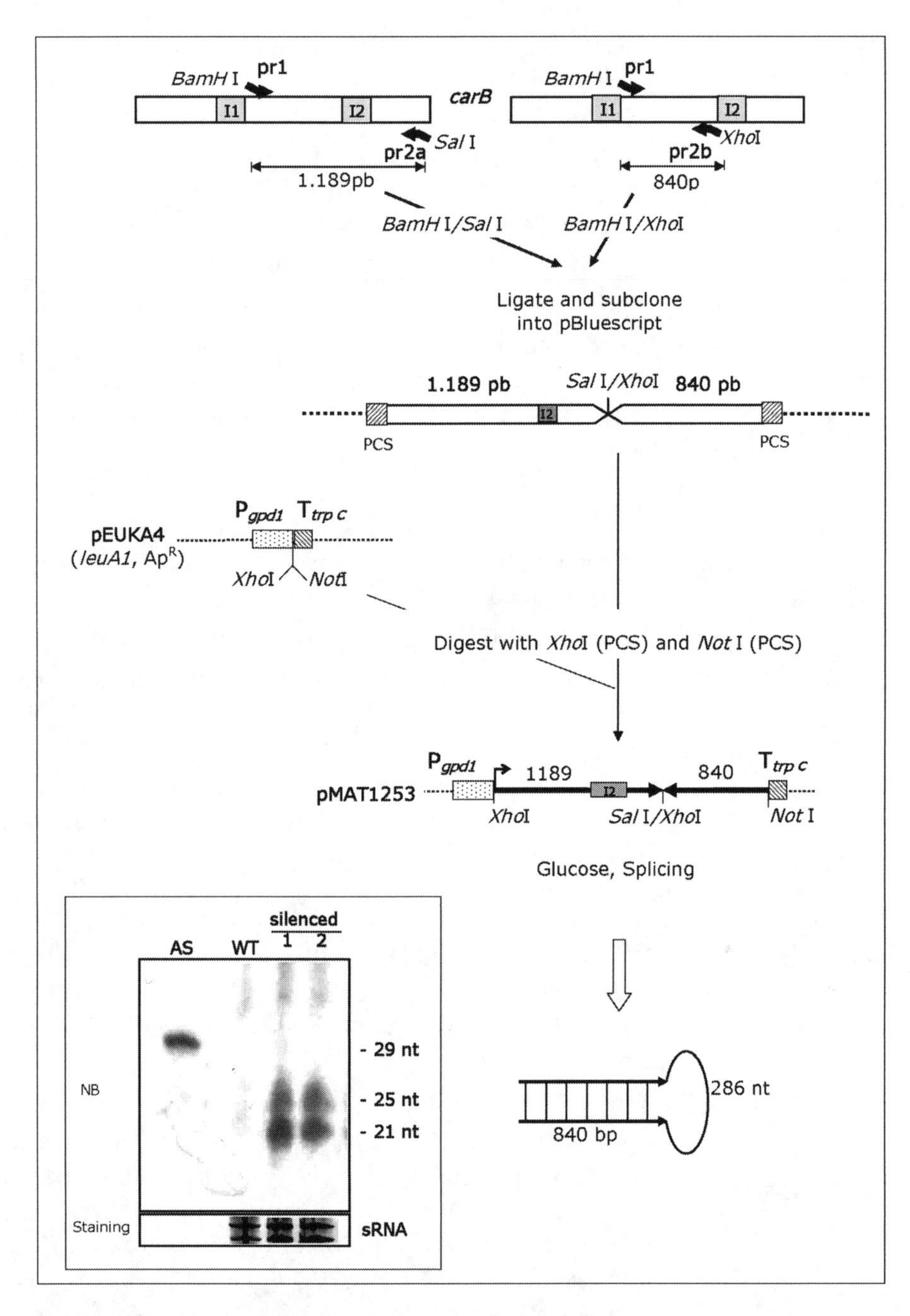

Fig. 2. Schematic representation of the silencing plasmid pMAT1253. This plasmid contains the inverted-repeat transgene for silencing the *carB* gene. *pr1*, *pr2a*, and *pr2b* are the primer used for PCR amplification of *carB* fragments included in the inverted repeat. Promoter (P), terminator (T), and introns (I1 and I2) are represented as *boxes*. The stem-and-loop region size of the hpRNA obtained upon transcription is shown. Only relevant restriction sites are indicated. Inset: Northern blot analysis of low molecular weight RNAs (50 μg) isolated from cultures of the wild-type strain (WT) and from two silenced transformants (1 and 2) carrying the hpRNA-expressing plasmid pMAT1253, grown for 24 h in liquid medium. Ten picomoles of 29-mer DNA oligonucleotide in antisense orientation (AS) was used as a size marker and to control the hybridization specificity. The RNA blot (NB) was hybridized with a hydrolyzed *carB* antisense-specific riboprobe, which corresponded to a 1,662-bp DNA fragment of the *carB* sequence that extends from position +863 to the end of the *carB* gene (6). The predominant RNA species in the small RNA samples (sRNA) were stained with ethidium bromide after size separation by agarose gel electrophoresis.

1. Use the PCR primers pr1 (forward) and pr2a (reverse) to amplify a 1,189-bp fragment, which contain *carB* sequences including intron 2.
2. Use the PCR primers pr1 (forward) and pr2b (reverse) to generate an 840-bp fragment, which contains *carB* sequences overlapping with those present in the 1,189-bp fragment.
3. Digest the 1,189-bp fragment with *Sal*I and *Bam*HI.
4. Digest the 840-bp fragment with *Xho*I and *Bam*HI.
5. Ligate both fragments with the pBlueScript vector restricted with *Bam*HI (*Sal*I and *Xho*I are compatible; this ligation produces, among others, a fragment that contains *carB* sequences with opposite orientations).
6. Select the correct construction by restriction analysis.
7. Digest this construction with *Xho*I and *Not*I (restriction sites present into the PCS sequence) to excise the 2-kb inverted repeat fragment.
8. Digest the pEUKA4 plasmid with *Xho*I and *Not*I.
9. Ligate the resulting fragment with the 2-kb fragment. The new plasmid contains the 2-kb inverted repeat under the *gpd1* promoter.
10. Use the silencing construction to transform the adequate recipient strain, following the protocols described in Subheading 3.1.

3.4. Low Molecular Weight RNAs (sRNA) Isolation

1. Incubate a 500-mL Erlenmeyer flask with 50 mL YNB medium pH 4.5 inoculated with 5×10^4 spores/mL, at 26°C for 48 h, shaking at 150 rpm.
2. Isolate the mycelium by filtration and dry well with filter paper.
3. Weigh about 100 mg of mycelium; add liquid nitrogen and grind it on a mortar and pestle to a fine powder.
4. Transfer immediately the powder to an ice cold 50-mL Falcon tube and add 1.5 mL Trizol. Vortex vigorously to resuspend the powder completely. Transfer the solution to a 2-mL centrifuge tube and centrifuge at $10,000 \times g$ for 10 min at 4°C.
5. Keep the supernatant at room temperature for 5 min.
6. Add 300 μL of chloroform.
7. Vortex vigorously for 15 s.
8. Keep at room temperature for 3 min.
9. Centrifuge at $10,000 \times g$ for 15 min at 4°C.
10. Take the aqueous phase and add 750 μL of isopropanol. Incubate for 2 h on ice.
11. Centrifuge at $10,000 \times g$ for 10 min at 4°C.

12. Wash the pellet with 1.5 mL of 70% ethanol, centrifuge at 10,000×*g* for 5 m at 4°C, and dry it for 15 min at 37°C.
13. Resuspend the dried pellet in 300 μL of DEPC-treated double distilled water and maintain it at 65°C for 10 min.
14. Add 50 μL of 4 M NaCl and 40 μL 50% PEG 8000. Mix well and incubate on ice for 30 min.
15. Centrifuge at 10,000×*g* for 10 min at 4°C. The supernatant contains the low molecular weight RNA (small RNAs). The pellet, which contains the rRNA and mRNA, can be resuspended in 50 μL of 50% formamide and electrophoresed on 1.5% agarose gel (at 150 V for 30 min), as a control of the RNA integrity.
16. To purify the low molecular weight RNA, add to the supernatant 3 volumes of 100% ethanol and 0.1 volume of 3 M sodium acetate pH 5, keep at −20°C overnight. After that, centrifuge at 10,000×*g* for 30 min at 4°C.
17. Wash the pellet with 2 volumes of 80% ethanol, centrifuge at 10,000×*g* for 10 min at 4°C, and dry it at 37°C for 15 min.
18. Resuspend the dried pellet in 20 μL of DEPC-treated water.
19. The concentration of the RNA preparation is quantified by spectrophotometric analysis.
20. Samples can be frozen for long-term storage at −80°C.

3.5. Electrophoresis of Small RNAs and Blotting

1. Mix the RNA sample (50–60 μg) with an equal volume of 2× formamide loading buffer, with a maximum final volume of 40 μL.
2. Incubate the samples at 65°C for 5 min (to open RNA secondary structures) and then place them on ice for 5 min.
3. Load the samples in a 40 mL 0.5× TBE/7 M urea/15% acrylamide gel.
4. Electrophorese at 300 V for 3 h in a Protean II Cell System, until the bromophenol blue (the faster-migrating dye) is about 1 cm of the end. If possible, use the central lanes, to avoid the "smiling" effect. The low molecular weight RNA localizes between the two dye bands.
5. After electrophoresis is complete, remove the upper glass. Mark positions of the markers.
6. Divide horizontally the gel in two parts, the cutting line being the upper dye band (xylene cyanol). The upper part of the gel contains the fraction of small RNAs (sRNA), as tRNA, 5S rRNAs, etc., which will be used as a loading control.
7. Incubate the upper part of the gel with 1× TBE and 5 μL of ethidium bromide, shaking at 60 rpm for 20 min. Wash with 1× TBE and incubate again at 60 rpm for 20 min. The gel is now ready to take a picture using UV light.

8. Cut six pieces of 3 MM Whatman paper, about 1 cm larger than the lower part of the gel.
9. Cut one piece of Hybond™-N+ nylon transfer membrane slightly bigger than the lower part of the gel.
10. Place the membrane on top of the gel, removing the bubbles by rolling with a sterile pipette.
11. Pre-wet three pieces of Whatman paper with 1× TBE and place on top of the membrane, removing again the bubbles.
12. Invert the plate carefully to remove the gel off the glass.
13. Place it on the bottom (anode) electrode plate of the semidry electroblotting unit. The order must be: bottom plate, paper, membrane, and gel. Now place the last three pieces of 1× TBE pre-wet filter paper on top of the gel and then, the upper (cathode) electrode plate on top of the stack.
14. Run at 3 mA/cm^2 gel for 1 h.
15. After transfer, fix the RNA by ultraviolet cross-linking (0.120 J/cm^2, 2 min) in a UV-Crosslinker (see Note 12).

3.6. Riboprobe Preparation

1. The riboprobe is synthesized using the in vitro transcription kit MAXIscript. The source of DNA template for the riboprobe synthesis can be either a T7 promoter-containing vector or a PCR product. In the first case, the DNA sequence to be transcribed must be cloned in the polylinker region, downstream the T7 polymerase promoter. The recombinant plasmid must be linearized by cutting on the polylinker region with the appropriate enzyme, downstream the sequence to be transcribed. Thus, the polymerase produces a transcription unit (riboprobe) that extends from the T7 promoter to the cleavage site. If the DNA template is a PCR product, the primers designed to amplify the DNA must include a 5′ T7 promoter sequence. Use 1 μg of DNA template for the transcription reaction. All the components needed for the transcription reaction and instructions are supplied in the kit.
2. Place the T7 RNA polymerase on ice (it is stored in glycerol and will not be frozen at −20°C), and vortex the 10× transcription buffer and ribonucleotide solutions until they are completely in solution. Once they are thawed, store the ribonucleotides on ice, but keep the 10× transcription buffer at room temperature (the spermidine in the 10× transcription buffer can coprecipitate the DNA template if the reaction is assembled on ice).
3. Assemble the transcription reaction at room temperature, in a RNase-free Eppendorf-like tube, by adding the DNA template (1 μg), nuclease-free water to a 20 μL final volume, 1 μL of 10 mM ATP, 1 μL of 10 mM CTP, 1 μL of 10 mM GTP, and 2 μL of 10× transcription buffer.

4. Then add 2 μL of the T7 polymerase, but do not mix it with the other components by keeping the liquid in the tube wall. Then add finally 5 μL of α-^{32}P UTP (800 Ci/mmol).
5. Gently flick the tube and then microfuge briefly to collect the reaction mixture at the bottom of the tube.
6. Incubate at 37°C for 10 min.
7. To remove the DNA template, add 1 μL of RNase-free DNaseI (included in the kit) and incubate at 37°C for 15 min.
8. Add 1 μL of 0.5 M EDTA to stop the reaction.
9. Add 300 μL of 1× alkaline buffer and incubate at 60°C for 3 h (see Note 13).
10. Add 20 μL of 3 M sodium acetate pH 5.0 to stop the hydrolysis reaction.
11. Prepare Sephadex G-50 spin columns: Resuspend and equilibrate Sephadex G-50 with 2 volumes of TE, then wash with several volumes of TE. Place the resuspended and washed resin in 1.5 volumes of TE in a glass bottle and autoclave. Store at 4°C until use. Rinse a 1–3 mL spin column thoroughly with DEPC-treated double-distilled water; frits may be pre-installed, or made by plugging the bottom of a 1-mL syringe with a support such as siliconized glass beads. Pipet 1–3 mL of the prepared, well mixed resin into the washed spin column. Place the column in a 15-mL plastic centrifuge tube and spin at 360 × *g* for 10 min in a centrifuge with a swinging bucket rotor. Place the end of the spin column containing the spun resin into an appropriate microfuge tube (typically 0.5 mL) and insert the assembly into a new 15-mL centrifuge tube. Load 20–100 μL of the sample onto the center of the resin bed (dilute sample with nuclease-free water or TE if necessary), and spin at 360 × *g* for 10 min. The eluate collected in the microfuge tube should be approximately the same volume as the sample loaded onto the column, and it will contain about 75% of the nucleic acid applied to the column.
12. The riboprobe solution can now be used for hybridization or stored at 4°C.

3.7. Riboprobe Hybridization

1. Membranes must be prehybridized, in a hybridization oven, with 20 mL of prehybridization/hybridization buffer containing 400 μL of denatured salmon sperm DNA (stock 5 mg/mL: denature by boiling at 100°C for 10 min, and then cooled on ice; final concentration: 100 μg/mL). Incubate at 30°C for 2–3 h.
2. Heat all the riboprobe solution at 95°C for 10 min and chill on ice. Then, add the riboprobe to the hybridization solution and incubate overnight at 30°C in the hybridization oven.

3. After hybridization, wash the membranes with the SDS/SSC solution at 50°C for 20 min. Repeat two more times.
4. Expose the membranes to the Kodak BioMax™ MS film (high sensitive) at −70°C.
5. If necessary, remove nonspecifically bound riboprobe by washing the membranes with 20 mL of RNase buffer containing 10 μg/mL of RNase A, at 37°C for 1 h. Wash vigorously the membranes using SDS/SSC solution at room temperature for 3 min to remove RNase.

3.8. Cell Lysate Preparation

Blue light regulates carotene biosynthesis in the fungus *M. circinelloides*, and this response has been demonstrated to be controlled by *crgA* and *mcwc-1c* gene. CrgA shows characteristics of ubiquitin ligases and represses carotenogenesis in the dark, whereas *mcwc-1c*, one of three *white collar 1*-like genes present in *M. circinelloides*, is required for its light induction. The effect of *crgA* on carotenogenesis is mediated by *mcwc-1b*, another *white collar 1*-like gene, which acts as a carotenogenesis activator. CrgA is involved in proteolysis independent mono- and di-ubiquitylation of MCWC-1b, which results in its inactivation. Proteolysis independent ubiquitylation is a novel and interesting mechanism of regulation of carotenogenesis. In this section, we describe the procedures to prepare cell lysates of *M. circinelloides*, to inmunoprecipitate proteins (MCWC-1b) and to the specific detection of ubiquitylated proteins (Fig. 3).

1. Inoculate 2.5×10^5 spores over cellophane paper in a 9-cm diameter plate with the proper agar medium at pH 4.5. Grow for about 18 h (inoculation of a high amount of spores per plate allows production of enough young mycelium in a short time).
2. Separate the mycelium from the cellophane paper and dry well with filter paper.
3. Weigh about 100–150 mg of mycelium and grind it with liquid nitrogen on a mortar and pestle to a fine powder.
4. Place the powder to a microcentrifuge tube and add 1 mL of ice cold lysis buffer (washing buffer 1) containing a 1:50 dilution (20 μL) of the Protease Inhibitor Cocktail, and 10 μL of 100 mM benzamidine. Keep 30 min on ice.
5. Centrifuge at $13{,}000 \times g$ for 30 min at 4°C.
6. Recover the supernatant using a pipette.
7. The supernatant contains the fraction of total proteins. At this point can be used for western blotting, or frozen for long-term storage at −80°C.

3.9. Immunoprecipitation

1. Transfer the supernatant to a microcentrifuge tube.
2. Add 1–5 μg of purified antibody against MCWC-1b (primary antibody) (see Notes 14 and 15).

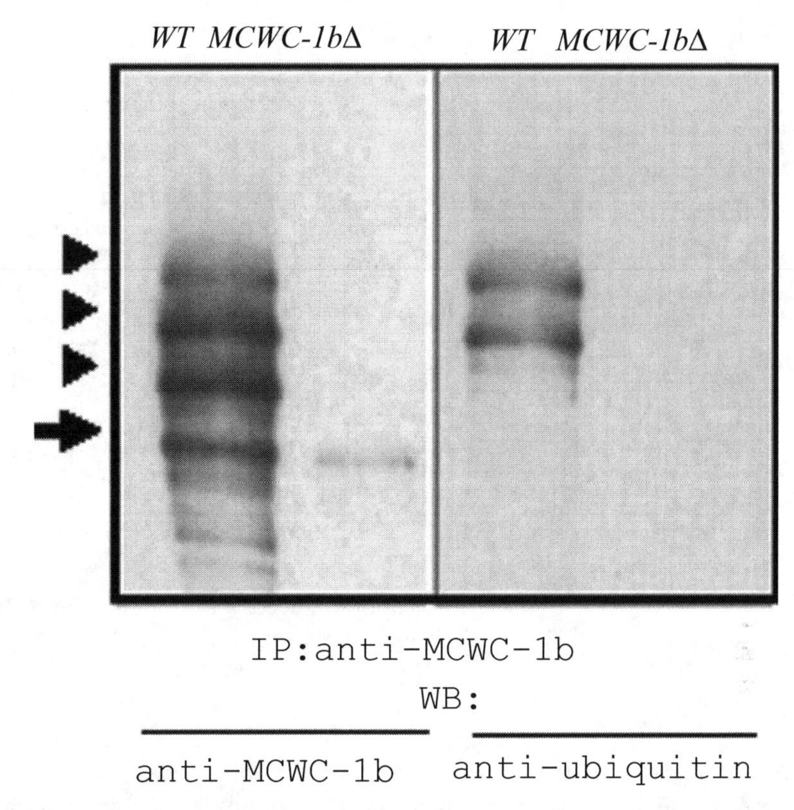

Fig. 3. Identical aliquots of crude lysates from mycelia grown in dark (18 h on YNB pH 4.5) of the wild-type strain of *M. circinelloides* (WT), and the null *mcwc-1b* mutant transformed with pLEU4 to complement its leucine auxotrophy (22), were immunoprecipitated using anti-MCWC-1b monospecific antibodies. Western blots assays were subsequently performed using the same antibody (*left*) or polyclonal antibody against human ubiquitin (*right*). The anti-MCWC-1b antibody detected an unspecific band as it also appeared in the *mcwc-1b*Δ mutant (*arrow*), and specific proteins with apparent molecular weights of 106, 115, and 125 kDa, respectively (*arrowheads*). The anti-ubiquitin antibody detected the ubiquitylated forms of the MCWC-1b protein.

3. Incubate overnight at 4°C on a rotating wheel.
4. Add 30 μL of 50% aqueous suspension Protein A-Agarose Fast Flow and keep at 4°C for 4 h. The Protein A-Agarose tube must be vortex vigorously, before taking the aliquot with a cutoff pipette tip.
5. Centrifuge at 13,000 × *g* at 4°C for 60 s.
6. Discard the supernatant and add 1 mL of cold washing buffer 1 to the pellet. Incubate at 4°C for 20 min on a rotating wheel. To discard supernatant in steps 6, 9, 12, and 14, use a 1-mL pipette, and avoid pipetting pellet protein A-agarose beads.
7. Centrifuge at 13,000 × *g* at 4°C for 60 s.
8. Repeat steps 6 and 7.

9. Discard the supernatant and add 1 mL of cold washing buffer 2 to the pellet. Incubate at 4°C for 20 min on a rotating wheel.
10. Centrifuge at 13,000 × *g* at 4°C for 60 s.
11. Repeat steps 9 and 10.
12. Discard the supernatant and add 1 mL of cold washing buffer 3 to the pellet. Incubate at 4°C for 20 min on a rotating wheel.
13. Centrifuge at 13,000 × *g* at 4°C for 60 s.
14. Discard the supernatant and add to the pellet 60 μL of cold washing buffer 3 and 15 μL of protein loading buffer. Shake well and incubate at 100°C for 5 min. After that, keep on ice.
15. Centrifuge at 13,000 × *g* at 4°C for 60 s.
16. Use a 100-μL pipette to remove the supernatant and transfer it to a clean centrifuge tube, avoid pipetting protein A-Agarose bead pellet.
17. The supernatant should contain the fraction of proteins reacting against the antibody. At this point can be used for SDS-PAGE and western blotting, or frozen for long-term storage at −80°C.

3.10. Western Blotting: Gel Assembly

1. Assemble the glass cassette and casting stand of the Mini-Protean 3 Cell, or any equivalent system.
2. The gel is divided into an upper "stacking" gel of low acrylamide percentage and low pH (6.8) and a resolving gel with a pH of 8.8.
3. Prepare the 7% resolving gel (see Note 16).
4. Take up the gel with a Pasteur pipette, put the pipette on the edge of glass, hold it at a 90° angle and begin filling to 2–2.5 cm below the top of the smallest glass plate.
5. Carefully, add 800 μL of double-distilled water on top of the acrylamide solution by using a 1,000-μL pipette.
6. Let sit for 30–45 min until the gel is polymerized.
7. Invert the apparatus and remove the water. Blot with Whatman paper strips to remove any water.
8. Prepare the stacking gel.
9. Pour in stacking gel as the resolving gel.
10. Insert comb into gel. Let sit for 10 min, checking polymerization.
11. Remove comb when stacking gel is polymerized.
12. Remove the gel cassette sandwich from the casting frame and place it into the electrode assembly with the short plate facing inward.

13. Slide the gel cassette sandwich and electrode assembly into the clamping frame.
14. Press down the electrode assembly while closing the two cam levers of the clamping frame.
15. Lower it into the mini tank and fill it with 1× running buffer.

3.11. Western Blotting: Samples Preparation and Running Gel

1. Total extract: add to a microcentrifuge tube 20 μL of total protein extract and 5 μL of 5× protein loading buffer.
2. Inmunoprecipitated proteins: add to a microcentrifuge tube 30 μL of total immunoprecipitation extract.
3. Marker: add to a microcentrifuge tube 7–10 μL of BenchMark prestained protein ladder.
4. Boil samples at 100°C during 5–10 min.
5. Quick spin (30 s) at 4°C.
6. Load them on gel.
7. Run gel at 200 V for 45–60, until the sample dye reaches the bottom of the gel.

3.12. Western Blotting: Transfer, Antibody Reaction, and Detection

1. Cut six pieces of 6×9 cm 3 MM Whatman paper.
2. Cut one piece of 5×9 cm Protran nitrocellulose membrane.
3. Pre-wet three pieces of Whatman paper with transfer buffer and place on the bottom (anode) electrode plate of the semi-dry electroblotter unit, removing bubbles by rolling with a pipette.
4. Place the membrane on top of the filter paper pack, removing bubbles carefully.
5. Gently nudge gel off plate with spatula, and place on top of the membrane (see Note 17). Remove bubbles carefully.
6. Cutoff stacking gel/wells with razor or spacers.
7. Pre-wet three pieces of Whatman paper with western blotting buffer and place them, one by one, on top of the gel, removing again the bubbles.
8. Place the upper (cathode) electrode plate on top of the stack.
9. Run at 200 mA for 45 min.
10. Use the prestained marker as control of transfer.
11. Block membrane with 10 mL blocking buffer at room temperature for 1–2 h or overnight at 4°C.
12. Wash with blocking buffer for 10 min at room temperature.
13. Add a 1/2,500 dilution of primary antibody (see Note 15) (3 μL in 7.5 mL of blocking buffer; dilution could vary,

depending of the antibody) and 1/200 dilution of anti-ubiquitin antibody.

14. Gently sake at room temperature for 1–2 h or overnight at 4°C.
15. Discard the blocking buffer and wash three times with 10 mL PBST for 10 min each.
16. Add 10 mL of blocking buffer with a 1/2,500 dilution of secondary antibody (supplied in the ECL kit).
17. Gently shake at room temperature for 1–2 h.
18. Discard the blocking buffer and wash three times with 10 mL of PBST for 10 min each.
19. Place the membrane in double-distilled water.
20. Remove the membrane and place it on a piece of plastic wrap.
21. Distribute a mix of 0.750 mL ECL Reagent 1 and 0.750 mL of Reagent 2 over the membrane. Cover with the plastic wrap and expose the membrane to a high performance chemiluminescence film into a cassette for 5 min. If necessary, expose to another film for a longer time.
22. Develop the film with a 1:10 dilution of multigrade paper developer, and fix with rapid fixer.

4. Notes

1. *M. circinelloides* R7B, the standard strain used as wild type, is a leucine auxotroph derived from *M. circinelloides* CBS277.49 (21).
2. For a normal growth using YPG, YPG-agar, YNB, YNB-agar, MMC, and MMC-agar, adjust pH to 4.5 with 1 M HCl before autoclaving. When colony growth is required (for viability counts, etc.) the media are adjusted to pH 3.2 with 1 M HCl before autoclaving. YPG and YNB are, respectively, the complete and minimal media for *Mucor*. To avoid hydrolysis of agar, double-strength solutions of agar and the other media components were autoclaved separately, and mixed after cooling at 50°C.
3. The Biotaq DNA polymerase is supplied with the 10× PCR buffer and the 50 mM $MgCl_2$ solution.
4. DEPC is a strong RNase inhibitor, and causes irritation to eyes, skin, and mucous membranes. It is a suspected carcinogen. Use it in the fume hood, wear gloves, and avoid getting DEPC on your skin.

5. When adding SSC and SDS make sure to add water first to dilute one before adding the other or a precipitate will be formed.
6. The cell wall digestion, and the appearance of protoplasts, is monitored with a light microscope (protoplasts are identified because the cell wall refringence is lost). This usually takes about 90 min.
7. Usually, each transformation mixture is spread on three plates.
8. Plasmids used for *Mucor* transformation are all self-replicative. For this reason, and because the initial transformants are usually heterokaryons due to the presence of several nuclei in the protoplasts, these transformants must be grown in selective medium for several vegetative cycles to increase the proportion of transformed nuclei. Since the two main selectable marker genes used in *M. circinelloides* are genes *leuA1* and *pyrG*, the proportion of the Leu$^+$ or PyrG$^+$ spores is used as an indicator of transformation or DNA integration. When PyrG$^+$ transformants must be selected, mainly when the recipient strain is a double mutant *pyrG$^-$ leu$^-$*, the selective medium must be MMC, because other way the mycelial growth is very poor and spore production scarce. When Leu$^+$ transformants have to be selected, the selective medium must be YNB. Once transformants have grown, spores from each transformant are plated on minimal medium YNB pH 3.2 with and without leucine (20 μg/mL) or MMC with and without uridine (200 μg/mL) to determine the proportion of Leu$^+$ or PyrG$^+$ spores, respectively. For each cycle, spores from each transformant are harvested from colonies grown under selective pressure, that is, without leucine or uridine. Integration (or efficient transformation) is denoted by the successive increase in the proportion of Leu$^+$ or PyrG$^+$ spores throughout the cell cycles. Usually, in case of DNA integration, after 2–4 cycles, homokaryotic transformants showing 100% stable Leu$^+$ or PyrG$^+$ spores can finally be obtained.
9. In our hands, gene replacement based on the RF homologous integration works better when the ends of the linear RF correspond to sequences of the target gene and not to vector sequences used for cloning. This can be achieved either by digesting the recombinant plasmid with restriction enzymes that cut appropriate restriction sequences present into the 5′ and 3′ ends of the RF, or by including the adequate restriction sequences in the design of primers used for amplification of the flanking regions of the target gene.
10. It is recommended to use the rapid PCR amplification procedure after the percentage of transformed spores is higher than 50%.
11. The highest frequency of gene silencing is obtained when using an inverted repeat transgene as a silencing trigger. This type of

construction produces, upon transcription, a hairpin RNA molecule (hpRNA), which contains sequences corresponding to the target gene. Because both the frequency and stability of gene silencing in *Mucor* are highly dependent on the level of hpRNA transcription, the strong *gpd1* promoter, which is active during the fungal vegetative growth and is regulated by the carbon source, can be used to drive transgene transcription. A spliceable intron must be included in the loop region, since the presence of introns increase the silencing efficiency in *M. circinelloides.* The hairpin stem length should be longer than 500–600 bp, and the loop after intron elimination around 300 bp.

12. It is not convenient to allow the membrane to dry completely prior to UV cross-linking. To avoid that, introduce the membrane on the top of the three wet pieces of filter paper, with the side of the blot with the bound RNA exposed to the UV light source. A 15-cm Petri dish can be used to place the membrane into the unit.
13. The alkaline buffer is used to hydrolyze the riboprobe to small RNA fragments (average size of 50 nt) to obtain a greater resolution in the hybridization.
14. This amount of antibody refers to purified antibody, and it should be calibrated according to each antibody and experimental condition. If nonpurified antibody is used (i.e., serum preparation) this calibration is necessary.
15. Monospecific antibodies against MCWC-1b are generated by immunization of rabbits with a 119-amino acid MCWC-1b peptide (from I193 to C311), which is expressed fused to a six histidine tag and purified by Nickel affinity chromatography (2).
16. Good resolution of MCWC-1b protein requires a 7% acrylamide resolving gel. Proteins with different molecular weights may require other acrylamide percentages.
17. Wet your finger with transfer buffer to manipulate the gel.

Acknowledgments

The work in our laboratory is supplied by Fundacion Séneca de la Comunidad Autónoma de la Región de Murcia, Spain (project 08802/PI/08), by D.G. de Investigación y Política Científica (Comunidad Autónoma de la Región de Murcia, Spain; Project Bio-BMC 07/01-0005), and Spanish Ministerio de Ciencia e Innovación (BFU2006-02408, BFU2009-07220).

References

1. Silva F, Torres-Martínez S, Garre V (2006) Distinct *white collar-1* genes control specific light responses in *Mucor circinelloides*. Mol Microbiol 61:1023–1037
2. Silva F, Navarro E, Peñaranda A, Murcia-Flores L, Torres-Martínez S, Garre V (2008) A RING-finger protein regulates carotenogenesis via proteolysis-independent ubiquitylation of a White Collar-1-like activator. Mol Microbiol 70:1026–1036
3. Idnurm A, Walton FJ, Floyd A, Heitman J (2008) Identification of the sex genes in an early diverged fungus. Nature 451:193–196
4. Lee SC, Corradi N, Byrnes EJ III, Torres-Martínez S, Dietrich FS, Keeling PJ, Heitman J (2008) Microsporidia evolved from ancestral sexual fungi. Curr Biol 18:1–5
5. Zemach A, McDaniel IE, Silva P, Zilberman D (2010) Genome-wide evolutionary analysis of eukaryotic DNA methylation. Science 328:916–919
6. Nicolás FE, Torres-Martínez S, Ruiz-Vázquez RM (2003) Two classes of small antisense RNAs in fungal RNA silencing triggered by non integrative transgenes. EMBO J 22:3983–3991
7. Nicolás FE, De Haro JP, Torres-Martínez S, Ruiz-Vázquez RM (2007) Mutants defective in a *Mucor circinelloides dicer*-like gene are not compromised in siRNA silencing but display developmental defects. Fungal Genet Biol 44:504–516
8. Nicolás FE, Torres-Martínez S, Ruiz-Vázquez RM (2009) Transcriptional activation increases RNA silencing efficiency and stability in the fungus *Mucor circinelloides*. J Biotechnol 142:123–126
9. De Haro JP, Calo S, Cervantes M, Nicolás FE, Torres-Martínez S, Ruiz-Vázquez RM (2009) A single gene required for efficient gene silencing associated with two classes of small antisense RNAs in *Mucor circinelloides*. Eukaryot Cell 8:1486–1497
10. Nicolás FE, Moxon S, De Haro JP, Calo S, Grigoriev IV, Torres-Martínez S, Moulton V, Ruiz-Vázquez RM, Dalmay T (2010) Endogenous short RNAs generated by Dicer 2 and RNA-dependent RNA polymerase 1 regulate mRNAs in the basal fungus *Mucor circinelloides*. Nucleic Acids Res 38:5535–5541
11. Rodríguez-Sáiz M, Paz B, De La Fuente JL, López-Nieto MJ, Cabri W, Barredo JL (2004) *Blakeslea trispora* genes for carotene biosynthesis. Appl Environ Microbiol 70:5589–5594
12. Ruiz-Hidalgo MJ, Benito EP, Sandmann G, Eslava AP (1997) The phytoene dehydrogenase gene of *Phycomyces*: regulation of its expression by blue light and vitamin A. Mol Gen Genet 253:734–744
13. Arrach N, Fernández-Martín R, Cerdá-Olmedo E, Ávalos J (2001) A single gene for lycopene cyclase, phytoene synthase and regulation of carotene biosynthesis in *Phycomyces*. Proc Natl Acad Sci USA 98:1687–1692
14. Velayos A, Eslava AP, Iturriaga EA (2000) A bifunctional enzyme with lycopene cyclase and phytoene synthase activities is encoded by the *carRP* gene of *Mucor circinelloides*. Eur J Biochem 267:5509–5519
15. Velayos A, Blasco JL, Álvarez MI, Iturriaga EA, Eslava AP (2000) Blue-light regulation of the phytoene dehydrogenase (carB) gene expression in *Mucor circinelloides*. Planta 210:938–946
16. van Heeswijck R, Roncero MIG (1984) High frequency transformation of *Mucor* with recombinant plasmid DNA. Carlsberg Res Commun 49:691–702
17. Gutiérrez A, López-García S, Garre V (2011) High reliability transformation of the basal fungus *Mucor circinelloides* by electroporation. J Microbiol Methods 84:442–446
18. Navarro E, Lorca-Pascual JM, Quiles-Rosillo MD, Nicolás FE, Garre V, Torres-Martínez S, Ruiz-Vázquez RM (2001) A negative regulator of light-inducible carotenogenesis in *Mucor circinelloides*. Mol Gen Genet 266:463–470
19. Nicolás FE, Calo S, Murcia-Flores L, Garre V, Ruiz-Vázquez RM, Torres-Martínez S (2008) A RING-finger photocarotenogenic repressor involved in asexual sporulation in *Mucor circinelloides*. FEMS Microbiol Lett 280:81–88
20. Roncero MIG (1984) Enrichment method for the isolation of auxotrophic mutants of *Mucor* using the polyenic antibiotic *N*-glycosylpolyfungin. Carlsberg Res Commun 49:685–690
21. Wolf AM, Arnau J (2002) Cloning of glyceraldehyde-3-phosphate dehydrogenase-encoding genes in *Mucor circinelloides* (syn. *racemosus*) and use of the *gpd1* promoter in recombinant protein production. Fungal Genet Biol 35:21–29
22. Roncero MIG, Jepsen LP, Strøman P, van Heeswijck R (1989) Characterization of a *leuA* gene and *ARS* element from *Mucor circinelloides*. Carlsberg Res Commun 84:335–343

Chapter 6

Gene Fusions for the Directed Modification of the Carotenoid Biosynthesis Pathway in *Mucor circinelloides*

Enrique A. Iturriaga, Tamás Papp, María Isabel Álvarez, and Arturo P. Eslava

Abstract

Several fungal species, particularly some included in the Mucorales, have been used to develop fermentation processes for the production of β-carotene. Oxygenated derivatives of β-carotene are more valuable products, and the preference by the market of carotenoids from biological sources has increased the research in different carotenoid-producing organisms. We currently use *Mucor circinelloides* as a model organism to develop strains able to produce new, more valuable, and with an increased content of carotenoids. In this chapter we describe part of our efforts to construct active gene fusions which could advance in the diversification of carotenoid production by this fungus. The main carotenoid accumulated by *M. circinelloides* is β-carotene, although it has some hydroxylase activity and produces low amounts of zeaxanthin. Two enzymatic activities are required for the production of astaxanthin from β-carotene: a hydroxylase and a ketolase. We used the *ctrW* gene of *Paracoccus* sp. N81106, encoding a bacterial β-carotene ketolase, to construct gene fusions with two fungal genes essential for the modification of the pathway in *M. circinelloides*. First we fused it to the *carRP* gene of *M. circinelloides*, which is responsible for the phytoene synthase and lycopene cyclase activities in this fungus. The expected activity of this fusion gene would be the accumulation by *M. circinelloides* of canthaxanthin and probably some astaxanthin. A second construction was the fusion of the *crtW* gene of *Paracoccus* sp. to the *crtS* gene of *Xanthophyllomyces dendrorhous*, responsible for the synthesis of astaxanthin from β-carotene in this fungus, but which was shown to have only hydroxylase activity in *M. circinelloides*. The expected result in *M. circinelloides* transformants was the accumulation of astaxanthin. Here we describe a detailed and empirically tested protocol for the construction of these gene fusions.

Key words: Astaxanthin, Xanthophylls, *Mucor*, Metabolic engineering, Gene fusions

1. Introduction

Carotenoids comprise a group of naturally occurring pigments with a broad range of biological functions such as a strong antioxidant activity (1), an enhancement of the immune response (2), or even they could be effective in preventing several types of cancer (3).

José-Luis Barredo (ed.), *Microbial Carotenoids From Fungi: Methods and Protocols*, Methods in Molecular Biology, vol. 898, DOI 10.1007/978-1-61779-918-1_6, © Springer Science+Business Media New York 2012

So, its application is emerging in the pharmaceutical and personal care industries. Many of them are also used in the food and feed industries as a supplement in poultry and aquaculture industries (4). Despite the availability of a variety of natural sources of carotenoids, synthetic ones are still cheaper and preferred by these industries, although this is changing due to some public concern over the safety of artificial food colorants.

The availability of genes involved in carotenoid biosynthesis, isolated from many different organisms, together with the development of improved culture conditions have been used in biotechnological research and metabolic engineering of carotenoid production (5–7) in naturally nonproducing and producing organisms, such as some bacteria, *Saccharomyces cerevisiae* (8), and other fungi, the yeast *Xanthophyllomyces dendrorhous,* and the algae *Haematococcus pluvialis*, the two major microorganisms able to synthesize natural astaxanthin (9, 10).

Carotenoid-producing fungi often accumulate a single main carotenoid. Among the Mucorales, β-carotene is the key carotenoid accumulated by *Blakeslea trispora*, *Mucor circinelloides*, and *Phycomyces blakesleeanus*, while astaxanthin predominates in the basidiomycetous yeast *X. dendrorhous.* The basic technology for the industrial production of β-carotene by fungi is already set up. However, the best known results for xanthophylls are still obtained at the laboratory and semi-industrial fermentation levels (9).

Several years ago we started to use *M. circinelloides* as a model organism to study the biosynthesis of β-carotene and its regulation. The result was the isolation of many mutants and the structural genes involved in the pathway (11–14). Later we took a preliminary applied research by improving strain and culture conditions for β-carotene production with this fungus. We obtained several β-carotene-overproducing strains, a number of them able to grow as yeasts aerobically (*M. circinelloides* is a dimorphic fungus that shifts from hyphae to yeast in anaerobiosis) (15). We also introduced in the wild-type strain, under the control of a strong promoter (*gpd1*), the genes responsible for the initial steps in the synthesis of β-carotene (IPP isomerase, FPP synthase, and GGPP synthase), getting an increase of up to four times in carotenoid content (16). Since astaxanthin is a more valuable carotenoid, we also checked the possibility of *M. circinelloides* to produce it by introducing in selected strains the *crtW* and *crtZ* genes from *Paracoccus* sp. N81106 (formerly *Agrobacterium aurantiacum*), which both together carry out the transformation of β-carotene into astaxanthin. Although we obtained *M. circinelloides* pink transformants which accumulate astaxanthin, its amount with respect to the total carotenoid content was only about 10% (17). This is not surprising, since the astaxanthin ratio with respect to the total carotenoid content in the native strain of *Paracoccus* sp. is about 19.5%. In general, it is assumed that in astaxanthin-producing organisms, the rate-limiting step for astaxanthin production is the activity of the *crtW* enzymes (the ketolases), which show different activities and substrate specificities (18, 19). In the case of *M. circinelloides*, the low amount of conversion of β-carotene to astaxanthin could also be due to other facts: first, it has been shown that there is a clear compartmentalization in the production of sterol, carotenoids, and other isoprenoids (20, 21), and second, but related to the first, the product of the exogenous genes introduced in *M. circinelloides* could be not working where β-carotene is produced. This is critical if we want to improve the production of astaxanthin in this fungus and compete with other astaxanthin-producing microorganisms. So we decided to construct fusion genes in which the *crtW* product would be there where the β-carotene or its derivatives are synthesized. To do this, we fused the *crtW* of *Paracoccus* sp. to the *carRP* gene of *M. circinelloides*, responsible for the first and third steps of β-carotene biosynthesis in this fungus (*M. circinelloides* was shown to have some hydroxylase activity) (17). We also fused the *crtW* gene to the *crtS* gene of *X. dendrorhous*, which was shown to have only hydroxylase activity in *M. circinelloides* (22). Here we describe an example and a detailed protocol for the construction of gene fusions of different genes for the conversion of β-carotene into astaxanthin in *M. circinelloides* that in our hands are currently working well and efficiently.

2. Materials

1. *M. circinelloides* CBS277.49 (Centraalbureau for Schimmelcultures, CBS-KNAW Fungal Biodiversity Centre, Utrecht, The Netherlands) (see Note 1).
2. *M. circinelloides* MS12 (23) (see Note 1).
3. Plasmid pPT43 (17) (see Note 2).
4. Plasmid pAK96K (Dr. N. Misawa, Central Laboratories for Key Technology, Kirin Brewery Co., Ltd., Kanagawa, Japan) (24) (see Note 3).
5. Plasmid pALMC11 (Dr. J.L. Barredo, Antibióticos SA, León, Spain) (22) (see Note 4).
6. YPG liquid medium: 0.3% yeast extract, 1% peptone, and 2% glucose, pH 4.5.
7. SSE buffer: 0.5 M sucrose, 0.1 M KCl, 10 mM EDTA, 1 mM spermine, 4 mM spermidine, 1 mM PMSF, and 14 mM β-mercaptoethanol.
8. Lysis buffer A: 50 mM Tris–HCl, 20 mM EDTA pH 8.0, and 1% sarkosyl.
9. Phenol–chloroform–isoamyl alcohol (25:24:1 v/v).

10. Ultracentrifuge and accessories.
11. Lysis buffer B: 4 M guanidine isothiocyanate, 30 mM sodium acetate pH 5.2, and 1 M β-mercaptoethanol.
12. Separating solution: 5.7 M CsCl and 2 mM EDTA.
13. Chloroform–butanol (4:1 v/v).
14. Diethyl pyrocarbonate (DEPC)-treated water.
15. OligodT column.
16. Superscript cDNA synthesis kit (Gibco BRL Spain, Invitrogen S.A., Barcelona, Spain).
17. Primer #1 (*crtSW01*): 5′-GGGTATCGATAATGTTCATCT-3′. Includes the *Cla*I recognition sequence for the directed cloning in pPT43, and the first nucleotides of the coding region of the *crtS* cDNA.
18. Primer #2 (*crtSW02*): 5′-CTCGCGGCCGCTCATGCGGTGT CCCCCTTG-3′. Includes the *Not*I recognition sequence for the directed cloning in pPT43, and the last nucleotides of the *crtW* gene.
19. Primer #3 (*crtSW03*): 5′-GGCATGTGCGCTCATTTCGACC GGCTTGACCTGC-3′.
20. Primer #4 (*crtSW04*): 5′-GCAGGTCAAGCCGGTCGAAA TGAGCGCACATGCC-3′.
21. Expand Long Template PCR System (Roche Applied Science, Barcelona, Spain).
22. Thermocycler.
23. Thin-walled 0.2-mL PCR tubes.
24. Nucleospin Extract II (Macherey-Nagel, Düren, Germany).
25. PGEMT-Easy cloning vector (Promega, Madison, WI, USA).
26. PS buffer: 0.01 M sodium phosphate buffer and 0.5 M sorbitol.
27. Chitosanase RD (US Biologicals, Swampscott, MA, USA).
28. LE enzymes (Sigma-Aldrich, St. Louis, MO, USA).
29. SMC buffer: 0.5 M sorbitol, 10 mM MOPS pH 6.3, and 50 mM $CaCl_2$.
30. PMC buffer: 40% PEG_{4000}, 0.4 M sorbitol, 10 mM MOPS pH 6.3, and 50 mM $CaCl_2$.

3. Methods

3.1. DNA Purification

High molecular weight pure DNA is always the best source to work with fungal genomic DNA. This simple procedure lets us obtain a high amount of ultrapure DNA from a single 1-week DNA extraction for years (see Notes 5 and 6).

1. Inoculate 200 mL of YPG liquid medium with 10^5 spores/mL of wild-type *M. circinelloides* CBS277.49 strain.
2. Grow the culture for 16–20 h at 28°C and 180 rpm. Harvest the mycelium by paper filtration and wash it several times with distilled water. Dry the mycelium between several filter paper towels.
3. Put the dried mycelium in an ice-cooled mortar and add 5–10 mL of SSE buffer, just to wet it. Freeze immediately adding liquid nitrogen.
4. Disrupt the mycelia (5–10 g) with a pestle in the mortar adding liquid nitrogen regularly until a fine powder is obtained.
5. Pass the powder to an appropriate baker that will maximize the interface in the phenol extraction (see below), and allow it to slowly reach room temperature. Then add 20 mL of lysis buffer A.
6. Incubate at 65°C for 15 min and gently homogenate the mix using a sterile glass rod.
7. Add an equal volume of phenol–chloroform–isoamyl alcohol and shake at 100–120 rpm for 3–4 h at 4°C in the baker.
8. Separate the two phases by spinning at 3,500 × *g* for 15 min and 4°C.
9. Collect the aqueous phases and add CsCl to a 49% w/v, and bisbenzimide to 0.5 mg/mL, final concentrations. Set up ultracentrifugation tubes and devices, and centrifuge at 20°C, for 48 h at 100,000 × *g*.
10. After centrifugation, extract the DNA from the ultracentrifuge tubes with a syringe and needle, and put it into dialysis tubing. Dialyze several times against distilled water and quantify. The final concentration of DNA should be about 200 ng/μL.

3.2. RNA Purification and cDNA Synthesis

Total RNA extraction from *M. circinelloides*, mRNA isolation, and cDNA synthesis were carried out following standard methods.

1. Get a liquid nitrogen-frozen fine powder (about 20 g) of *M. circinelloides* mycelium as in Subheading 3.1.
2. Add 50 mL of lysis buffer B. Vortex for 5–10 min at full speed.
3. Centrifuge at 28,500 × *g* for 30 min at 4°C to eliminate cellular debris.
4. Overlay 5 mL of the supernatant onto previously prepared ultracentrifuge tubes filled with 20 mL of the separating solution, and centrifuge in a swinging bucket rotor at 25,000 × *g*, 20°C, for 12–18 h.
5. Resuspend the precipitates in 2–3 mL of sterile DEPC-treated water.

6. Extract three to four times with chloroform–butanol to eliminate salts (see Note 7).
7. Collect the aqueous phases and precipitate the total RNA with sodium acetate pH 5.2 (to 0.3 M) and ethanol (2.5 volumes).
8. Wash with the same volume of 70% ethanol and store the tubes at –80°C until needed. When required, desiccate the tubes under vacuum and resuspend in DEPC-treated water. Once resuspended, make aliquots, and store at –80°C.
9. Purify the mRNA passing the total RNA through an oligodT column.
10. To obtain single-stranded cDNA, use the Superscript cDNA synthesis kit or equivalent. Use the population of single-stranded cDNA molecules as the DNA source for PCR amplification of the desired cDNA.

3.3. Construction of Gene Fusions

A simple two-step PCR strategy was used to get a single gene from two different ones. The key point consists in the construction of two partially complementary molecules (PCR products A and B in Fig. 1b) that would be used as a template for a second amplification reaction (see Note 8). Primers #1 and #2 (Fig. 1) correspond to the 5′ end of the first gene and the 3′ end of the second gene, respectively, and are normally about 20 nt long (see Note 9).

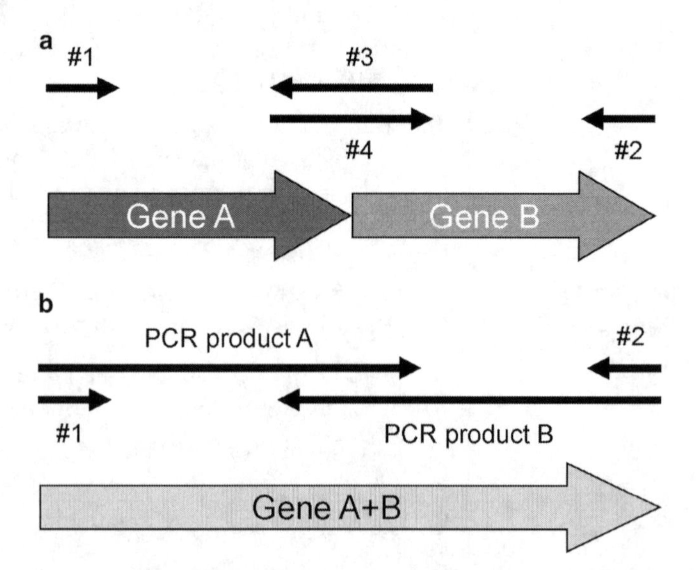

Fig. 1. (**a**) PCR strategy for gene fusions. Four oligonucleotides (#1, #2, #3, and #4) are needed to create a partially complementary molecule. Oligonucleotides #3 and #4 are complementary and designed in an opposite orientation. They are used in the first step of the PCR amplification. (**b**) When PCR products A and B are isolated and purified, they are used as DNA templates in the second step to create the fusion gene using oligonucleotides #1 and #2 in PCR amplification.

Primers #3 and #4 (Fig. 1a) are usually complementary (see Note 10), and include the 3′ end region of the first gene and the 5′ end region of the second gene. These primers must be longer (about 40 nt) and will be used in the first step to create the partially complementary molecules (see Notes 11 and 12). The system we have used (Expand Long Template PCR System) consists of a mix of thermostable DNA polymerases (*Taq* polymerase and *Tgo* polymerase; the latter with proofreading activity), with the availability of three different buffers to efficiently amplify DNA up to 20 kb (see Note 13). The optimal conditions (template DNA, incubation times, temperatures, Mg^{2+} concentration) depend on the sizes of the DNAs to be amplified and the thermal block cycler, and must be determined empirically.

1. Thaw and equilibrate all buffers at 37–56°C before use. If crystals are visible, incubate the 10× buffers until they are dissolved (see Note 14).
2. Prewarm the PCR thermocycler.
3. To a sterile 0.2-mL thin-walled PCR tube add (on ice) the following components in strict order: (1) Sterile distilled water (up to 50 μL); (2) dNTP nucleotide mix (final concentration 350–500 μM each, depending on the buffer used) (see Note 15); (3) primers (final concentration 300 nM); (4) 5 μL 10× PCR buffer (see Note 15); and (5) Template DNA (see Note 16).
4. Mix well (do not vortex).
5. Spin down and add 0.75 μL of the enzyme mix (5 U/μL).
6. Start the PCR reactions immediately. Do not delay or store the reactions.
7. Standard PCR conditions (see Note 17): (1) Initial denaturation step: 92–94°C for 2 min (see Note 18); (2) ten cycles of denaturation (92–94°C for 10 s), annealing (45–65°C for 30 s), and elongation (68°C for 45 s–30 min) (see Note 19); (3) 15–25 cycles of denaturation (92–94°C for 10 s), annealing (45–65°C for 30 s), and elongation (68°C for 45 s–30 min + 20 s elongation time for each successive cycle) (see Note 20); and (4) a final elongation step at 68°C for 7 min.
8. Second set of PCR conditions (see Note 17): (1) Initial denaturation step: 94°C for 2 min; (2) ten cycles of denaturation (94°C for 10 s), annealing (62°C for 30 s), and elongation (68°C for 6 min); (3) ten cycles of denaturation (94°C fort 10 s), annealing (62°C for 30 s), and elongation (68°C for 6.5 min); (4) ten cycles of denaturation (94°C for 10 s), annealing (62°C for 30 s), and elongation (68°C for 7 min); and (5) a final elongation step at 68°C for 7 min.

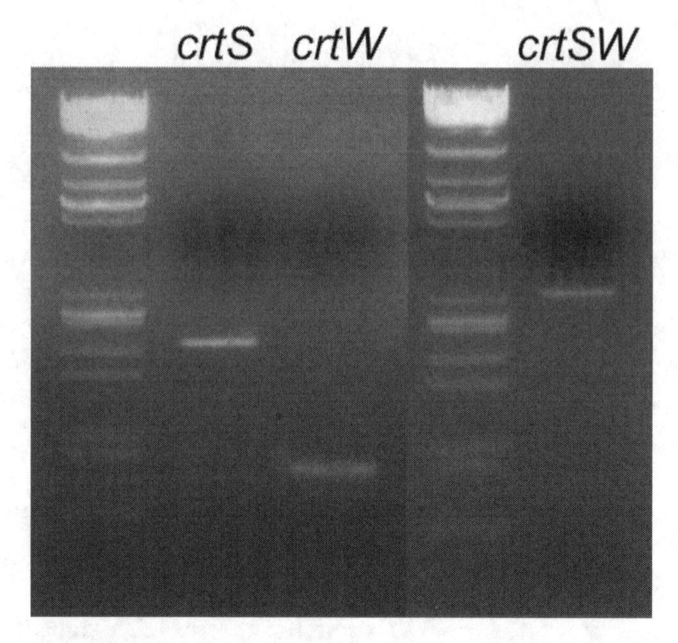

Fig. 2. First and second steps of the gene fusions. First step: amplification of *crtS* (*lane 2*) and *crtW* (*lane 3*). Second step: amplification of the fusion product *crtSW* (*lane 5*). *Lanes 1* and *4* show a molecular weight marker prepared by the mix of λ DNA digested with *Hind*III and *Eco*RI + *Hind*III.

3.4. Construction of Gene Fusions

1. After checking the PCR reactions by gel electrophoresis, isolate the first-step PCR products from preparative gels using a silica-based purification kit (see Note 21).
2. Use an appropriate amount (10–50 ng) of the two first-step PCR products to set up the second-step PCR (see Note 22).
3. Check and purify the second-step PCR product (Fig. 2) as in step 1.
4. Digest pPT43 with *Cla*I and *Not*I in two consecutive steps. Gel-purify the DNA fragment (see Note 23).
5. Ligate the pPT43-digested vector (about 10–20 ng) to the purified second-step PCR product in a 1:10 ratio. Leave the ligation at room temperature for at least 3 h.
6. Transform any competent *Escherichia coli* strain (we use DH5α), by chemical standard methods or electroporation.
7. Analyze several clones and select one containing the fusion gene under the control of the *gpd1* gene promoter and terminator sequences. Check by PCR using primers #1 and #2, and comparison to the PCR products in step 3 (Fig. 2).

3.5. Transformation of M. circinelloides

Transformation of *M. circinelloides* was carried out as previously described (25, 26) with minor modifications. A method for the electroporation of *M. circinelloides* protoplasts has been recently developed (27).

1. Inoculate 10^8 spores of MS12 in 10 mL of YPG appropriately supplemented (i.e., +leucine and uracil).
2. Leave at room temperature without agitation for at least 2–4 h to let the spores swallow. This step helps in a synchronized germination of the spores. Store at 4°C overnight.
3. Next day, incubate the spores at 180 rpm, 28°C, for 3.5–4 h. Monitor the germination of spores using a light microscope. When the germinating tubes are approximately twice the size of the spores, harvest the cells at 200 × *g* for 5 min and wash them twice with PS.
4. To remove the cell wall, resuspend the cells in a total volume of 10 mL PS buffer with chitosanase (15 μg/mL) and LE enzymes (0.5 mg/mL). Incubate at 30°C and 60 rpm for at least 1.5 h. Use a phase contrast microscope to observe the loss of refringence associated with the cell wall removal.
5. Wash twice with SMC buffer and resuspend in 0.8 mL.
6. Use 0.2 mL of protoplasts per transformation experiment. Mix them in a 10-mL tube with 10 μL of plasmid DNA (0.5–10 μg) and 20 μL of PMC buffer. Incubate for 30 min on ice.
7. Add 2.5 mL PMC and incubate for 25 min at room temperature.
8. Wash twice with 0.5 M sorbitol to remove the PMC buffer. Resuspend in 1 mL of 0.5 M sorbitol.
9. Add soft agar and plate on an appropriate selective medium.

Here we present a fast and efficient method to get gene fusions based on a simple PCR strategy. Preliminary analyses of the transformants show an increase of up to four times in the relative content of astaxanthin when using the *crtSW* fusion gene (with reference to transformants with the *crtS* and *crtW* is separate plasmids) (17).

4. Notes

1. *M. circinelloides* CBS277.49 (wild type) was used as the source of DNA and mRNA. *M. circinelloides* MS12 mutant (*pyrG*⁻, *leu*⁻) was used as the recipient strain in transformation experiments.
2. Plasmid pPT43 was constructed by inserting a polylinker cloning site between the strong promoter and terminator sequences of the *gpd1* gene (EMBL Accession No. AJ293012) of *M. circinelloides* and was used to clone the fusion genes (17). The *pyrG* gene of *M. circinelloides* (EMBL Accession No. M69112) was then added to these constructions for selection in transformations experiments.

3. The genes of *A. aurantiacum* (*crt* gene cluster, EMBL Accession No. D58420) were amplified from the plasmid pAK96K (24).
4. pALMC11 includes the *X. dendrorhous crtS* (EMBL Accession No. DQ202402) cDNA (22).
5. At the beginning of this work we had plasmids with the *carRP* gene of *M. circinelloides*, its corresponding cDNA, and also with the cDNA of the *carS* gene of *X. dendrorhous.* Whenever possible, cDNA was preferred as the source of DNA to be PCR amplified because it maintains the length of the gene fusions as short as possible. It all depends on the sizes of the genes to be fused, the availability of cDNA copies, and the organism where we want them to be expressed. For instance, the *crtS* gene of *X. dendrorhous* contains 17 introns and the final destination in our case is *M. circinelloides.* It could happen that this zygomycete does not process efficiently all of the introns, so the DNA source of election will be then the cDNA. We checked using total genomic DNA instead of the cloned cDNA of the *carRP* gene of *M. circinelloides* as a starting material for PCR amplifications, and it worked equally well in the transformation experiments.
6. When a plasmid (containing either the genomic sequences or the cDNAs) is used as a source of DNA amplification by PCR, the region to be amplified is always previously isolated from the bacterial sequences in the plasmid. This reduces the possible background during the amplifications.
7. Extractions with chloroform–butanol reduce the volume of the aqueous phase containing the RNA. Be aware of adding more DEPC-treated water to maintain the volume of the RNA solution. If too much water dissolves in the butanol solution, the RNA gets stuck to the walls of the tube and it is almost impossible to resuspend it.
8. Since the final use of the fused genes will be its expression in *M. circinelloides* transformants, the use of thermostable DNA polymerases with proofreading activity is strongly recommended because it minimizes the risk of getting point mutations in the fused gene. Several commercial long-range, thermostable DNA polymerases with proofreading activity were used in preliminary experiments, although we chose the Expand Long Template PCR System on the basis of its consistent results. In any case, it is advised to sequence the final PCR amplification product. We have found less than one nucleotide substitution per kilobytes with this system.
9. The length of the oligonucleotide primers must be enough to get a G+C content of at least 50%. It is opportune to introduce in these oligonucleotides recognition sequences for restriction enzymes that will facilitate the cloning of the gene fusion into

the vectors used for the transformation experiments. Never put this recognition sequence at the 5′ end. Insert at least 3–4 nt between the 5′ end and the recognition sequence, since many enzymes are unable to digest 5′ ended targets.

10. In fact, they do not have to be necessarily fully complementary, but should maintain a complementarity of at least 30 nt. Since they are not used at the same time in first-step PCR reactions, there are no problems with possible primer dimer reactions.
11. A common mistake when designing these primers is to maintain in their sequence the stop codon of the first gene. The stop codon must be released from the sequence. Be aware of it.
12. The spatial structure of the fused protein must be taken into account. It could happen that the tertiary and quaternary structures would be nonfunctional because of amino acid interactions at the fusion point. If problems arise, a good solution is to introduce four to five neutral amino acid codons between the last codon of the first gene and the initiation codon of the second.
13. Although the system is prepared for the amplification of genomic DNA fragments, it works efficiently using cloned genes.
14. Do not vortex, but mix by inversion (specially buffer 3). It contains detergents and bubbles will be formed.
15. Three buffers are available from the system. They differ in $MgCl_2$ concentration and buffer 3 also contains detergents. They are prepared to fit different ranges of amplification: from 0.5 to 9 kb (buffer 1), from 9 to 12 kb (buffer 2), and for >12 kb (buffer 3). The suppliers recommend increasing the concentration of Mg^{2+} when increasing the dNTP concentration. This Mg^{2+} increase is already included in the buffers, so only in particular circumstances should be taken into account. We used buffer 2. In our hands (amplifying DNA fragments from 0.7 to 6.0 kb), buffer 1 gave always high nonspecific backgrounds, but buffers 2 and 3 gave us clear bands (see Fig. 2).
16. Keep the template DNA as minimum as possible. When using fungal total genomic DNA never exceed 100 ng. 1–10 ng of DNA isolated from plasmids (see Note 6) give the best, clear results. It is preferable to have a relatively low yield than getting a high background.
17. We used two sets of PCR conditions: the standard conditions suggested by the suppliers and a serendipity one that worked really well with the amplification of our DNA fragments (0.7–6.0 kb). The standard conditions gave us a high amplification yield, but with some background. The second set of conditions gave us cleaner amplification reactions but to a lower yield.

18. The suppliers recommend keeping denaturation steps as short as possible and denaturation temperatures as low as possible. We checked these parameters and it is really true. Maintaining short and low temperature denaturation steps minimizes background in the reactions.
19. The annealing temperature depends on the primers used. Start using a high annealing temperature. If there is no amplification, repeat the assay lowering it by 2°C. The elongation times depend on the fragment length, say: 2 min for up to 3 kb, 4 min for up to 6 kb, 8 min for up to 10 kb, etc.
20. For instance, cycle number 11 is 20 s longer than cycle number 10, cycle number 12 is 40 s longer than cycle number 10, and so on.
21. Isolating the PCR products from a preparative gel instead of purifying the whole reaction ensures the elimination of primers, template DNA, and background DNA that could affect negatively the second-step reaction.
22. In the first cycles of the second-step PCR reaction, we need that the two partially complementary molecules form a single molecular structure (i.e., that they complement each other). Half of these molecules will be good templates for the polymerases, because of their 3′OH ends. But it could happen, if we add the two end primers (#1 and #2) at the beginning, that they would be amplifying only each first-step PCR product. We check running five cycles of PCR before adding these oligonucleotides, to obtain more complete molecules of the fusion product before the amplification. A higher yield was obtained, but with a higher background.
23. *Cla*I and *Not*I are relatively close in the polylinker. It is always preferable to digest the vector first with one enzyme, check out the complete digestion, and add the second enzyme. A single double digestion could result in plasmids only digested by either one or another, which will religate easily in the ligation reaction, reducing the cloning efficiency. Purification of the plasmid vectors from preparative gels is also advised to eliminate the polylinker.

Acknowledgments

This work was supported by a grant of the Junta de Castilla y León (Spain) (GR64), and in part for the Hungarian Scientific Research Fund and the National Office for Research and Technology (OTKA CK80188).

References

1. Nishikawa Y, Minenaka Y, Ichimura M, Tatsumi K, Nadamoto T, Urabe K (2005) Effects of astaxanthin and vitamin C on the prevention of gastric ulcerations in stressed rats. J Nutr Sci Vitaminol 51:135–141
2. Chew BP, Park JS (2004) Carotenoid action on the immune response. J Nutr 134:257–261
3. Vainio H, Rautalahti M (1998) An international evaluation of the cancer preventive potential of carotenoids. Cancer Epidemiol Biomarkers Prev 7:725–728
4. Johnson EA, Schroeder WA (1995) Microbial carotenoids. In: Fiechter A (ed) Advances in biochemical engineering and biotechnology. Springer, Berlin
5. Misawa N, Shimada H (1997) Metabolic engineering for the production of carotenoids in non-carotenogenic bacteria and yeasts. J Biotechnol 59:169–181
6. Schmidt-Dannert C, Umeno D, Arnold FH (2000) Molecular breeding of carotenoid biosynthesis pathways. Nat Biotechnol 18:750–753
7. Umeno D, Tobias AV, Arnold FH (2005) Diversifying carotenoid biosynthetic pathways by directed evolution. Microbiol Mol Biol Rev 69:51–78
8. Ukibe K, Hashida K, Yoshida N, Takagi H (2009) Metabolic engineering of *Saccharomyces cerevisiae* for astaxanthin production and oxidative stress tolerance. Appl Environ Microbiol 75:7205–7211
9. de la Fuente JL, Rodríguez-Sáiz M, Schleissner C, Díez B, Peiro E, Barredo JL (2010) High-titer production of astaxanthin by the semi-industrial fermentation of *Xanthophyllomyces dendrorhous*. J Biotechnol 148:144–146
10. Domínguez-Bocanegra AR, Ponce-Noyola T, Torres-Muñoz JA (2007) Astaxanthin production by *Phaffia rhodozyma* and *Haematococcus pluvialis*: a comparative study. Appl Microbiol Biotechnol 75:783–791
11. Iturriaga EA, Velayos A, Eslava AP (2000) The structure and function of the genes involved in the biosynthesis of carotenoids in the Mucorales. Biotechnol Bioprocess Eng 5:263–274
12. Velayos A, Eslava AP, Iturriaga EA (2000) A bifunctional enzyme with lycopene cyclase and phytoene synthase activities is encoded by the *carRP* gene of *Mucor circinelloides*. Eur J Biochem 267:5509–5519
13. Iturriaga EA, Velayos A, Eslava AP, Álvarez MI (2001) The genetics and molecular biology of carotenoid biosynthesis in *Mucor*. In: Pandalai SG (ed) Recent research developments in current genetics. Research Signpost, Trivandrum
14. Velayos A, Papp T, Aguilar-Elena R, Fuentes-Vicente M, Eslava AP, Iturriaga EA, Álvarez MI (2003) Expression of the *carG* gene, encoding geranylgeranyl pyrophosphate synthase, is up-regulated by blue light in *Mucor circinelloides*. Curr Genet 43:112–120
15. Iturriaga EA, Papp T, Breum J, Arnau J, Eslava AP (2005) Strain and culture conditions improvement for β-carotene production with *Mucor*. In: Barredo JL (ed) Microbial processes and products. Humana Press, Totowa
16. Csernetics A, Nagy G, Iturriaga EA, Szekeres A, Eslava AP, Vágvölgyi C, Papp T (2011) Expression of three isoprenoid biosynthesis genes and their effects on the carotenoid production of the zygomycete *Mucor circinelloides*. Fungal Genet Biol 48(7):696–703. doi:10.1016/j.fgb.2011.03.006
17. Papp T, Velayos A, Bartók T, Eslava AP, Vágvölgyi C, Iturriaga EA (2006) Heterologous expression of astaxanthin biosynthesis genes in *Mucor circinelloides*. Appl Microbiol Biotechnol 69:526–531
18. Tao L, Wilczek J, Odom JM, Cheng Q (2006) Engineering a β-carotene ketolase for astaxanthin production. Metab Eng 8:523–531
19. Fraser PD, Miura Y, Misawa N (1997) In vitro characterization of astaxanthin biosynthetic enzymes. J Biol Chem 272:6128–6135
20. Domenech C, Giordano W, Avalos FJ, Cerdá-Olmedo E (1996) Separate compartments for the production of sterols, carotenoids and gibberellins in *Gibberella fujikuroi*. Eur J Biochem 239:720–725
21. Kuzina V, Domenech C, Cerdá-Olmedo E (2006) Relationships among the biosynthesis of ubiquinone, carotene, sterols, and triacylglycerols in Zygomycetes. Arch Microbiol 186:485–493
22. Álvarez V, Rodríguez-Sáiz M, de la Fuente JL, Gudiña EJ, Godio RP, Martín JF, Barredo JL (2006) The *crtS* gene of *Xanthophyllomyces dendrorhous* encodes a novel cytochrome-P450 hydroxylase involved in the conversion of β-carotene into astaxanthin and other xanthophylls. Fungal Genet Biol 43:261–272
23. Velayos A, López-Matas MA, Ruiz-Hidalgo MJ, Eslava AP (1997) Complementation analysis of carotenogenic mutants of *Mucor circinelloides*. Fungal Genet Biol 22:19–27
24. Misawa N, Satomi Y, Kondo K, Yokoyama A, Kajiwara S, Saito T, Ohtani T, Miki W (1995)

Structure and functional analysis of a marine bacterial carotenoid biosynthesis gene cluster and astaxanthin biosynthetic pathway proposed at the gene level. J Bacteriol 177: 6575–6584

25. Van Heeswijk R, Roncero MIG (1984) High frequency transformation of *Mucor* with recombinant plasmid DNA. Carlsberg Res Commun 49:691–702

26. Iturriaga EA, Díaz-Mínguez JM, Benito EP, Álvarez MI, Eslava AP (1992) Heterologous transformation of *Mucor circinelloides* with the *Phycomyces blakesleeanus leu1* gene. Curr Genet 21:215–223

27. Gutiérrez A, López-García S, Garre V (2011) High reliability transformation of the basal fungus *Mucor circinelloides* by electroporation. J Microbiol Methods 84:442–446

Chapter 7

Integration of a Bacterial β-Carotene Ketolase Gene into the *Mucor circinelloides* Genome by the *Agrobacterium tumefaciens*-Mediated Transformation Method

Tamás Papp, Árpád Csernetics, Ildikó Nyilasi, Csaba Vágvölgyi, and Enrique A. Iturriaga

Abstract

Plasmids introduced in *Mucor circinelloides* (and most transformable Mucorales) tend to replicate autonomously, and hardly ever integrate in the genome. This is critical if we want to express exogenous genes, because plasmids are easily lost during vegetative growth, and the ratio of plasmid molecules/nuclei is invariably low. Linearized molecules of DNA have been used to get their genomic integration but the transformation efficiency drops extremely. We have developed and highly optimized an efficient *Agrobacterium*-mediated transformation system for *M. circinelloides* to facilitate the integration of transforming DNA in the genome of the recipient strain that could also be used for other Mucorales.

Key words: Astaxanthin, Xanthophylls, *Mucor circinelloides*, Metabolic engineering, *Agrobacterium*-mediated transformation

1. Introduction

Mucor circinelloides is a zygomycetous fungus that accumulates β-carotene and has been used as a model system for the study of the carotenoid biosynthesis pathway. A part of our research in the last years was devoted to modify strains of this fungus to obtain β-carotene overproducing mutants (1), and other strains able to transform the β-carotene into astaxanthin, a more valuable carotenoid, by introducing plasmids with the *crtW* (ketolase) and *crtZ* (hydroxylase) genes of the marine algae *Paracoccus* sp. N81106 (formerly *Agrobacterium aurantiacum*) in selected strains of

José-Luis Barredo (ed.), *Microbial Carotenoids From Fungi: Methods and Protocols*, Methods in Molecular Biology vol. 898, DOI 10.1007/978-1-61779-918-1_7,

M. circinelloides (2). The amount of β-carotene converted to astaxanthin in all the transformants invariably reached at best only about 10%. This could be due to the fact that plasmids replicate autonomously in this fungus. So, having in mind that *M. circinelloides* is a coenocytic fungus, the ratio of plasmids/nuclei is usually low, and that the presence of a small number of copies of the selectable auxotrophic gene is usually enough to let the transformants grow, the expression of the *crtW* and *crtZ* genes would be also lower than expected. We then decided that integration of the transforming DNA would be desirable to ensure a 1:1 ratio with the carotenoid structural genes. Former attempts to integrate exogenous DNA in *M. circinelloides* failed because the transformation efficiency drops dramatically, or it had to be directed to specific loci by gene replacement.

Agrobacterium tumefaciens is a plant pathogenic bacterium responsible for the formation of the crown gall tumor. This bacterium is able to transfer part of its DNA, the T-DNA, flanked by two short direct repeats and located on a tumor-inducing plasmid (Ti plasmid), into the infected cells. The transfer depends on the expression of virulence proteins encoded in the *vir* region of the Ti plasmid; these *vir* genes can be induced by secreted compounds of wounded plant cells, such as acetosyringone.

Agrobacterium tumefaciens-mediated transformation (ATMT) is commonly used to transfer genes to a wide variety of plants, and it has successfully been adapted for the transformation of several fungal species, including several zygomycetes such as *Rhizomucor miehei* (3), *Rhizopus oryzae* (4, 5), *M. circinelloides* (6), and *Backusella lamprospora* (7). The method has the advantages that it leads to the integration of the transferred DNA into the host genome and generally allows the transfer of relatively high molecular weight DNA (8). Currently used ATMT systems are the so-called binary vector systems, where separate plasmids harbor the T-DNA (binary plasmid) and the *vir* region (helper plasmid). In this system, the binary plasmid can be easily manipulated with standard molecular biological techniques in *Escherichia coli* and then it can be transformed into an *A. tumefaciens* strain harboring the helper plasmid with the *vir* genes.

In practice, ATMT is performed through the cocultivation of the fungus with an *A. tumefaciens* strain harboring the appropriately modified Ti plasmid, where it is important to balance the bacterial and the fungal growth. Therefore, conditions of cocultivation, especially the quality and quantity of the fungal starting material, the length of the cocultivation period, and the incubation temperature, are crucial factors of an efficient transformation. Various fungal starting materials can be used for ATMT (4, 9). Among zygomycetes, germlings and sporangiospores have been successfully used for the transformation of *R. miehei* (3) and *M. circinelloides* (6), respectively. In the case of *R. oryzae*, only protoplasts could be transformed with *A. tumefaciens* (4) and for *B. lamprospora*, the use of protoplasts also proved to be much more efficient than that of sporangiospores directly (7).

Here, we describe an optimized method for the ATMT of *M. circinelloides.* The protocol is a modification and optimization of methods previously described (6, 7, 10).

2. Materials

1. *M. circinelloides* f. *lusitanicus* MS12 (11) (see Note 1).
2. *A. tumefaciens* GV3101 (12) (see Note 2).
3. Luria Broth (LB): 10 g/L NaCl, 10 g/L tryptone, and 5 g/L yeast extract (pH 7.0). Autoclave at 121°C for 15 min.
4. LB agar: LB and 20 g/L agar.
5. LB-rifampicin agar plates: Autoclave LB-agar and when the solution cools down to 70–80°C, add rifampicin to 100 μg/mL. Pour 20–25 mL into each Petri dish (see Note 3).
6. LB-gentamycin agar plates: Autoclave LB-agar and when the solution cools down to 70–80°C, add gentamycin to 25 μg/mL. Pour 20–25 mL into each Petri dish (see Note 3).
7. LB-kanamycin agar plates: Autoclave LB-agar and when the solution cools down to 70–80°C, add kanamycin to 40 μg/mL. Pour 20–25 mL into each Petri dish (see Note 3).
8. SOC: 20 g/L tryptone, 5 g/L yeast extract, 0.5 g/L NaCl, 10 mL/L of 250 mM KCl, and 5 mL/L of 2 M $MgCl_2$. Adjust pH to 7.0 with about 0.2 mL of 5 N NaCl and autoclave at 121°C for 15 min. Add 20 mL of 1 M glucose just before use.
9. Induction medium (IM): 1× MM salt solution, 10 mM glucose, 0.5% (v/v) glycerol, and 15 g/L agar for solid medium. After autoclaving and cooling the medium, add 40 mM MES and 200 μM acetosyringone, the latter just before use (13).
10. MM salt solution (2.5×): 26.6 mM KH_2PO_4, 29.4 mM K_2HPO_4, 6.4 mM NaCl, 5.1 mM $MgSO_4{\cdot}7H_2O$, 1.1 mM $CaCl_2{\cdot}2H_2O$, 22.3 μM $FeSO_4{\cdot}7H_2O$, and 9.5 mM $(NH_4)_2SO_4$. Sterilize by filtration.
11. MES [2-(*N*-morpholino) ethanesulfonic acid] stock solution: 1 M MES. Adjust to pH 5.3 with 5 M KOH. Sterilize by filtration (see Note 4).
12. Acetosyringone (3,5-dimethoxy-4-hydroxyacetophenone) stock: 25 mM acetosyringone. Adjust to pH 8 with 5 M KOH (see Note 5).

13. YNB: 10 g/L glucose, 0.5 g/L Yeast Nitrogen Base without amino acids, 1.5 g/L $(NH_4)_2SO_4$, 1.5 g/L glutamic acid, and 20 g/L agar for solid medium. Supplement with 500 μg/mL leucine and 300 μg/mL cefotaxime (see Note 6).
14. Cefotaxime stock solution: 50 mM (23.8 mg/mL) cefotaxime (Merck, Darmstadt, Germany) solution in sterile distilled water.
15. YPG: 3 g/L yeast extract, 10 g/L peptone, 20 g/L glucose, and 20 g/L agar for solid medium.
16. 10% Glycerol in sterile distilled water.
17. Electroporation cuvette with 1 mm gap.
18. Electro Cell Manipulator ECM-600 (BTX Instrument Division, Holliston, MA, USA).
19. Falcon tubes.
20. Erlenmeyer flasks.
21. Sterile cellophane sheets (see Note 7).
22. Binary vector pCA15 (Fig. 1) (see Note 8).
23. 1-kb DNA ladder.

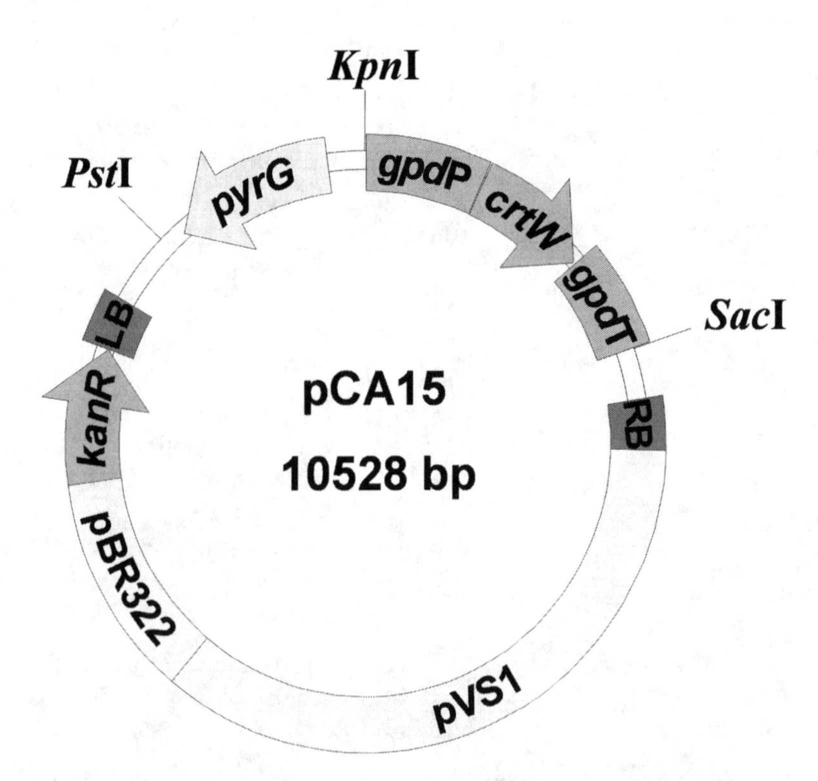

Fig. 1. Plasmid pCA15 used for ATMT of *Mucor circinelloides*. It contains the *pyrG* gene and the bacterial *crtW* gene under the control of the *M. circinelloides gpd1* regulator sequences (*gpd*P and *gpd*T, respectively). *LB* left T-DNA border, *RB* right T-DNA border.

3. Methods

3.1. Preparation of Electrocompetent Cells from A. tumefaciens

1. Culture *A. tumefaciens* in liquid LB medium supplemented with the proper antibiotics (corresponding to the resistance genes in their genome and on the helper plasmid) for 24 h at 28°C.
2. To obtain a fresh culture, inoculate 3 mL of 1-day-old culture into 300 mL prewarmed LB supplemented with antibiotics and cultivate it at 28°C by shaking at 200 rpm. When the OD_{600} of the culture reaches 0.5 (see Note 9), cool it on ice for 15 min.
3. Spin down the cooled culture at 2,100 × *g* for 15 min at 4°C. Discard the supernatant and wash the pellet with 10 mL of distilled water.
4. After a subsequent centrifugation (at 2,100 × *g* for 15 min at 4°C), pour off the water phase and wash the cells in 10 mL of 10% glycerol.
5. After a final centrifugation (at 2,100 × *g* for 15 min at 4°C), discard the supernatant and resuspend the cells in 3 mL of 10% glycerol.
6. Split the resuspended cells into 200 μL aliquots in Eppendorf tubes, freeze them in liquid nitrogen and store at –70°C.

3.2. Transformation of A. tumefaciens Cells with DNA to Be Transformed into M. circinelloides

1. Thaw an aliquot of the competent cells on ice.
2. Add approximately 100 ng of the plasmid (binary vector) to 40 μL of the bacterial suspension and place them into an electroporation cuvette with 1 mm gap previously prechilled on ice. Perform the electroporation using 400 Ω, 25 μF, and 2.5 kV.
3. Add 1 mL of SOC solution to the cuvette and, after gently pipetting it up and down, transfer the electroporated suspension into a sterile Falcon tube or flask.
4. After shaking the electroporated cells at 28°C for 3 h, plate them on LB plates supplemented with the appropriate antibiotics for colony selection and cultivate them for 1.5–2 days at 28°C.

3.3. A. tumefaciens-Mediated Transformation of Sporangiospores of M. circinelloides (See Note 10)

1. Grow *A. tumefaciens* carrying the DNA to be transferred in LB medium supplemented with the appropriate antibiotics overnight at 28°C.
2. Dilute 1 mL of the culture with 20 mL of fresh liquid IM (see Note 11).
3. Incubate the resulting bacterial solution under shaking (200 rpm) for approximately 4–5 h at 28°C, then, measure the

optical density (OD_{600}) of the culture; it should be in the range from 0.6 to 0.8.

4. Harvest sporangiospores from YPG plates after growing *M. circinelloides* for 4 days at 24°C (see Note 12).
5. For the coincubation, mix 100 μL of the *Agrobacterium* cells (OD_{600} between 0.6 and 0.8) and an equal volume of the sporangiospore suspension (10^5 sporangiospores/mL), and plate them onto sterile cellophane sheets placed previously on the surface of solid IM (see Note 13). Incubate the mixture for 1–3 days at 28°C (see Note 14).
6. After the coincubation, transfer the cellophane sheets onto the selection medium and incubate them at 24°C for another 4–5 days, until colonies appear. Putative transformants can be then transferred to fresh selection plates, and monosporic cultures (from single colony isolations) can be obtained (see Note 15).

Using the described method, the *crtW* gene of the marine, astaxanthin-producing bacterium, *Paracoccus* sp. N81106, encoding a β-carotene ketolase can be transferred into the *M. circinelloides* genome. The transformation frequency is about 7–10 transformant colonies per experiment. The presence of the *crtW* gene can be detected in the transformants by PCR (Fig. 2). Southern hybridization analysis of the transformants using the *crtW* gene as a probe suggests single copy integrations at different loci of the host genome. Due to the expression of the bacterial gene, transformants are able to produce canthaxanthin, the double-ketolated form of β-carotene.

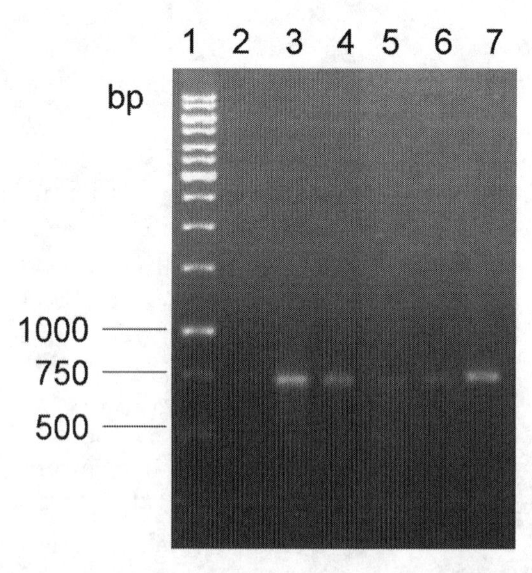

Fig. 2. Amplification pattern obtained with *crtW* specific primers from the *M. circinelloides* transformants. *Lane 1*: size standards (1 kb DNA ladder). *Lanes 2–6*: transformants MS12/A1–A5. *Lane 7*: pCA15 plasmid. The size of the detected amplification product is 728 bp.

4. Notes

1. *M. circinelloides* f. *lusitanicus* MS12 is a leucine, uracil double-auxotroph mutant derived from the wild-type strain CBS 277.49 (11).
2. *A. tumefaciens* GV3101 harbors the helper plasmid pMP90. The strain GV3101 includes the rifampicin resistance gene, and pMP90 harbors the gentamycin resistance gene as selection marker. Other *A. tumefaciens* strains can also be used in the ATMT experiments. In this case, the media should contain the appropriate antibiotics corresponding to the selection markers (resistance genes) in the bacterial genome and on the helper plasmid. For *M. circinelloides*, the *A. tumefaciens* strain GV2260 harboring the pTiB6S3ΔT helper plasmid (14) and the strain AGL1 (15) harboring the pTiBo542ΔT helper plasmid (16) are also successfully used (6).
3. For the selection of the bacterial cells, the helper plasmid and the binary vector, the medium should be supplemented with the appropriate antibiotics corresponding to the resistance gene on the bacterial chromosome and the applied plasmids.
4. If the salt precipitates, it can be dissolved incubating the solution at 65°C and by vortexing. The stock can be stored at –20°C until further usage.
5. Acetosyringone can be dissolved by stirring the solution for 1 h. The solution should be sterilized by filtration. The stock can be stored at –20°C until further usage.
6. Cefotaxime is necessary to eliminate the *Agrobacterium* cells after the transformation; it should be added after autoclaving and cooling the medium. Leucine supplementation of the medium is necessary because a leucine, uracil double-auxotroph *M. circinelloides* strain (MS12) is used as the recipient in the transformation experiments and the transferred binary plasmid contained the *pyrG* gene complementing only the uracil auxotrophy. Of course, if a single-auxotroph strain is transformed (e.g., the leucine auxotroph strain ATCC90608, which is also often used in transformation experiments), preparation of YNB supplemented with only cefotaxime is enough.
7. Sterile cellophane sheets are for the cocultivation of the bacterial cells and the fungal spores. Nitrocellulose membrane can also be used for this purpose, although the transparent cellophane disks are cheaper and monitoring of putative transformants is easier than in the case of nitrocellulose membranes.
8. pCA15 derives from the pPK2 *A. tumefaciens* binary vector (10) by the replacement of the original *hph* (hygromycin B phosphotransferase) cassette with the orotidine-5-phosphate

decarboxylase gene (*pyrG* EMBL Accession No. M69112) of *M. circinelloides* and the *crtW* gene encoding the β-carotene ketolase of *Paracoccus* sp. N81106. To increase the expression of the *crtW* gene in the fungus, it was located between the promoter and terminator sequences of the strongly expressed glyceraldehyde-3-phosphate dehydrogenase 1 gene (*gpd1*; EMBL Accession No. AJ293012) (Fig. 1). Construction of the transforming DNA can be performed using standard DNA manipulation procedures in *E. coli* strains. We use *E. coli* TOP10F$^-$ strains to construct and amplify the plasmid pCA15. Both pPK2 *A. tumefaciens* binary vector (10) and pBHt2, a derivative of pCAMBIA (*Cambia*, Australia; (17)), can be used to construct vectors for ATMT of *M. circinelloides* by the replacement of the DNA segments, which were originally built between the left and right borders of the plasmids with newly constructed cassettes.

9. OD_{600} of the culture should not be above 0.8.
10. In the case of some Mucorales fungi, such as *R. oryzae* (4) or *B. lamprospora* (7), transformants are only obtained when protoplasts are used as starting material in the ATMT experiments. In our tests, sporangiospores of *M. circinelloides* could be transformed by ATMT as effectively as protoplasts. Moreover, handling of the fungal starting material and optimization of the cocultivation were easier with the usage of spores than that of protoplasts. If protoplasts are to be transformed, the *A. tumefaciens* cells containing the DNA to be transferred and the protoplasts can be mixed in a ratio of 1:5 for the cocultivation, where the concentration of the protoplasts is approximatly 10^5 cells/mL (see step 5, Subheading 3.3). Protoplast production methods for *M. circinelloides* are well established and are described by several authors, such as in ref. (18).
11. The optimal OD_{600} of the diluted bacterial solution is 0.1 and the final volume of the solution should be 20 mL.
12. It is important to use always a freshly harvested spore suspension.
13. Alternatively, coincubation can also be performed in liquid medium that may result in higher transformation frequency in the case of some *M. circinelloides* strains. The same amounts of the fungal spores and the bacterial cells are inoculated into 50 mL of liquid IM as described in the method and the culture is shaken for 16 h at 25°C. After collecting the cells by centrifugation (at 2,100 × *g* for 15 min at 4°C), they are spread on the selective medium and incubated until transformant colonies appear (4–12 days).
14. Application of temperatures higher to 28°C is not recommended for the cocultivation because it may inactivate the

vir system, of which activity is necessary to the DNA transfer process (19). Cocultivation of the strain MS12 with the *A. tumefaciens* cells can be performed at 28°C, which is optimal for the bacterial growth, because the germination of this strain is restricted on the IM due to the lack of leucine (thus transformant colonies appear only after the spores have been transferred onto the selection medium). However, other *M. circinelloides* strains may grow very rapidly on the IM at this temperature which can cause an overgrowth of the fungus and an inappropriate fungus/bacterium ratio. Moreover, too much fungal growth makes the subsequent isolation of the transformants difficult. In this case, temperature of the cocultivation can be decreased even to 15°C (6). It is important to take into account that *A. tumefaciens* will also grow slowly at the decreased temperature, and the cocultivation period should be extended to 4–6 days.

15. *Mucor* species generally produce multinucleate spores and form coenocytic mycelia; primary integrative transformants are therefore heterokaryotic to the transferred DNA. Attainment of the homokaryotic stage frequently requires some consecutive cultivation cycles on the selection medium.

Acknowledgments

This work was supported by a grant of the Hungarian Scientific Research Fund and the National Office for Research and Technology (OTKA CK80188), and in part by a grant of the Junta de Castilla y León (Spain) (GR64).

References

1. Iturriaga EA, Papp T, Breum J, Arnau J, Eslava AP (2005) Strain and culture conditions improvement for β-carotene production with *Mucor*. In: Barredo JL (ed) Microbial processes and products. Humana Press, Totowa
2. Papp T, Velayos A, Bartók T, Eslava AP, Vágvölgyi C, Iturriaga EA (2006) Heterologous expression of astaxanthin biosynthesis genes in *Mucor circinelloides*. Appl Microbiol Biotechnol 69:526–531
3. Monfort A, Cordero L, Maicas S, Polaina J (2003) Transformation of *Mucor miehei* results in plasmid deletion and phenotypic instability. FEMS Microbiol Lett 224:101–106
4. Michielse CB, Salim K, Ragas P, Ram AFJ, Kudla B, Jarry B, Punt J, van den Hondel CAMJJ (2004) Development of a system for integrative and stable transformation of the zygomycete *Rhizopus oryzae* by *Agrobacterium*-mediated DNA transfer. Mol Genet Genomics 271:499–510
5. Ibrahim AS, Skory CD (2007) Genetic manipulation of zygomycetes. In: Kavanagh K (ed) Medical mycology. Wiley, New York, pp 305–326
6. Nyilasi I, Ács K, Papp T, Vágvölgyi C (2005) *Agrobacterium tumefaciens*-mediated transformation of *Mucor circinelloides*. Folia Microbiol 50:415–420
7. Nyilasi I, Papp T, Csernetics Á, Vágvölgyi C (2008) *Agrobacterium tumefaciens*-mediated transformation of the zygomycete fungus,

Backusella lamprospora. J Basic Microbiol 48:59–64

8. Hamilton CM, Frary A, Lewis C, Tanksley S (1996) Stable transfer of intact high molecular weight DNA into plant chromosomes. Proc Natl Acad Sci USA 93:9975–9979
9. de Groot MJA, Bundock P, Hooykaas PJJ, Beijersbergen AGM (1998) *Agrobacterium tumefaciens*-mediated transformation of filamentous fungi. Nat Biotechnol 16:839–842
10. Covert SF, Kapoor P, Lee M, Briley A, Nairn CJ (2001) *Agrobacterium tumefaciens*-mediated transformation of *Fusarium circinatum*. Mycol Res 105:259–264
11. Velayos A, López-Matas MA, Ruiz-Hidalgo MJ, Eslava AP (1997) Complementation analysis of carotenogenic mutants of *Mucor circinelloides*. Fungal Genet Biol 22:19–27
12. Koncz C, Schell J (1986) The promoter of the T_L-DNA gene 5 controls the tissue-specific expression of chimeric genes carried by a novel type of *Agrobacterium* binary vector. Mol Gen Genet 204:383–396
13. Bundock P, Hooykaas PJJ (1996) Integration of *Agrobacterium tumefaciens* T-DNA in the *Saccharomyces cerevisiae* genome by illegitimate recombination. Proc Natl Acad Sci USA 93:15272–15275
14. McBride KE, Summerfelt KR (1990) Improved binary vectors for *Agrobacterium*-mediated plant transformation. Plant Mol Biol 14:269–276
15. Lazo GR, Stein PA, Ludwig RA (1991) A DNA transformation-competent *Arabidopsis* genomic library in *Agrobacterium*. Nat Biotechnol 9:963–967
16. Hood EE, Helmer GL, Fraley RT, Chilton MD (1986) The hypervirulence of *Agrobacterium tumefaciens* A281 is encoded in the region pTiBo542 outside the T-DNA. J Bacteriol 168:1291–1301
17. Mullins ED, Chen X, Romaine P, Raina R, Geiser DM, Kang S (2001) *Agrobacterium*-mediated transformation of *Fusarium oxysporum*: an efficient tool for insertional mutagenesis and gene transfer. Phytopathology 91:173–180
18. Iturriaga EA, Díaz-Mínguez JM, Benito EP, Álvarez MI, Eslava AP (1992) Heterologous transformation of *Mucor circinelloides* with the *Phycomyces blakesleeanus leu1* gene. Curr Genet 21:215–223
19. Fullner KJ, Nester EW (1996) Temperature affects the T-DNA transfer machinery of *Agrobacterium tumefaciens*. J Bacteriol 178:1498–1504

Chapter 8

Metabolic Engineering of *Mucor circinelloides* for Zeaxanthin Production

Marta Rodríguez-Sáiz, Juan-Luis de la Fuente, and José-Luis Barredo

Abstract

Mucor circinelloides is a β-carotene producing zygomycete amenable to metabolic engineering using molecular tools. The *crtS* gene of the heterobasidiomycetous yeast *Xanthophyllomyces dendrorhous* encodes the enzymatic activities β-carotene hydroxylase and ketolase, allowing this yeast to produce the xanthophyll called astaxanthin. Here we describe the fermentation of *X. dendrorhous* in astaxanthin producing conditions to purify mRNA for the cloning of the cDNA from the *crtS* gene by RT-PCR. Further construction of an expression plasmid and transformation of *M. circinelloides* protoplasts allow the heterologous expression of the *crtS* cDNA in *M. circinelloides* to obtain β-cryptoxanthin and zeaxanthin overproducing transformants. These two xanthophylls are hydroxylated compounds from β-carotene. These results show that the *crtS* gene is involved in the conversion of β-carotene into xanthophylls, being potentially useful to engineer carotenoid pathways.

Key words: Zeaxanthin, Astaxanthin, β-Carotene, β-Cryptoxanthin, Xanthophyll, Fermentation, *Xanthophyllomyces dendrorhous*, *Mucor circinelloides*

1. Introduction

Astaxanthin (3,3′-dihydroxy-β,β-carotene-4-4′-dione; $C_{40}H_{52}O_4$) is a red xanthophyll (oxygenated carotenoid) with large importance in the aquaculture, pharmaceutical, and food industries. It is widely used as a feed additive to pigment flesh in salmon and trout aquaculture (1) and also conferring a characteristic coloration to crustaceans and some birds. It belongs to the carotenoids, a family of yellow to orange-red terpenoid pigments synthesized by photosynthetic organisms and by many bacteria and fungi (2) that are used as colorants, feed supplements, and nutraceuticals in the food, medical, and cosmetic industries (3).

José-Luis Barredo (ed.), *Microbial Carotenoids From Fungi: Methods and Protocols*, Methods in Molecular Biology, vol. 898, DOI 10.1007/978-1-61779-918-1_8, © Springer Science+Business Media New York 2012

Despite the availability of a variety of natural and synthetic carotenoids, there is currently renewed interest in microbial sources (4). The green alga *Haematococcus pluvialis* and the heterobasidiomycetous yeast *Xanthophyllomyces dendrorhous*, the teleomorphic state of *Phaffia rhodozyma* (5–8), are currently known as the main microorganisms useful for astaxanthin production at the industrial scale. The improvement of astaxanthin titer by microbial fermentation is a requirement to be competitive with the synthetic manufacture by chemical procedures, which at present is the major source in the market.

The biosynthetic pathway of astaxanthin (Fig. 1) has been studied in *X. dendrorhous* (9–12). The *idi* gene encodes isopentenyl pyrophosphate (IPP) isomerase, which catalyzes the isomerization of IPP to dimethylallyl pyrophosphate (DMAPP) (13). The conversion of these isoprenoid precursors into β-carotene is catalyzed by four enzymatic activities: (i) geranylgeranyl pyrophosphate (GGPP) synthase (encoded by the *crtE* gene), which catalyzes the sequential addition of three IPP molecules to DMAPP to give the C20-precursor GGPP (14); (ii) phytoene synthase (encoded by the *crtYB* gene), which links two molecules of GGPP to form phytoene (15); (iii) phytoene desaturase (encoded by the *crtI* gene), which introduces four double bonds in the phytoene molecule to yield lycopene (16); and (iv) lycopene cyclase (also encoded by the *crtYB* gene), which sequentially converts the ψ acyclic ends of lycopene to β rings to form γ- and β-carotenes (15). Two additional enzymatic activities convert β-carotene into astaxanthin through several biosynthetic intermediates: a ketolase which incorporates two 4-keto groups in the molecule of β-carotene, and a hydroxylase which introduces two 3-hydroxy groups. In *X. dendrorhous*, both activities could be together as a single enzyme (astaxanthin synthetase; CrtS) encoded by the *crtS* gene, which sequentially catalyzes the 4-ketolation of β-carotene followed by the 3-hydroxylation (17). However, the expression of the *crtS* gene in a β-carotene producing strain of *Mucor circinelloides* revealed the lack of astaxanthin and ketolated intermediates (18, 19), which does not support that hypothesis. In contrast, two independent genes have been described in other astaxanthin producing microorganisms. Additionally, a cytochrome P450 reductase, encoded by the *crtR* gene, has been shown to have an auxiliary role to CrtS in *X. dendrorhous*, providing with the necessary electrons for substrate oxygenation (20). Expression of the *crtS* gene in a β-carotene producing strain of *Saccharomyces cerevisiae* did not lead to astaxanthin production, but coexpression of *crtS* and *crtR* genes resulted in the accumulation of a small amount of astaxanthin (21). The existence of a monocyclic pathway diverging from the dicyclic pathway at neurosporene and proceeding through β-zeacarotene, 3,4-didehydrolycopene, torulene, 3-hydroxy-3′,4′-didehydro-β, ψ-carotene-4-one (HDCO) to the end product 3,3′-dihydroxy-β, ψ-carotene-4,4′-dione (DCD) (Fig. 1) was also proposed (22).

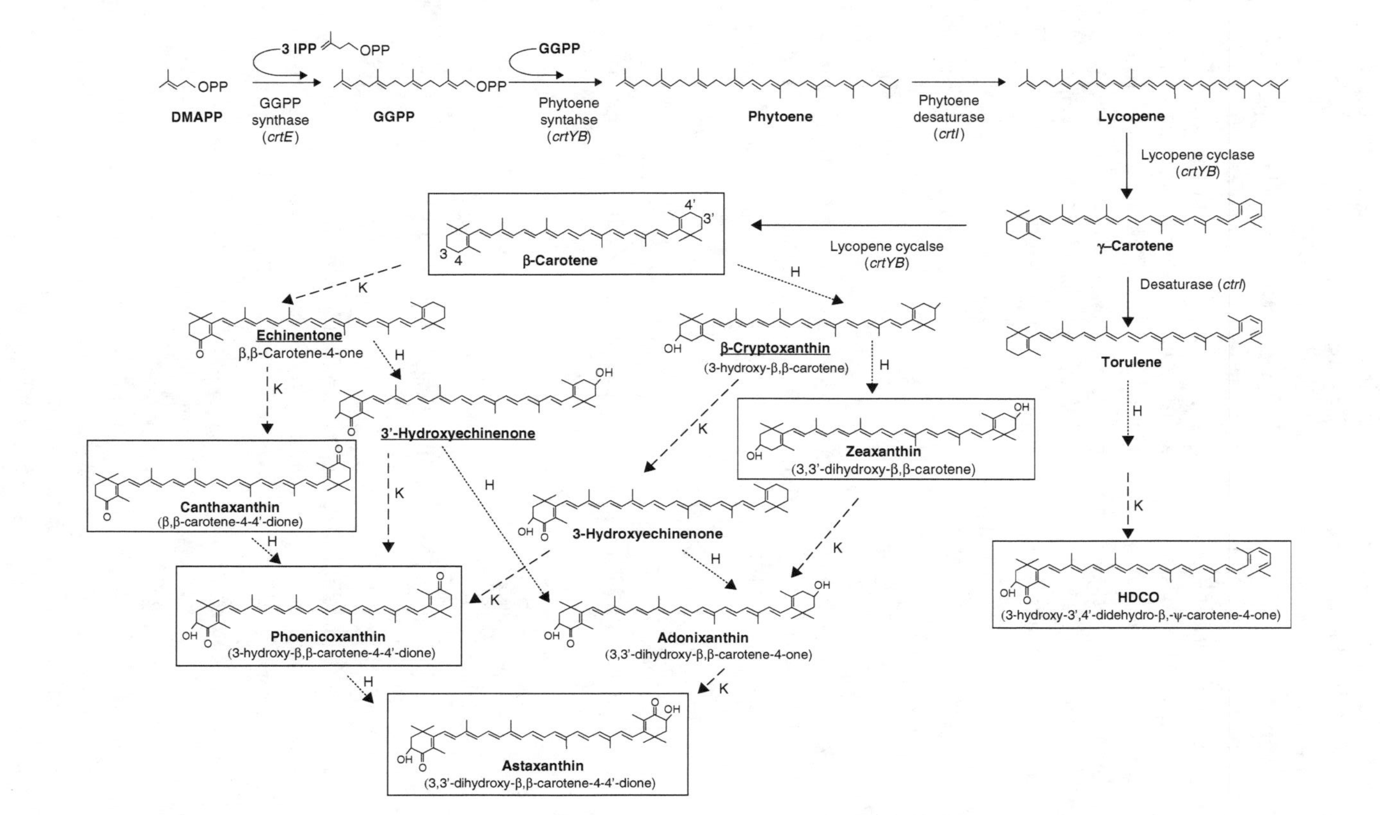

Fig. 1. Biosynthetic pathway leading to astaxanthin in *Xanthophyllomyces dendrorhous.* One molecule of dimethylallyl pyrophosphate (DMAPP) and three molecules of isopentenyl pyrophosphate (IPP) are coupled to form geranylgeranyl pyrophosphate (GGPP) by the GGPP synthase. Subsequently, two molecules of GGPP are joined by the phytoene synthase, encoded by the *crtYB* gene, to form phytoene. Afterwards, the phytoene desaturase, encoded by the *crtI* gene, introduces four double bonds in the molecule of phytoene to yield lycopene. Finally, the lycopene cyclase, encoded also by the *crtYB* gene, sequentially converts one of the ψ acyclic ends of lycopene as β-ring to form γ-carotene, and subsequently the other to generate β-carotene. Potential biosynthetic steps for the conversion of β- and γ-carotene into xanthophylls in *X. dendrorhous* include the incorporation of two 4-keto groups in the molecule of β-carotene by the ketolase (K) activity, and the introduction of two 3-hydroxy groups by the hydroxylase (H) activity. Both activities (K and H) seem to be included in a single enzyme (astaxanthin synthetase; CrtS) encoded by the *crtS* gene. The cytochrome P450 reductase encoded by the *crtR* gene has an auxiliary role to CrtS, providing with the necessary electrons for substrate oxygenation. The existence of a monocyclic pathway to HDCO has also been proposed (22). Main carotenoids detected in *X. dendrorhous* broths are shown inside a *rectangle.* This figure has been modified from Álvarez et al. (18), with permission.

Attempts to increase astaxanthin content in *X. dendrorhous* involved optimization of cultivation methods and searching for astaxanthin-hyperproducing mutants by screening of classically mutagenized *X. dendrorhous* strains. Additionally, carotenogenic genes have been cloned and transformation systems have been developed, allowing the directed genetic modification of the astaxanthin pathway. However, the knowledge about carotenogenesis regulation is still limited in comparison to other carotenogenic fungi. The cloning and characterization of the genes involved in the astaxanthin biosynthetic pathway of *X. dendrorhous* have provided valuable opportunities to construct improved strains by metabolic pathway engineering. Here, we describe the expression of the cDNA from the *crtS* gene of *X. dendrorhous* in *M. circinelloides* to obtain zeaxanthin overproducing transformants.

2. Materials

This section describes the specific material used in the procedures described in the Subheading 3. Those of general use are only described the first time.

2.1. Flask Fermentation of X. dendrorhous

1. *X. dendrorhous* Y2476 (23).
2. YEPD solid medium (YEPDA): 20 g/L glucose, 10 g/L yeast extract, 20 g/L bacto-peptone, and 20 g/L agar. Adjust pH to 6.0 with HCl. Sterilize at 121°C for 20 min (24) (see Note 1).
3. Saline solution: 0.9 g of NaCl and 100 mL of deionized water. Sterilize at 121°C for 20 min.
4. Seed medium: 70 g/L glucose, 6.2 g/L yeast extract, 5.5 g/L cotton meal, 2 g/L KH_2PO_4, 0.4 g/L K_2HPO_4, 1.5 g/L $MgSO_4·7H_2O$, 2 g/L $(NH_4)_2SO_4$, 0.2 g/L NaCl, 0.2 g/L $CaCl_2$, 50 µg/L biotin, 500 µg/L thiamine HCl, and 2 mg/L calcium pantothenate. Adjust to pH 5.8–6.0 with HCl. Distribute 25 mL aliquots in 500-mL flasks. Sterilize at 121°C for 20 min (25).
5. Fermentation medium: 200 g/L glucose, 6.2 g/L yeast extract, 5.5 g/L cotton meal, 2 g/L KH_2PO_4, 0.4 g/L K_2HPO_4, 1.5 g/L $MgSO_4·7H_2O$, 2 g/L $(NH_4)_2SO_4$, 0.2 g/L NaCl, 0.2 g/L $CaCl_2$, 1.6 g/L $CaCO_3$, 50 µg/L biotin, 500 µg/L thiamine HCl, and 2 mg/L calcium pantothenate. Distribute 25 mL aliquots in 500-mL flasks. Adjust pH to 5.8–6.0 with HCl. Sterilize at 121°C for 20 min (25) (see Note 2).
6. Erlenmeyer flask of 500 mL.
7. Autoclave.

8. Flow laminar cabinet.
9. Freezer −86°C.
10. Milli-Q Integral water purification system (Millipore, Billerica, MA, USA).
11. Elix water purification system (Millipore, Billerica, MA, USA).
12. pH meter.
13. Orbital shaker (see Note 3).

2.2. Cloning of cDNA of crtS Gene from *X. dendrorhous* by RT-PCR

To avoid RNA degradation: (i) all the material used in this procedure must be RNase-free, (ii) disposable gloves must be used, and (iii) material must be washed with water-DEPC (treated with 0.2% diethyl pyrocarbonate (DEPC) to eliminate RNases) and autoclaved.

2.2.1. Purification of RNA from *X. dendrorhous*

1. Liquid nitrogen (see Note 4).
2. Mortar and pestle.
3. Centrifuge.
4. Refrigerated centrifuge.
5. RNase-free microtubes.
6. RNeasy plant mini kit (QIAGEN, Valencia, CA, USA).
7. Micropipettes.
8. Agarose gel electrophoresis equipment.
9. TAE (50×) stock solution: Dissolve 242 g of Tris base (MW = 121.14) in approximately 750 mL of ultrapure water. Add 57.1 mL of glacial acid and 100 mL of 0.5 M EDTA, and adjust the solution to a final volume of 1 L with ultrapure water. Sterilize at 121°C for 30 min (see Note 5).
10. TAE buffer (1×) working solution: Dilute the stock solution by 50-fold with sterile ultrapure water. Final concentrations are 40 mM Tris acetate and 1 mM EDTA.
11. Agarose gel (1%): 0.5 g of agarose and 50 mL of 1× TAE buffer (see Note 6).
12. Ethidium bromide (0.5 μg/mL) (see Note 7).

2.2.2. DNA Amplification from RNA by RT-PCR

1. Primer #126 (CTC CCC ATG GTC ATC TTG GTC TTG CTC) (see Note 8).
2. Primer #127 (GAA TCA ACT CAT TCG ACC GGC) (see Note 8).
3. Titan one tube RT-PCR system (Roche Molecular Biochemicals, Mannheim, Germany).
4. PCR thermocycler.
5. Molecular imager gel doc (Bio-Rad, Hercules, CA, USA).
6. QIAex gel extraction kit (QIAGEN, Valencia, CA, USA).

7. pGEM-T (Promega, Madison, WI, USA) (see Note 9).
8. Restriction enzymes.
9. T4-DNA ligase.
10. DNA polymerase I large (Klenow) fragment.
11. *Escherichia coli* DH5α (ATCC, Manassas, VA, USA).
12. Gene pulser II (Bio-Rad, Hercules, CA, USA) (see Note 10).
13. Luria-Bertani (LB) medium: 10 g/L bacto-tryptone, 5 g/L bacto-yeast extract, 10 g/L NaCl, and 20 g/L agar. Adjust to pH 7.0 with 5 N NaOH. Sterilize at 121°C for 15 min.
14. Ampicillin stock solution 100 mg/mL (see Note 11).
15. Kanamycin sulfate stock solution 50 mg/mL (see Note 11).
16. Phenol–chloroform–isoamyl alcohol (25:24:1) (v/v).
17. TE buffer: 10 mM Tris–HCl, pH 8.0, and 1 mM EDTA.
18. IPTG (isopropyl-β-D-1-thiogalactopyranoside).
19. X-Gal (5-bromo-4-chloro-3-indolyl-β-D-galactopyranoside).

2.3. Construction of the Expression Plasmid

1. Plasmid pLeu4 (26) (see Note 12).
2. P*carRA* (27) (see Note 13).

2.4. Transformation of M. circinelloides Protoplasts

2.4.1. Isolation of Cell Walls of Blakeslea trispora

1. *B. trispora* F-816 (+) (28) (see Note 14).
2. YPDA: 20 g/L glucose, 20 g/L Bacto-peptone, 10 g/L yeast extract, and 15 g/L agar. Adjust pH to 6.0 with HCl. Sterilize at 121°C for 20 min.
3. Nytal filter 30-μm pore (Swiss silk bolting cloth fabric, Zurich, Switzerland).
4. Microscope counting chamber (hemocytometer).
5. Ball mill with 0.5-mm grinding beads.
6. Saline solution: 9 g/L NaCl.
7. Sieve (50 mesh, 0.3 mm nominal opening).

2.4.2. Obtaining Chitosanase (Streptozyme) from Streptomyces sp.

1. *Streptomyces* sp. n°6 (29) (see Note 15).
2. Slant tubes (200 × 30 mm).
3. MEY sporulation medium: 4 g/L yeast extract, 10 g/L maltose, 0.5 mL/L $CoCl_2$ (1%), and 20 g/L agar. Sterilize at 121°C for 20 min (30).
4. Jeniaux medium: 5 g/L yeast extract, 3 g/L glucose, 0.5 g/L NaCl, 1 g/L KH_2PO_4, 0.5 g/L $MgSO_4 \cdot 7H_2O$, and 0.01 g/L $FeSO_4 \cdot 7H_2O$. Adjust to pH 7.0 with NaOH 30%. Sterilize at 121°C for 20 min (31).
5. Induction medium: 20 g/L cell walls from *B. trispora*, 0.5 g/L NaCl, 1 g/L KH_2PO_4, 0.5 g/L $MgSO_4 \cdot 7H_2O$, and 0.01 g/L $FeSO_4 \cdot 7H_2O$. Sterilize at 121°C for 20 min.

6. Orbital shaker (see Note 16).
7. Bench centrifuge.
8. PG buffer: 4.82 g/L $NaH_2PO_4 \cdot H_2O$ and 4.03 g/L of $Na_2HPO_4 \cdot 7H_2O$ in a 1 L of deionized water. Add 50 g glycerol and mix until homogenization. Check the pH before sterilize at 121°C for 20 min.
9. PD-10 desalting columns (GE Healthcare, Miami, FL, USA) (see Note 17).
10. Bio-Rad Protein Assay (Bio-Rad, Hercules, CA, USA).

2.4.3. Obtaining and Transformation of Protoplasts from M. circinelloides

1. *M. circinelloides* MS12 (−) (*leuA1*, *pyrG4*) (32) (see Note 18).
2. YPG: 20 g/L glucose, 3 g/L yeast extract, 10 g/L Bacto-peptone, and 6 g/L malt extract. Adjust to pH 4.5. Sterilize at 121°C for 20 min (33).
3. Uracil (St Louis, MO, USA) (see Note 19).
4. Leucine (St Louis, MO, USA) (see Note 19).
5. Sodium phosphate buffer 10 mM pH 6.5: 0.964 g of $NaH_2PO_4 \cdot H_2O$ and 0.807 g of $Na_2HPO_4 \cdot 7H_2O$ in 1 L of deionized water. Check the pH and sterilize at 121°C for 20 min.
6. Caylase C3 (Cayla, Toulouse, France) (see Note 20).
7. YNB: 0.5 g/L yeast nitrogen base without amino acids, 1.5 g/L ammonium sulfate, 10 g/L glucose, 1.5 g/L glutamic acid, and 20 g/L agar. Sterilize at 121°C for 20 min (34) (see Note 21).
8. Soft YNB: YNB with 10 g/L agar (instead of 20 g/L). Sterilize at 121°C for 20 min (see Note 21).
9. SMC buffer: 0.55 M sorbitol, 10 mM MOPS pH 6.3, and 50 mM $CaCl_2$. Sterilize at 121°C for 20 min.
10. PMC buffer: 40% PEG 4000, 10 mM MOPS pH 6.3, 0.4 M sorbitol, and 50 mM $CaCl_2$. Sterilize at 121°C for 20 min.

2.5. Carotenoid Analysis

1. Dichloromethane:methanol (50:50).
2. Alliance high performance liquid chromatography (HPLC) equipment, including the 2960, separation module and 996 PAD (Waters, Milford, MA, USA).
3. Nucleosil 100 NH_2 column (resin particle diameter, 5 μm; 250×4.6 mm).
4. Mobile phase: Hexane:ethyl acetate (50:50).
5. Zeaxanthin standard solution (Sigma, St Louis, MO, USA) (see Note 22).

3. Methods

This section describes the heterologous expression of *crtS* gene from *X. dendrorhous* in *M. circinelloides* to produce xanthophylls. The methods outline the flask fermentation of *X. dendrorhous* to produce astaxanthin, the cloning of the cDNA of *crtS* gene by RT-PCR from RNA of *X. dendrorhous* obtained in astaxanthin production conditions the construction of an expression plasmid of this cDNA under the control of the P*carRA* promoter of *B. trispora*, the selection of *M. circinelloides* transformants, and the carotenoid profile of these transformants. DNA manipulations are performed according to standard procedures (35).

3.1. Flask Fermentation of X. dendrorhous

Fermentation of *X. dendrorhous* is carried out as previously described (25), with some modifications.

1. Inoculate 1 mL of frozen cells of *X. dendrorhous* Y2476 in a Petri plate (15-cm diameter) with 75 mL of YEPDA and incubate for 120 h at 17–20°C.
2. Resuspend the cells in 1.5 mL of sterile saline solution and seed a 500-mL flask with 25 mL of seed medium (see Note 23).
3. Incubate at 17–20°C and 250 rpm for 48 h.
4. Seed 2.5 mL of seed culture into a 500-mL flask with 25 mL of fermentation medium. Incubate for 120–168 h at 17–20°C and 250 rpm under illumination.

3.2. Cloning of the crtS cDNA from X. dendrorhous by RT-PCR

One of the limitations of the heterologous expression of eukaryotic genes is the capability of the heterologous host for the proper processing of introns. The *crtS* gene from *X. dendrorhous* has 17 introns, which largely difficult their correct processing by the host. The expression of the cDNA solves this problem as it has spliced introns. This section describes the purification of total RNA from *X. dendrorhous* in astaxanthin production conditions to maximize the *crtS* expression (Subheading 3.2.1), and cDNA amplification by RT-PCR for further expression (Subheading 3.2.2).

3.2.1. Purification of RNA from X. dendrorhous

RNA from *X. dendrorhous* is isolated using the RNeasy plant mini kit, a technology that combines the selective binding properties of a silica gel-based membrane with the speed of microspin technology. This procedure isolates high-quality RNA molecules longer than 200 nucleotides, showing a size distribution of the isolated RNA comparable to that obtained by centrifugation through a CsCl cushion.

1. Centrifuge the *X. dendrorhous* culture at 4,000 × *g* for 10 min at 4°C and carefully recover the cells.

2. Freeze immediately the cells in liquid nitrogen, and store at –70°C until use (see Note 4).
3. Disrupt the cells in a mortar under liquid nitrogen to obtain a fine powder (see Note 24).
4. Transfer a quantity of powder of about 500 μL to a liquid nitrogen-cooled sterile microtube (2.2 mL) with the help of a frozen spatula (see Note 25).
5. Purify RNA using RNeasy plant mini kit.
6. Analyze 3–5 μL of RNA sample onto 1% agarose gel and perform electrophoresis at 120 V for 20 min (see Note 26).

3.2.2. RT-PCR

This procedure allows synthesis of cDNA from mRNA and subsequent amplification of the DNA. In this case, we use the "Titan One Tube RT-PCR System" according to the instructions of the manufacturer. All the material must be RNase-free to avoid sample degradation. Disposable gloves must be used and material must be autoclaved twice. Solutions must be done with water treated with 0.2% diethyl pyrocarbonate (water-DEPC) to eliminate RNases.

1. Thaw the RNA samples in an ice water bath.
2. Prepare the RT reactions with the reactives from the "Titan One Tube RT-PCR System."
3. Incubate RNA with dNTPs, primer #126, primer #127, AMV reverse transcriptase, and DNA polymerase at 50°C for 30 min.
4. Transfer the samples to a thermocycler and do 35 cycles of amplification reactions as follows: 94°C, 10 s; 58°C, 30 s; 68°C, 60 s (extending 5 s per cycle after 11th).
5. Load the sample onto 0.7% agarose gel and perform electrophoresis at 100 V for 40 min (see Note 27).
6. Gel purify the DNA fragment using "Qiaex II Gel Extraction Kit," and resuspend the DNA in 10 μL of sterile ultrapure water.
7. Check 1 μL DNA by 0.7% agarose gel electrophoresis (100 V for 40 min).
8. Digest pGEM-T with *Not*I and fill the edges with DNA polymerase (Klenow). Then, digest with *Nco*I and ligate to the PCR fragment previously digested with *Nco*I.
9. Check the cloned fragment in pGEM-T by sequencing analysis to confirm the absence of mutations that may have occurred during the DNA amplification from RNA.

3.3. Construction of an Expression Plasmid

The autonomous replicating plasmid pALMC11 is constructed to express the cDNA of the *crtS* gene of *X. dendrorhous* in *M. circinelloides* MS12. This plasmid includes the *crtS* cDNA (1.6 kb *Nco*I–*Not*I) expressed under the control of the *carRA* promoter of *B. trispora* (611 pb *Nco*I) (27), and the *leuA* gene as an auxotrophic marker (26)

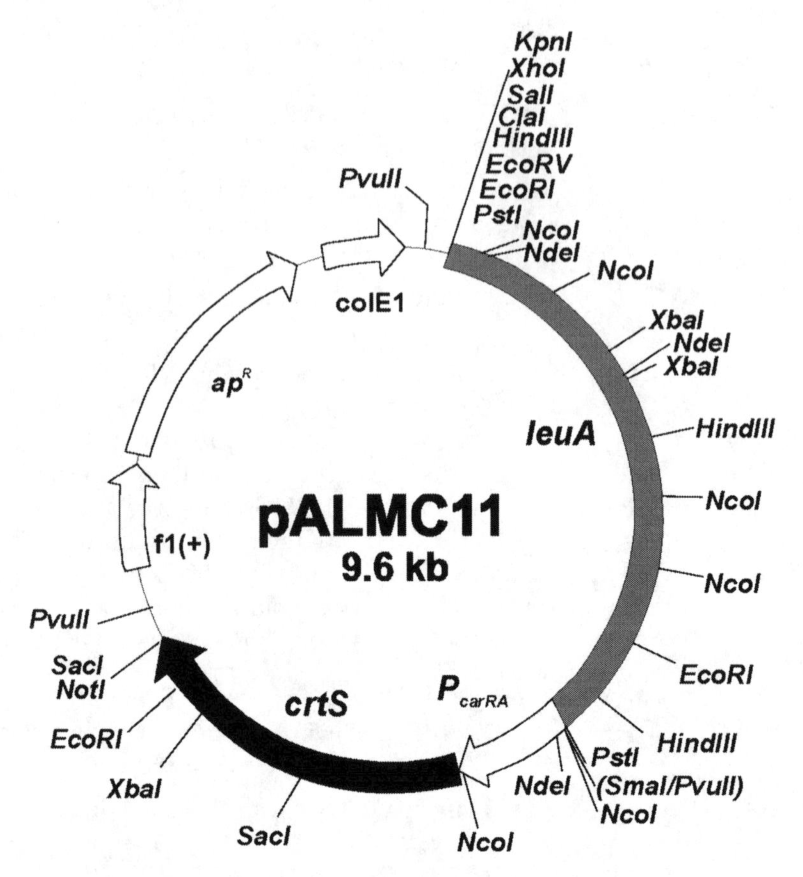

Fig. 2. Restriction map of the plasmid pALMC11 used for the expression of the cDNA from the *crtS* gene of *X. dendrorhous* in *M. circinelloides* MS12. This figure has been reproduced from Álvarez et al. (18), with permission.

(Fig. 2). General procedures for plasmid DNA purification, cloning, and transformation of *E. coli* are those of Sambrook et al. (35).

1. To digest the DNA, mix 1–3 μg of DNA, 5 μL of 10× buffer, and 25–40 U of enzyme adding ddH_2O to a final volume of 50 μL. Incubate at 37°C for 3 h.
2. Check the digestion products by 0.7% agarose gel electrophoresis (120 V for 20–30 min).
3. Purify the digestion products with the "Qiaex II Gel Extraction Kit."
4. Resuspend the purified DNA in 10 μL of sterile ultrapure water for further reactions.
5. Prepare standard ligation reactions (35).

3.4. Transformation of M. circinelloides Protoplasts

One of the most efficient methods to express exogenous DNA in zygomycete fungi is the polyethylene glycol-based transformation method. This procedure requires the previous formation of protoplasts to allow introducing the exogenous DNA through the cell membrane. The cellular wall of Zygomycetes fungi as *M. circinelloides* is mainly composed of complex polysaccharides such as chitin and chitosan that confers a strong rigidity to the cell. The treatment with lytic enzymes able to digest this type of structure results in the formation of protoplasts, susceptible then for DNA transformation. Here we describe the protocol to obtain Streptozyme (31), a lytic enzymatic mixture synthesized by *Streptomyces* sp. n°6 in the presence of cellular walls of fungi, able to degrade chitosan. In this procedure, we use cellular walls from *B. trispora*, a β-carotene producing Mucoral fungus, to induce the synthesis of the lytic enzymes (Subheadings 3.4.1 and 3.4.2). Transformation of *M. circinelloides* is carried out as previously described (36), with some modifications (Subheading 3.4.3).

3.4.1. Isolation of Cell Walls of B. trispora

1. Seed about 10^7 spores of *B. trispora* F-816 (+) on Petri plates with 75 mL of YPDA medium.
2. Incubate for 3–5 days at 25°C until the aerial mycelium covers the surface of the plates.
3. Recover carefully the mycelium with a sterile spatula and wash three times with sterile water by centrifugation at 3,000 × *g* and room temperature (see Note 28).
4. Freeze the mycelium at –20°C for several hours and then thaw to room temperature (see Note 29).
5. Disrupt the mycelium using a ball mill by ten cycles of 2 min at 4°C, leaving 30 s among each cycle (see Note 30).
6. Filter through 50-mesh sieve to recover the grinding beads (see Note 31).
7. Centrifuge the filtrate at 7,000 × *g* for 15 min.
8. Wash the pellet with 100 mL of sterile water and centrifuge again in the same conditions.
9. Repeat the washing steps until the supernatant is transparent.
10. Transfer the pellet to a tube and weigh it. This extract is used at final concentration of 2% to prepare the induction medium.

3.4.2. Obtaining of Chitosanase Enzyme (Streptozyme) from Streptomyces sp.

1. Prepare slant tubes with 25 mL of MEY (see Note 32).
2. Seed the slant tubes with 0.1 mL of frozen spores of *Streptomyces* sp. n°6, and incubate for 7 days at 28°C until sporulation.
3. Resuspend the spores of each slant in 10 mL of sterile water, briefly scraping with a sterile pipette.
4. Seed 0.25 mL of the spore solution in 500-mL flasks with 75 mL of Jeniaux medium. Incubate at 28°C and 250 rpm for 48 h.

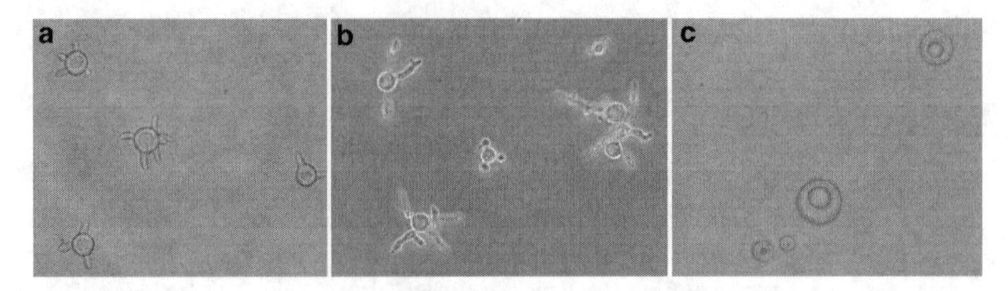

Fig. 3. Microscopic observation of the formation of *M. circinelloides* protoplasts (×40). (**a**) Early germination of spores. (**b**) Germinated spores ready for lytic enzyme treatment. (**c**) Protoplasts.

5. Harvest the mycelium by centrifugation at 4,000 × *g* for 5 min in sterile conditions. Wash the pellet with sterile water and centrifuge in the same conditions.
6. Resuspend the pellet in 1/10 volume of sterile water and seed 5 mL in 500-mL flasks with 75 mL of induction medium. Incubate at 28°C and 250 rpm for 18–24 h.
7. Centrifuge the culture at 7,000 × *g* for 30 min at 4°C and recover the supernatant.
8. Precipitate with ammonium sulfate using 50 g of salt for each 100 mL supernatant (see Note 33).
9. Centrifuge at 10,000 × *g* for 30 min at 4°C. Discard the supernatant and resuspend the pellet in the minimum volume of PG buffer.
10. Equilibrate PD-10 desalting columns with 4 volumes of PG buffer.
11. Add 2 mL of sample and elute with PG buffer, harvesting 2 mL aliquots in sterile tubes (see Note 34).
12. Add cold glycerol to a final concentration of 20% and freeze the aliquots at −20°C until their quantification.
13. Check the protein content by the Bradford method (37) using the "Bio-Rad Protein Assay."

3.4.3. Transformation of M. circinelloides

1. Inoculate *M. circinelloides* MS12 at about 10^7–10^9 spores/mL in flasks of 100 mL with 10 mL of YPG medium supplemented with 0.2 g/L of leucine and 0.4 g/L of uracil.
2. Incubate for 2 h at room temperature without agitation and 12 h at 4°C. Then, incubate at 28°C and 200 rpm to favor spore germination.
3. Take samples after 2.5–3 h incubation and check the germination rate by microscopic examination (Fig. 3).
4. Once germinated, centrifuge the culture at 1,000 × *g* for 5 min at 4°C in sterile conditions. Wash the pellet with 10 mM phosphate buffer pH 6.5 and centrifuge again.

5. Resuspend the pellet in 10 mL of 10 mM phosphate buffer pH 6.5 supplemented with 0.55 M sorbitol.
6. Add the lytic enzymes: 1.5 mg/mL of caylase C3 and 0.5 mL of streptozyme (Subheading 3.4.2).
7. Mix homogenously and incubate in a 100-mL flask at 22°C for 2–3 h with gentle agitation (no more than 65 rpm).
8. Check the protoplast release by microscopic examination (see Note 35).
9. Centrifuge the protoplast suspension at 4,000 × *g* for 5 min at 4°C.
10. Discard the supernatant and carefully wash the protoplast pellet twice with SMC buffer, by centrifugation at 4,000 × *g* for 5 min at 4°C.
11. Suspend the pellet in 1–2 mL of SMC buffer and check the protoplast concentration by microscopic observation.
12. Mix 200 μL of protoplast suspension with 0.5–5 μg of the expression plasmid pALMC11 and 20 μL of PMC buffer in a 10-mL sterile polypropylene tube.
13. Incubate for 30 min in a water-ice bath and then add 2.5 mL of PMC buffer, mixing gently by inversion until homogenization. Incubate at room temperature for 25 min.
14. Wash with 10 mL of SMC buffer and centrifuge at 4,000 × *g* for 5 min at 4°C.
15. Resuspend in 5 mL of YPG medium supplemented with 0.55 M sorbitol pH 4.5 and incubate with gentle agitation (50–65 rpm) for 30 min at room temperature.
16. Wash twice with 10 mL of SMC buffer as indicated in step 10 and finally resuspend in 3 mL.
17. Add 6 mL of soft YNB medium, mix gently inverting the tube, and spread over Petri plates with YNB medium supplemented with 0.4 g/L uracil (see Note 36).
18. Incubate for 4–5 days at 22°C until the colonies appear, and transfer them to individual fresh plates for storage and further analysis.

3.5. Verification of the Transformants

1. Confirm the stability of the transformants by successive growth cycles in YNB medium supplemented with 0.4 g/L of uracil (see Note 37).
2. Inoculate the selected colonies into 10 mL of liquid YNB medium supplemented with 0.4 g/L of uracil. Grow at 30°C for 3–5 days with 250 rpm agitation.
3. Mix 5 mL of culture with 5 mL of 50% glycerol, and aliquot in cryogenic vials for their conservation. Store at –70°C until use.

4. Use the remaining culture to purify plasmid DNA.
5. Mix 2 μL of isolated plasmid with 50 μL of *E. coli* DH5α cells and put the mixture into a prechilled electroporation cuvette.
6. Electroporate the cells at 2.5 kV, and quickly add 1 mL of LB medium.
7. Incubate at 37°C and 250 rpm for 1 h, and then spread the cells on a LB plate supplemented with 100 μg/mL ampicillin.
8. Incubate the Petri plates at 37°C for 16–18 h until colonies appear.
9. Inoculate single colonies in 3 mL of LB supplemented with 100 μg/mL ampicillin, and grow at 37°C for 12–16 h.
10. Purify the plasmids from each culture.
11. Verify the integrity and lack of rearrangements of pALMC11.

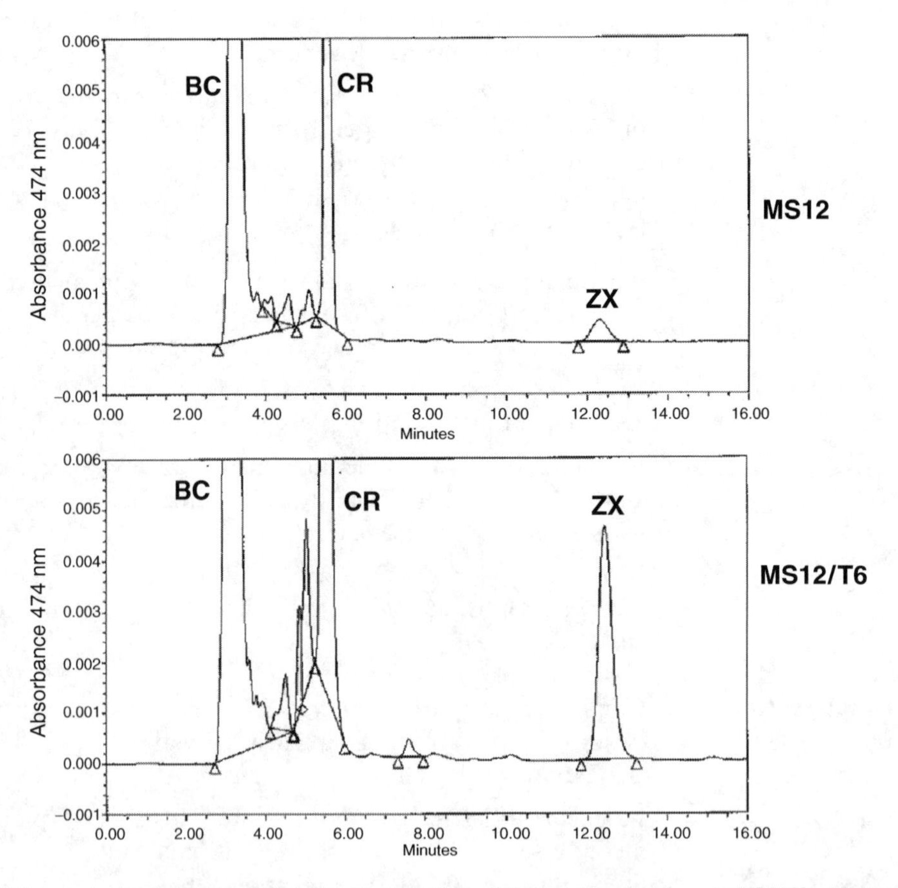

Fig. 4. HPLC analysis of carotenoids produced by *M. circinelloides* MS12 and the transformant MS12/T6. *M. circinelloides* carotenoid production was tested using as starting material the same quantity of mycelium (0.5 g) grown on YNB plates. β-carotene (BC), β-cryptoxanthin (CR), and zeaxanthin (ZX). This figure has been reproduced from Álvarez et al. (18), with permission.

3.6. Zeaxanthin Analysis

1. Place 1 g of total broth in a 100-mL flask and extract with methanol/dichloromethane (50/50, v/v) for 1 h under magnetic stirring.
2. Filter through nylon—0.2 μm and dilute with acetone if necessary.
3. Perform HPLC analysis at 474 nm using Nucleosil 100 NH_2 column and hexane/ethyl acetate (50/50 ratio) at 1 mL/min flow rate as mobile phase (Fig. 4).

4. Notes

1. To favor the complete dissolution of the agar, prepare the medium with hot water and under stirring.
2. Glucose must be autoclaved separately to avoid Schiff base reaction during the sterilization.
3. For optimal astaxanthin productivity, the orbital shaker should work at 250 rpm with 5 cm stroke under illumination. Astaxanthin biosynthesis is induced by white or UV light in *X. dendrorhous* (23, 28). Best results are obtained with three-baffled, high collar 500-mL flasks, and using milk filters instead foam plugs to allow maximum aeration.
4. Liquid nitrogen must be manipulated with caution. Skin contact may lead to a frostbite burn. The release of large quantities of liquid nitrogen in a confined space displaces air and can lead to asphyxiation.
5. This stock solution can be stored at room temperature. The pH of this buffer is not adjusted and should be about 8.5.
6. Microwave until agarose is completely melted. Cool the solution to approximately 70–80°C. Wash the electrophoresis equipment and the comb with 10% hydrogen peroxide and rinse with ultrapure water. Pour 25–30 mL of solution onto an agarose gel rack with appropriate 2- or 8-well combs.
7. Ethidium bromide is a carcinogen and must be manipulated with caution. Wear gloves and discard the contaminated tips and gels in specific containers for ethidium bromide solid waste disposal.
8. Primers are designed from nucleotide sequence of the *crtS* gene (39). Primer #126 includes a target site for the restriction endonuclease *Nco*I (underlined) to facilitate the subsequent cloning of the amplified DNA in different expression vectors.
9. pGEM-T is used for subcloning of PCR products. It has a size of 3,006 bp and confers resistance to ampicillin.

10. Gene pulser II is used to transform plasmids into *E. coli* by electroporation.
11. Ampicillin and kanamycin are antibiotics used for the selection of plasmid transformants. They are dissolved in water and sterilized by filtration. Final volumes of the stock solutions added to the culture medium are 100 μL/mL ampicillin and 50 μL/mL kanamycin.
12. Plasmid pLeu4 is derived from pUC13, which includes a 4.4-kb *Pst*I fragment with the *leuA1* gene for the complementation of the leucine auxotrophy (leu-). It also has the *ampR* gene that confers resistance to ampicillin.
13. It is a 611-bp *Nco*I fragment that includes the *carRA* promoter of *B. trispora*.
14. *B. trispora* F-816 (+) is a β-carotene overproducing strain. Its cell walls are used to induce chitosan lytic enzymes in *Streptomyces* sp. n°6.
15. This *Streptomyces* strain is used to obtain lytic enzymes able to degrade the chitosan of the *M. circinelloides* cell walls.
16. Optimal working conditions are 250 rpm and 5 cm stroke.
17. PD10 desalting columns containing 8.3 mL Sephadex™ G-25 resin for separation of high (Mr > 5,000) and low molecular weight substances (Mr < 1,000) by desalting and buffer exchange.
18. *M. circinelloides* MS12 (–) is an auxotrophic strain for leucine and uracil used for heterologous expression of *crtS* gene of *X. dendrorhous.*
19. Uracil and leucine are amino acids used to complement strains auxotrophies. Uracil is used at a final concentration of 400 mg/L of culture medium, and leucine at 200 mg/L.
20. Caylase C3 is an enzymatic mixture from *Tolypocladium geodes.*
21. Adjust YNB to pH 3.5 to grow isolated colonies or to pH 4.5 for a spread growth.
22. Dissolve 10 mg of zeaxanthin in 500 mL of hexane:ethyl acetate (50:50). Take 5 mL and dilute to 50 mL with acetone. Aliquot in vials and freeze at –20°C. To check the zeaxanthin concentration, measure OD_{452} with a specific extinction coefficient of 2,340 (100 mL/g/cm) (40).
23. Use crystal pipettes because *X. dendrorhous* is readily adsorbed on polystyrene surfaces.
24. Before to put the frozen cells in the mortar, clean the mortar with ethanol and cool it with liquid nitrogen.
25. Allow to evaporate the liquid nitrogen avoiding thawing the sample. Spatula must be cooled.

26. TAE 1× must be prepared with RNase-free water and then autoclaved to eliminate RNases. Electrophoresis equipment must be cleaned with diluted H_2O_2.
27. The size of amplified fragment must match the predicted by sequence. In this case, the amplified fragment of 1674 bp corresponds to the size of the *crtS* gene after introns processing.
28. Filtration through Nytal filter (30-μm pore) can be done instead the washing steps.
29. The freezing and thawing cycle promotes the cell wall disruption process.
30. To avoid overheating, refrigerate previously the mill by circulating cold water for 15 min. Use a grinding beads/mycelium ratio of 7/10 (v/v).
31. The glass grinding beads can be recycled by washing with water, 24 h treatment with concentrated sulfuric acid, and washing again with water to neutral pH.
32. The slants must be incubated before seeding for 2 days at 28°C. This dryness of the medium surface promotes *Streptomyces* sporulation.
33. Finely mill ammonium sulfate in a mortar to favor its dissolution. Mix at 4°C for 5 min maximum, controlling the pH and avoiding foaming. Then, stand at 4°C for 2–3 h before centrifugation.
34. If the sample volume after suspension is high, the desalting step can be achieved by overnight dialysis through a 5,000 Da

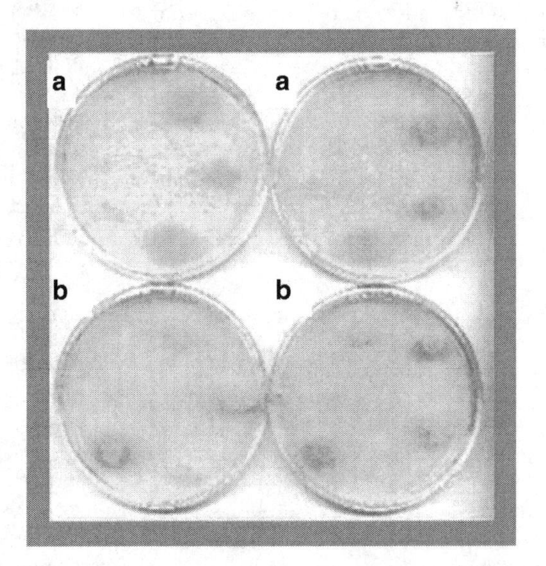

Fig. 5. Isolated transformants of *M. circinelloides* MS12. (**a**) Transformants including pALMC11. (**b**) Transformants including the expression plasmid without *crtS* gene (negative control).

cutoff membrane in PG buffer. In this case the cocktail activity expected will be lower due to higher volume and autolysis.

35. The protoplast yield after 3 h incubation should be more than 70%. If the protoplast yield is low, try to reduce the seeding, to reduce the germination time, to increase the washing volume, or to increase a bit the lytic enzyme concentration. Do not increase the lysis time (burst danger).
36. The growth of *M. circinelloides* in solid medium is very fast (3–6 cm/day). Lowering pH allows growing *M. circinelloides* as isolated colonies.
37. The transformants expressing pALMC11 show more intense yellow pigmented mycelium than the parental strain without the *crtS* cDNA (Fig. 5).

References

1. Johnson EA, Conklin D, Lewis MJ (1977) The yeast *Phaffia rhodozyma* as a dietary pigment source for salmonids and crustaceans. J Fish Res Board Can 34:2417–2421
2. Britton G (1998) Overview of carotenoid biosynthesis. In: Britton G, Liaaen-Jensen S, Pfander H (eds) Carotenoids: biosynthesis and metabolism. Birkhäuser Verlag, Basel, pp 13–147
3. Bauernfeind JC (1981) Carotenoids as colorants and vitamin A precursors: technical and nutritional applications. Academic, New York
4. Bhosale P (2004) Environmental and cultural stimulants in the production of carotenoids from microorganisms. Appl Microbiol Biotechnol 63:351–361
5. Phaff HJ, Miller MW, Yoneyama M, Soneda M (1972) A comparative study of the yeast florae associated with trees on the Japanese Islands and on the west coast of North America. In: Terui G (ed) Fermentation technology today. Society of Fermentation Technology, Japan, pp 759–774
6. Johnson EA, Lewis MJ (1979) Astaxanthin formation in the yeast *Phaffia rhodozyma*. J Gen Microbiol 115:173–183
7. Golubev WI (1995) Perfect state of *Rhodomyces dendrorhous* (*Phaffia rhodozyma*). Yeast 11:101–110
8. Baeza M, Sanhueza M, Flores O, Oviedo V, Libkind D, Cifuentes V (2009) Polymorphism of viral dsRNA in *Xanthophyllomyces dendrorhous* strains isolated from different geographic areas. Virol J 6:160
9. Andrewes AG, Phaff HJ, Starr MP (1976) Carotenoids of *P. rhodozyma*, a red pigmented fermenting yeast. Phytochemistry 15:1003–1007
10. Ducrey Sanpietro LM, Kula MR (1998) Studies of astaxanthin biosynthesis in *Xanthophyllomyces dendrorhous* (*Phaffia rhodozyma*). Effects of inhibitors and low temperature. Yeast 14: 1007–1016
11. Verdoes JC, Sandmann G, Visser H, Diaz M, van Mossel M, van Ooyen AJJ (2003) Metabolic engineering of the carotenoid biosynthetic pathway in the yeast *Xanthophyllomyces dendrorhous* (*Phaffia rhodozyma*). Appl Environ Microbiol 69:3728–3738
12. Visser H, van Ooyen AJJ, Verdoes JC (2003) Metabolic engineering of the astaxanthin-biosynthetic pathway of *Xanthophyllomyces dendrorhous*. FEMS Yeast Res 4:221–231
13. Kajiwara P, Fraser PD, Kondo K, Misawa N (1997) Expression of an exogenous isopentenyl diphosphate isomerase gene enhances isoprenoid biosynthesis in *Escherichia coli*. Biochem J 324:421–426
14. Niklitschek M, Alcaíno J, Barahona S, Sepúlveda D, Lozano C, Carmona M, Marcoleta A, Martínez C, Lodato P, Baeza M, Cifuentes V (2008) Genomic organization of the structural genes controlling the astaxanthin biosynthesis pathway of *Xanthophyllomyces dendrorhous*. Biol Res 41:93–108
15. Verdoes JC, Krubasik KP, Sandmann G, van Ooyen AJ (1999) Isolation and functional characterisation of a novel type of carotenoid biosynthetic gene from *Xanthophyllomyces dendrorhous*. Mol Gen Genet 262:453–461
16. Verdoes JC, Misawa N, van Ooyen AJ (1999) Cloning and characterization of the astaxanthin biosynthetic gene encoding phytoene desaturase of *Xanthophyllomyces dendrorhous*. Biotechnol Bioeng 63:750–755

17. Ojima K, Breitenbach J, Visser H, Setoguchi Y, Tabata K, Hoshino T, van den Berg J, Sandmann G (2006) Cloning of the astaxanthin synthase gene from *Xanthophyllomyces dendrorhous* (*Phaffia rhodozyma*) and its assignment as a beta-carotene-3-hydroxylase/4-ketolase. Mol Genet Genomics 275:148–158
18. Álvarez V, Rodríguez-Sáiz M, de la Fuente JL, Gudiña EJ, Godio RP, Martín JF, Barredo JL (2006) The *crtS* gene of *Xanthophyllomyces dendrorhous* encodes a novel cytochrome-P450 hydroxylase involved in the conversion of β-carotene into astaxanthin and other xanthophylls. Fungal Genet Biol 43:261–272
19. Martín JF, Gudiña E, Barredo JL (2008) Conversion of beta-carotene into astaxanthin: two separate enzymes or a bifunctional hydroxylase-ketolase protein? Microb Cell Fact 7:3
20. Alcaíno J, Barahona S, Carmona M, Lozano C, Marcoleta A, Niklitschek M, Sepulveda D, Baeza M, Cifuentes V (2008) Cloning of the cytochrome p450 reductase (*crtR*) gene and its involvement in the astaxanthin biosynthesis of *Xanthophyllomyces dendrorhous.* BMC Microbiol 8:169–181
21. Ukibe K, Hashida K, Yoshida N, Takagi H (2009) Metabolic engineering of *Saccharomyces cerevisiae* for astaxanthin production and oxidative stress tolerance. Appl Environ Microbiol 75:7205–7211
22. An GH, Cho MH, Johnson EA (1999) Monocyclic carotenoid biosynthetic pathway in the yeast *Phaffia rhodozyma* (*Xanthophyllomyces dendrorhous*). J Biosci Bioeng 88:189–193
23. De la Fuente JL, Rodríguez-Sáiz M, Schleissner C, Díez B, Peiro E, Barredo JL (2010) High-titer production of astaxanthin by the semi-industrial fermentation of *Xanthophyllomyces dendrorhous.* J Biotechnol 148:144–146
24. Howlett NG, Avery SV (1997) Relationship between cadmium sensitivity and degree of plasma membrane fatty acid unsaturation in *Saccharomyces cerevisiae.* Appl Microbiol Biotechnol 48:539–545
25. de la Fuente JL, Peiro E, Díez B, Marcos AT, Schleissner C, Rodríguez-Sáiz M, Rodríguez-Otero C, Cabri W, Barredo JL (2004) Method of producing astaxanthin by fermenting selected strains of *Xanthophyllomyces dendrorhous.* International Patent WO 03/066875
26. Roncero MIG, Jepsen LP, Strøman P, van Heeswijck R (1989) Characterization of a *leuA* gene and ARS element from *Mucor circinelloides.* Gene 84:335–343
27. Rodríguez-Sáiz M, Paz B, de la Fuente JL, López-Nieto MJ, Cabri W, Barredo JL (2004) Genes for carotene biosynthesis from *Blakeslea trispora.* Appl Environ Microbiol 70: 5589–5594
28. López-Nieto MJ, Costa J, Peiro E, Méndez E, Rodríguez-Sáiz M, de la Fuente JL, Cabri W, Barredo JL (2004) Biotechnological lycopene production by mated fermentation of *Blakeslea trispora.* Appl Microbiol Biotechnol 66:153–159
29. Jones D, Webley DM (1968) A new enrichment technique for studying lysis of fungal cell walls in soil. Plant Soil 28:147–157
30. Hopwood DA, Bibb MJ, Chater KF, Kieser T, Bruton CJ, Kieser H, Lydiate DJ, Smith CP, Ward JM, Schrempf H (1985) Genetic manipulation of streptomyces: a laboratory manual. The John Innes Foundation, Norwich
31. Jeniaux C (1966) Chitinases. Methods Enzymol 8:644
32. Ruiz-Hidalgo MJ, Eslava AP, Álvarez MI, Benito EP (1999) Heterologous expression of the *Phycomyces blakesleeanus* phytoene dehydrogenase gene (*carB*) in *Mucor circinelloides.* Curr Microbiol 39:259–264
33. Bartnicki-García S, Nickerson WJ (1962) Assimilation of carbon dioxide and morphogenesis *Mucor rouxii.* Biochim Biophys Acta 64:548–551
34. Lasker BA, Borgia PT (1980) High-frequency heterokaryon formation by *Mucor racemosus.* J Bacteriol 141:565–569
35. Sambrook J, Fritsch EF, Maniatis T (1989) Molecular cloning. A laboratory manual, 2nd edn. Cold Spring Harbor Laboratory Press, Cold Spring Harbor, New York
36. van Heeswijk R, Roncero MIG (1984) High frequency transformation of *Mucor* with recombinant plasmid DNA. Carlsberg Res Commun 49:691–702
37. Bradford MM (1976) A rapid and sensitive method for the quantitation of microgram quantities of protein utilizing the principle of protein-dye binding. Anal Biochem 72:248–254
38. Rodríguez-Sáiz M, de la Fuente JL, Barredo JL (2010) *Xanthophyllomyces dendrorhous* for the industrial production of astaxanthin. Appl Microbiol Biotechnol 88:645–658
39. Hoshino T, Ojiva K, Stoguchi Y (2000) Astaxanthin synthase. European Patent Application EP 1035206
40. Aasen AJ, Jensen SL (1966) Carotenoids of two further pigment types. Acta Chem Scand 20:2322–2324

Chapter 9

Bioengineering of Oleaginous Yeast *Yarrowia lipolytica* for Lycopene Production

Rick W. Ye, Pamela L. Sharpe, and Quinn Zhu

Abstract

Oleaginous yeast *Yarrowia lipolytica* is capable of accumulating large amount of lipids. There is a growing interest to engineer this organism to produce lipid-derived compounds for a variety of applications. In addition, biosynthesis of value-added products such as carotenoid and its derivatives have been explored. In this chapter, we describe methods to integrate genes involved in lycopene biosynthesis in *Yarrowia*. Each bacterial gene involved in lycopene biosynthesis, *crtE*, *crtB*, and *crtI*, will be assembled with yeast promoters and terminators and subsequently transformed into *Yarrowia* through random integration. The engineered strain can produce lycopene under lipid accumulation conditions.

Key words: Yeast, *Yarrowia*, Lipid accumulation, Carotenoid, Lycopene

1. Introduction

Yarrowia lipolytica has been traditionally used as an industrial organism to produce organic acids (citric acid and succinic acid), enzymes (lipase and esterase), single-cell protein, and single-cell oil (1). Through metabolic engineering, *Yarrowia* is being developed into a versatile production platform to convert lipids, fats, and oils into value-added products (2, 3).

As a value-added product, carotenoids have been widely used commercially (4). Biosynthesis of carotenoids is derived from the isoprenoid pathway (Fig. 1). In non-carotenogenic hosts such as *Escherichia coli* or *Saccharomyces cerevisiae*, introduction of bacterial carotenoid genes leads to production of various carotenoid compounds (5). Lycopene, the first color carotenoid in the pathway, can be made by coordinated expression of three genes, *crtE*, *crtY*, and *crtI*. The *crtE* gene encodes the geranylgeranyl diphosphate (GGPP) synthase, while *crtY* and *crtI* encode the phytoene

José-Luis Barredo (ed.), *Microbial Carotenoids From Fungi: Methods and Protocols*, Methods in Molecular Biology, vol. 898, DOI 10.1007/978-1-61779-918-1_9, © Springer Science+Business Media New York 2012

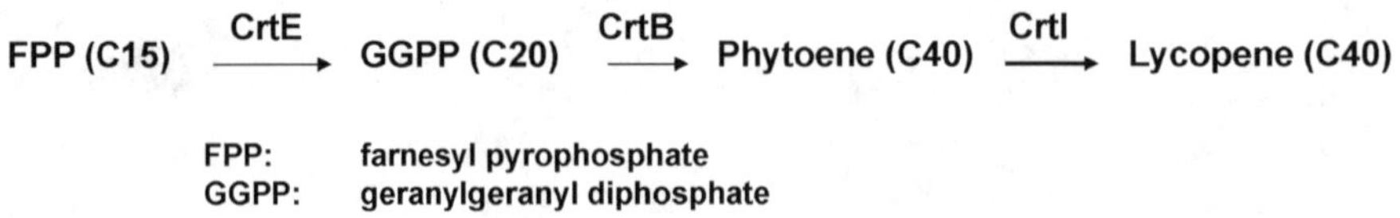

FPP (C15) —CrtE→ GGPP (C20) —CrtB→ Phytoene (C40) —CrtI→ Lycopene (C40)

FPP: farnesyl pyrophosphate
GGPP: geranylgeranyl diphosphate

Fig. 1. Lycopene biosynthetic pathway.

synthase and desaturase, respectively. Here in this chapter, we describe a method to produce lycopene in *Y. lipolytica* through expression of *crtE*, *crtY*, and *crtI* genes integrated in chromosomes. Chromosomal integration in *Yarrowia* is predominantly at random (6). Screening of transformants based on color formation is a convenient way to identify strains that produce a large amount of lycopene.

2. Materials

2.1. Strains, Plasmids, Genes, Promoters, and Terminators

1. *Y. lipolytica strain* Y2224 (Ura-, derivative of ATCC20362) (ATCC, Manassas, VA, USA).
2. *E. coli* XL2-Blue ultracompetent cells (Stratagene, La Jolla, CA, USA).
3. Ligase and rapid ligase buffer (Promega, Madison, WI, USA).
4. pZKleuN-6EP (GenBank accession number: GM621387).
5. pEXPGUS1-P (GenBank accession number: GM621389).
6. pZP34R (GenBank accession number: GM621390).
7. *crtE* codon optimized for expression in *Y. lipolytica*, derived from *Pantoea stewartii* DC413 (GenBank accession number: GM621381).
8. *crtB* codon optimized for expression in *Y. lipolytica*, derived from *P. stewartii* DC413 (GenBank accession number: GM621383).
9. *crtI* codon optimized for expression in *Y. lipolytica*, derived from *P. stewartii* DC413 (GenBank accession number: GM621385).
10. FBAIN promoter (2) (see Note 1).
11. *LIP1* terminator (2) (see Note 2).
12. GPDIN promoter (2) (see Note 3).
13. *LIP2* terminator (2) (see Note 4).
14. *EXP1* promoter (2) (see Note 5).
15. *OCT1* terminator (2) (see Note 6).

16. Ligation kit (Promega, Madison, WI, USA).
17. Lycopene (CaroteNature GmbH, Lupsingen, Switzerland).

2.2. Growth Medium

1. LB (Luria Bertani): Add 10 g of bacto-tryptone, 5 g of yeast extract, 10 g of NaCl, and dH_2O up to 1 L. Autoclave at 121°C for 15 min.
2. LB-agar: LB and 15 g/L agar. Autoclave at 121°C for 15 min.
3. Ampicillin stock solution (100 mg/mL).
4. YPD agar medium: Add 10 g of yeast extract, 20 g of peptone, and 20 g of glucose into dH_2O up to 1 L. Autoclave at 121°C for 15 min.
5. Minimal media (MM): Add 20 g of glucose, 1.7 g of yeast nitrogen base (YNB) without amino acids, 1.0 g of proline, and dH_2O up to 1 L, pH 6.1 (do not need to adjust). Autoclave at 121°C for 15 min.
6. Minimal Media + Uracil: Add 20 g of glucose, 1.7 g of YNB without amino acids, 1.0 g of proline, 0.1 g of uracil, 0.1 g of uridine, and dH_2O up to 1 L, pH 6.1 (do not need to adjust). Autoclave at 121°C for 15 min.
7. 5-FOA (5-fluoroorotic acid) (Zymo Research Corp., Orange, CA, USA).
8. Minimal media + 5-fluroorotic acid (MM + 5-FOA): Add 20 g of glucose, 1.7 g of YNB without amino acids, 1.0 g proline, and dH_2O up to 1 L, pH 6.1 (do not need to adjust). Autoclave at 121°C for 15 min. After autoclave, add 300 mg/L of sterile 5-FOA.
9. Fermentation medium (FM): Add 6.7 g of YNB, without amino acids and without ammonium sulfate, 6.0 g of KH_2PO_4, 2.0 g of K_2HPO_4, 1.5 g of $MgSO_4 \cdot 7H_2O$, 1.5 mg of thiamine hydrochloride, 20 g of glucose, 5.0 g of yeast extract, and dH_2O up to 1 L. Autoclave at 121°C for 15 min.
10. FM without yeast extract (YE): Add 6.7 g of YNB, without amino acids and without ammonium sulfate, 6.0 g of KH_2PO_4, 2.0 g of K_2HPO_4, 1.5 g of $MgSO_4 \cdot 7H_2O$, 1.5 mg of thiamine hydrochloride, 20 g of glucose, and dH_2O up to 1 L. Autoclave at 121°C for 15 min.
11. High glucose media (HGM): 80 g of glucose, 2.58 g of KH_2PO_4, 5.36 g of K_2HPO_4, pH 7.5 (do not need to adjust), and dH_2O up to 1 L. Sterilize by filtering through a 0.2-μm filter.
12. Transformation medium: 2.25 mL of 50% PEG (average MW 3350), 0.125 mL of 2 M lithium acetate (pH 6.0), and 0.125 mL of 2 M dithiothreitol (DTT) (see Note 7).

3. Methods

3.1. Cloning of crt Genes in the Integration Vector

1. Follow the specific steps of construction of integration vector for lycopene production outlined in Fig. 2. Set up a four-way ligation mixture containing the promoter, *crt* coding region, terminator, and the vector to clone each *crt* genes to (see Note 8).
2. Digest the plasmid containing codon-optimized *crt*E coding region with *Nco*I and *Not*I, and then gel-purify the fragment (see Note 9).
3. Digest the plasmid pZKleuN-6EP with *Bgl*II and *Nco*I, and then gel-purify the promoter FBAIN fragment.
4. Digest the plasmid pZKleuN-6EP with *Not*I and *Swa*I, and then gel-purify the *LIP1* terminator fragment.
5. Digest the plasmid pZKleuN-6EP with *Swa*I and *Bgl*II, and then gel-purify the vector fragment.
6. Set up a ligation reaction with above four fragments in ligation buffer with T-4 ligase, incubate the ligation reaction for 2 h, and then transform the ligation mixture into *E. coli* competent cells (see Note 10).
7. Name the resulting construct containing the FBAINr::*crtE*::LIP1 chimeric gene pYCRTE.
8. To clone the *crt*B coding region into pYCRTE, digest the plasmid containing synthetic *crt*B coding region with *Nco*I and *Not*I, and then gel-purify the fragment.

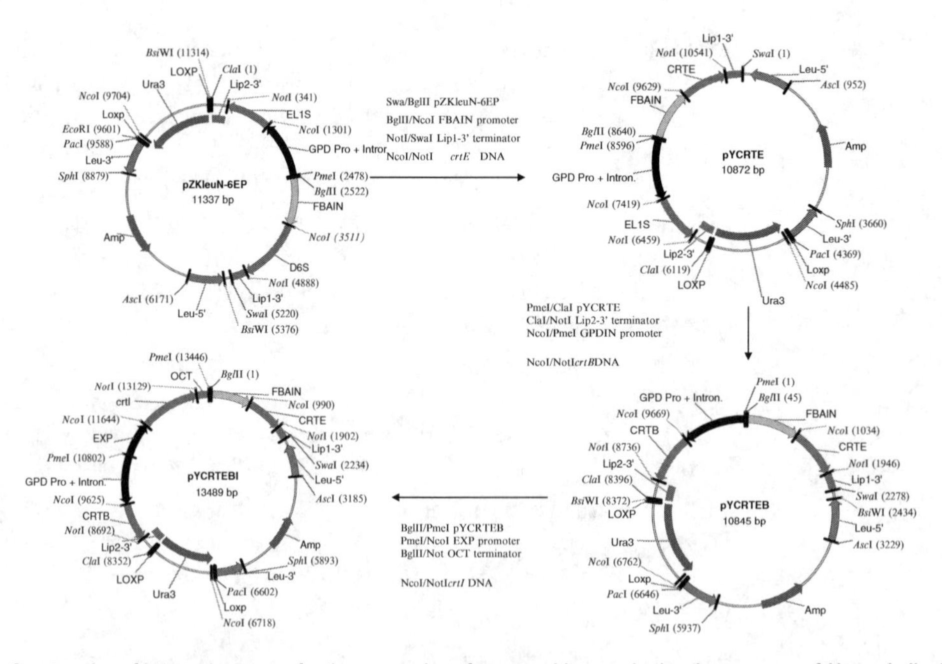

Fig. 2. Construction of integration vector for the expression of carotenoid genes in the chromosome of *Yarrowia lipolytica.*

9. Digest pZKleuN-6EP with *Pme*I and *Nco*I, and then gel-purify GPDIN promoter fragment.
10. Digest pZKleuN-6EP with *Cla*I and *Not*I, and then gel-purify the *LIP2* terminator fragment.
11. Digest pYCRTE with *Pme*I and *Cla*I, and then gel-purify the vector fragment.
12. Set up a ligation reaction with these four purified fragments in ligation buffer with T-4 ligase, incubate the ligation reaction for 2 h, and then transform the ligation mixture into *E. coli* competent cells (see Note 10).
13. Name the resulting construct containing FBAINr::*crtE*::LIP1 and GPDIN::CrtB::LIP2 chimeric genes pYCRTEB, which will be used to clone the *crtI* gene.
14. Digest the plasmid containing the synthetic *crtI* coding region with *Nco*I and *Not*I, and then gel-purify the fragment.
15. Digest the pEXPGUS1-P vector with *Pme*I and *Nco*I, and then gel-purify the EXP1 promoter fragment.
16. Digest pZP34R with *Bgl*II and *Not*I, and then gel-purify the OCT terminator fragment.
17. Digest pYCRTEB with *Bgl*II and *Pme*I, and then gel-purify the vector fragment.
18. Set up a ligation with all four fragments in ligation buffer with T-4 ligase, incubate the ligation reaction for 2 h, and transform the ligation mixture into *E. coli* competent cells (see Note 10).
19. Name the resulting construct containing FBAINr::crtE::LIP1, GPDIN::crtB::LIP2 and EXP::crtI::OCT three chimeric, pYCRTEBI.
20. Use pYCRTEBI digested with *Asc*I and *Sph*I to linearize the integration region from the vector backbone in the DNA transformations (see Note 11).

3.2. Integration of Three Chimeric crt Genes into Yarrowia Genome and Analysis of Lycopene Production

1. Streak *Y. lipolytica* Y2224 (Ura-) cells onto YPD agar plates 1 day prior to transformation.
2. Incubate the cultures at 30°C.
3. Transfer three large loops of cells from the YPD plates to 1 mL of transformation medium.
4. Mix 100–500 ng of digested plasmid DNA with 100 μL of the resuspended *Y. lipolytica* Y2224 (Ura-) cells.
5. Incubate the cell mixture at 39°C for 1 h. Mix the cell every 15 min using a Vortex mixer.
6. Spread the transformation mixture onto MM agar plates and incubate the plates at 30°C for 3–5 days (see Note 12).
7. Pick pink to red colonies from the plates.

8. Glow lycopene producing strain in 50 mL FM medium in a 250-mL flask and place the flask at 30°C in at shaker set at 250 rpm.
9. After 2 days of incubation, harvest 1 mL of culture by centrifugation at 12,000 × *g*.
10. Extract lycopene and analyze lycopene by HPLC (7). Commercial lycopene compound is used as standard.

3.3. Production of Lycopene Under Oil Accumulation Conditions

Oleaginous *Y. lipolytica* strain accumulates significant amounts of oil when grown in high glucose under nitrogen-limiting conditions.

1. Grow the lycopene producing *Yarrowia* strain in 100 mL FM medium in a 500-mL flask in 30°C shaker set at 225 rpm for 2 days.
2. Harvest 30 mL culture by centrifugation at 4,000 × *g*.
3. Resuspend the culture with 30 mL with HGM medium in a 125-mL flask.
4. Incubate the flask at a 30°C shaker (set at 225 rpm) for up to 4 days.
5. Analyze 1 mL of *Yarrowia* culture using HPLC analysis to determine the amount of lycopene produced (7).
6. Use rest of the cells for dry weight and oil analysis (8, 9).

4. Notes

1. *Bgl*II/*Nco*I fragment from pZKleuN-6EP; promoter region plus a portion of 5′ coding region that has an intron of the *FBA1* gene encoding for a fructose-bisphosphate aldolase enzyme (E.C. 4.1.2.13) of *Y. lipolytica*.
2. *Not*I/*Swa*I fragment from pZKleuN-6EP; terminator region of Lipase 1 gene of *Y. lipolytica*.
3. *Pme*I/*Nco*I fragment from pZKleuN-6EP; promoter region plus a portion of 5′ coding region that has an intron of the *GPD* gene encoding glyceraldehyde-3-phosphate-dehydrogenase of *Y. lipolytica*.
4. *Cla*I/*Not*I fragment from pZKleuN-6EP; terminator region of Lipase 2 gene of *Y. lipolytica*.
5. *Pme*I/*Nco*I fragment from pEXGUS1-P; promoter of the *EXP1* gene encoding EXP1 protein.
6. *Bgl*II/*Not*I fragment from pZP34R; terminator region of *OCT* gene encoding for 3-oxoacyl-coA thiolase of *Y. lipolytica*.
7. Prepare the DTT solution fresh prior to each use.
8. Instead of cloning, the each cassette containing the *crt* gene, promoter and terminator (Fig. 2) can be synthesized before

cloning into the integration vector. Alternatively, the entire integration region with all three genes and the necessary promoters and terminators can be synthesized to void multiple cloning steps.

9. All three synthetic *crt* genes contain an *Nco*I site around the start codon ATG and an *Not*I site after the translational stop codon.
10. We use the ligation kit from Promega. The ligation reaction has a final volume of 11 μL that contained 5.25 μL of 2× rapid ligation buffer, 0.75 μL of vector, 1.33 μL of each insert, and 1 μL of ligase. The reaction is allowed to proceed for 2 h at room temperature before transformation. After ligation, 2 μL of the reaction mixture is transformed into *E. coli* XL2-Blue ultracompetent cells. The transformed *E. coli* is spread on LB plates with ampicillin (100 μg/mL). Positive clones from each cloning step should be confirmed by restriction enzyme digestion and sequencing.
11. When transforming with integration plasmids, the DNA is linearized by digestion with appropriate restriction enzymes. There is no need to purify the fragment.
12. The integration fragment contains the *URA3* marker which confers growth selection for the integrated strain on MM plates.

References

1. Bankar A, Kumar A, Zinjarde S (2009) Environmental and industrial applications of *Yarrowia lipolytica*. Appl Microbiol Biotechnol 84:847–865
2. Zhu Q, Xue Z, Yadav N, Damude H, Pollak DW, Rupert R, Seip J, Hollerback D, Macool D, Zhang H, Bledsoe S, Short D, Tyreus B, Kinney A, Picataggio S (2010) Metabolic engineering of an oleaginous yeast for the production of omega-3 fatty acids. In: Cohen Z, Ratledge C (eds) Single cell oil, 2nd edn. ACOS Press, Urbana, pp 51–73
3. Sabirova J, Haddouche R, Van Bogaert I, Mulaa F, Verstraete W, Timmis K, Schmidt-Dannert C, Nicaud J, Soetaert W (2010) The "LipoYeasts" project: using the oleaginous yeast *Yarrowia lipolytica* in combination with specific bacterial genes for the bioconversion of lipids, fats and oils into high-value products. Microb Biotechnol 4: 47–54
4. Bhosale P, Bernstein P (2005) Microbial xanthophylls. Appl Microbiol Biotechnol 68:445–455
5. Misawa N, Shimada H (1997) Metabolic engineering for the production of carotenoids in non-carotenogenic bacteria and yeasts. J Biotechnol 59:169–181
6. Mauersberger S, Wang H, Gaillardin C, Barth G, Nicaud J (2001) Insertional mutagenesis in the n-alkane-assimilating yeast *Yarrowia lipolytica*: generation of tagged mutations in genes involved in hydrophobic substrate utilization. J Bacteriol 183:5102–5109
7. Ye R, Stead K, Yao H, He H (2006) Mutational and functional analysis of the beta-carotene ketolase involved in the production of canthaxanthin and astaxanthin. Appl Environ Microbiol 72:5829–5837
8. O'Fallon JV, Busboom JR, Nelson ML, Gaskins CT (2007) A direct method for fatty acid methyl ester synthesis: application to wet meat tissues, oils, and feedstuffs. J Anim Sci 85: 1511–1521
9. Yu X, Zheng Y, Dorgan KM, Chen S (2011) Oil production by oleaginous yeasts using the hydrolysate from pretreatment of wheat straw with dilute sulfuric acid. Bioresour Technol 102(10): 6134–6140

Chapter 10

Peroxisome Targeting of Lycopene Pathway Enzymes in *Pichia pastoris*

Pyung Cheon Lee

Abstract

Cellular targeting of biosynthetic pathway enzymes is an invaluable technique in metabolic engineering to modify metabolic fluxes towards metabolite of interest. Especially, recombinant carotenoid biosynthesis in yeasts should be balanced with a precursor pathway present in a specific cellular location because yeasts, being eukaryotes, have more defined intracellular location. Here, peroxisomal targeting of lycopene pathway enzymes, CrtE, CrtB, and CrtI, by fusing to peroxisomal targeting sequence 1 (PTS1) in *Pichia pastoris* X-33 is described.

Key words: Cellular targeting, Yeast, Peroxisomal targeting sequence, Lycopene, Metabolic engineering

1. Introduction

In most cases, metabolic pathway engineering of prokaryotic microorganisms, such as *Escherichia coli*, does not require a targeting strategy to send heterologously expressed pathway enzymes into specific cellular location (1–4). However, targeting strategies must be considered when engineering pathways in eukaryotic microorganisms. Eukaryotes such as yeasts have several cellular organelles such as mitochondria (5) and peroxisomes (6–8). The organelles, where metabolic activities occur, are physically separated from other cellular components by membrane structures. Therefore, the proper cellular location of heterologously expressed pathway enzymes is important, and both cellular and cytoplasmic membranes can be putative locations for membrane-bound enzymes to settle in. Carotenoid pathway enzymes are known to be membrane-associated or membrane-bound (9). When carotenogenic enzymes are heterologously

José-Luis Barredo (ed.), *Microbial Carotenoids From Fungi: Methods and Protocols*, Methods in Molecular Biology, vol. 898,
DOI 10.1007/978-1-61779-918-1_10,

expressed without a proper targeting system they can be randomly located on the membranes of cellular organelles as well as cytoplasmic membranes. The yeast *Pichia pastoris* has a cellular organelle peroxisome where farnesyl diphosphate, an important precursor for carotenoid biosynthesis (10), is known to be generated and present in large amounts (11).

Therefore, to make carotenoid biosynthesis balance with a precursor pathway present in peroxisomes in *P. pastoris*, the proper cellular targeting of carotenoid biosynthetic enzymes is very important. In this study, peroxisomal targeting of lycopene pathway enzymes, CrtE, CrtB, and CrtI, is investigated by fusing to enhanced green fluorescence protein (EGFP) with or without the peroxisomal targeting sequence 1 (PTS1) (12–14) in *P. pastoris* X-33.

2. Materials

Prepare all solution using double-distilled water (DDW) and analytical grade or molecular biology grade reagents. Prepare and store all reagents at 4°C (unless indicated otherwise).

2.1. Plasmids, Strains, and Molecular Biology Grade Chemicals

1. pEGFP (Clonetech, Mountain View, CA, USA).
2. pGAPZB (Invitrogen Corporation, Carlsbad, CA, USA) (see Note 1).
3. *E. coli* JM109 (New England Labs, Beverly, MA, USA).
4. *P. pastoris* X-33 strain (Invitrogen Corporation, Carlsbad, CA, USA) (see Note 2).
5. 100 mg/mL Zeocin™ (Invitrogen Corporation, Carlsbad, CA, USA) (see Note 3).
6. pUCM_CrtE (see Note 4).
7. pUCM_CrtB (see Note 4).
8. pUCM_CrtI (see Note 4).
9. Yeast lytic enzyme (ICN, Solon, OH, USA) (see Note 5).
10. Restriction enzymes *Eco*RI, *Xba*I, and *Avr*II.
11. T4 DNA ligase.
12. Vent DNA polymerases.
13. QIAquick Gel Extraction kit (Qiagen, Valencia, CA, USA).
14. QIAGEN Plasmid Mini kit (Qiagen, Valencia, CA, USA).
15. DDW (Qiagen, Valencia, CA, USA).
16. 0.1 mM TE (pH 8.0) (Qiagen, Valencia, CA, USA).

2.2. Culture Media

1. LB: 5 g/L yeast extract, 10 g/L tryptone, and 5 g/L NaCl. Adjust pH to 7.5 with 1 N NaOH. Autoclave on liquid cycle at 15 psi and 121°C for 20 min. Store at 4°C.
2. LB solid: LB and 15 g/L agar before autoclaving. Autoclave on liquid cycle at 15 psi and 121°C for 20 min. Store at 4°C.
3. LB–zeocin: LB and 50 μg/mL zeocin. Add 2,000× zeocin (100 mg/mL) into LB and use it immediately.
4. YPD: 10 g/L yeast extract, 20 g/L peptone, and 20 g/L dextrose. Autoclave on liquid cycle at 15 psi and 121°C for 20 min. Store at 4°C.
5. YPD solid: YPD and 20 g/L agar before autoclaving. Autoclave on liquid cycle at 15 psi and 121°C for 20 min. Store at 4°C.
6. YPD–zeocin: YPD and 100 μg/mL zeocin. Add 1,000× zeocin (100 mg/mL) into YPD and use it immediately.
7. YPDS: YPD and 1 M sorbitol. Store at 4°C.
8. YPDS–zeocin: YPDS and 100 μg/mL zeocin. Add 1,000× zeocin (100 mg/mL) into YPDS and use it immediately.

2.3. PCR Primers

The complete list of PCR primers designed for this study is shown in Table 1.

1. Forward primer F, e.g., CrtE-F, contains an *Eco*RI site and a Kozak consensus sequence (ATGG) at its 5′ end.
2. Reverse primer R, e.g., CrtE-R, contains a *Xba*I site and a stop codon at its 5′ end.
3. Reverse primer PTS-R, e.g., CrtE-PTS-R, contains a restriction enzyme site, a stop codon, and additional "CAACTTAGA" sequence (peroxisomal targeting sequence 1, PTS1) at its 5′ end. The PTS1 sequence is underlined.
4. Reverse fusion PCR primer I3, e.g., fEEGFP-I3, covers the C terminus of the carotenogenic enzyme with an extra six amino acids (Gly-Gly-Ala-Gly-Gly-Gly: GGT-GGA-GCA-GGA-GGT-GGC) as a spacer.

2.4. PCR

1. PCR machine.
2. 200-μL PCR tube.
3. QIAquick PCR purification kit (Qiagen, Valencia, CA, USA).
4. 10 mM dNTPs mixture.

2.5. Transformation

1. *P. pastoris* X-33 competent cells (Invitrogen Corporation, Carlsbad, CA, USA).
2. Gene Pulser Electroporation system (Bio-Rad Laboratories, Hercules, CA, USA).

Table 1
Primers used in this study. PTS1 sequences are underlined and spacers for fusion proteins are bold

Primers	Sequences (5′–3′)
CrtE-F	GGAATTCAAAATGGCAGTCTGCGCAAAA
CrtE-R	GCTCTAGAGCTTAACTGACGGCAGCG
CrtE-PTS-R	GCTCTAGAGCTTACAACTTAGAAC TGACGGCAGCGAG
CrtB-F	GGAATTCAAAATGGCAGTTGGCTCG
CrtB-R	GCTCTAGAGCCTAGAGCGGGCGCTG
CrtB-PTS-R	GCTCTAGAGCCTACAACTTAGAGAGCGGGCGCTGCC
CrtI-F	GGAATTCAAAATGGCACCAACTACGG
CrtI-R	GCTCTAGAGCTCAAATCAGATCCTCCAG
CrtI-PTS-R	GCTCTAGAGCTCACAACTTAGAAATCAGATCCTCCAGC
EGFP-F	CCGGAATTCAAAATGGTGAGCAAGGGCG
EGFP-R	GCTCTAGATTACTTGTACAGCTCGTCC
EGFP-PTS-R	GCTCTAGATTACAACTTAGACTTGTACAGCTCGTCC
fEEGFP-I3	**GCCACCTCCTGCTCCACC**ACTGACGGCAGCGAG
fEGFP-I5	**GGTGGAGCAGGAGGTGGC**ATGGTGAGCAAGGGCG
fBEGFP-I3	**GCCACCTCCTGCTCCACC**GAGCGGGCGCTGCC
fEGFP-I5	**GGTGGAGCAGGAGGTGGC**ATGGTGAGCAAGGGCG
fIEGFP-I3	**GCCACCTCCTGCTCCACC**AATCAGATCCTCCAGC
fEGFP-I5	**GGTGGAGCAGGAGGTGGC**ATGGTGAGCAAGGGCG

3. 1-mm Gap width disposable electroporation cuvettes (Bio-Rad Laboratories, Hercules, CA, USA).
4. 4-mm Gap width disposable electroporation cuvettes (Bio-Rad Laboratories, Hercules, CA, USA).

2.6. Fluorescence Microscope

1. Leica fluorescence microscope (Leica microsystem, Wetzlar, Germany) with an AxioCamMR camera (Carl Zeiss, Gőttingen, Germany), and MRGrab™ software version 1.0.0.4 (Carl Zeiss, Gőttingen, Germany).
2. Microscope slides and cover slips.
3. SCE buffer (pH 7.0): 0.6 M sorbitol, 0.3 M mannitol, 20 mM K_2HPO_4, 20 mM citric acid, 1 mM EDTA, and 0.1 M β-mercaptoethanol. Store at 4°C.

3. Methods

Carry out all procedures at room temperature unless otherwise specified.

3.1. Construction of EGFP Systems Targeted to Peroxisomes

1. Amplify the gene encoding fluorescent EGFP_pts from 0.1 to 0.2 μg pEGFP, as a template DNA, using a forward PCR primer EGFP-F and a reverse PCR primer EGFP-PTS-R under these PCR conditions: 95°C/5 min/1 cycle for initial denaturation, 94°C/30 s/29 cycles for denaturation, 50°C/1 min/29 cycles for annealing, 72°C/1 min/29 cycles for extension, 72°C/7 min/1 cycle for final extension. The PCR mixture (50 μL) is as follows: 37.5 μL of DDW, 5 μL of 10× PCR buffer, 5 μL of dNTPs, 1 μL of primer mix, 0.5 μL of DNA polymerase, and 1 μL of template DNA.
2. To amplify the second gene encoding fluorescent EGFP_n, repeat step 1 with primers EGFP-F and EGFP-R.
3. Clean up the PCR products with a QIAquick PCR purification kit (see Note 6), and then elute them with 50 μL DDW (see Note 7).
4. Digest both 1–2 μg PCR products and 2–3 μg pGAPZB in 100 μL reaction volumes with 5 U of *Eco*RI and *Xba*I at 37°C for 5 h.
5. Clean up the enzyme reaction mixtures with a QIAquick Gel Extraction kit, and then elute with ~30 μL DDW.
6. Ligate the digested PCR products into the *Eco*RI and *Xba*I sites of pGAPZB in 20 μL reaction volumes using 1 U of T4 DNA ligase at 16°C for 6 h (see Note 8).
7. Transform 1–4 μL of the ligation mixture into 50 μL *E. coli* competent cells in 1-mm electroporation cuvettes, precooled on ice, by Gene Pulser electroporation system.
8. After transformation, allow the transformants to recover in 1 mL LB at 37°C for 1 h.
9. Plate 25–100 μL the cells on LB–zeocin, prewarmed at 37°C for 1 h, and then incubate the plates at 37°C for 16 h.
10. Inoculate a single colony in 3 mL LB–zeocin and incubate overnight at 37°C.
11. After plasmid miniprep with QIAGEN Plasmid Mini kit, confirm the correct clones by restriction enzyme digestion and DNA sequencing.

3.2. Construction of Carotenogenic Enzyme_EGFP Systems Targeted to Peroxisomes

1. To obtain the first part of fusion proteins CrtX_EGFP_pts or CrtX_EGFP_n, amplify the gene *crt*X with a forward PCR primer CrtX-F and a reverse fusion PCR primer fXEGFP-I3 under the PCR conditions shown in Subheading 3.1, step 1 (see Note 9).

2. To obtain the second part of fusion protein, CrtX_EGFP_pts, which is targeted to peroxisomes, amplify the gene encoding EGFP using a forward fusion PCR primer fEGFP-I5 and a reverse PCR primer EGFP-PTS-R under the same PCR conditions.
3. To obtain the second part of fusion protein, CrtX_EGFP_n, which is not targeted to peroxisomes, repeat step 2 with fEGFP-I5 and EGFP-R.
4. Clean up the PCR products with a QIAquick PCR purification kit and then elute them with 50 μL DDW.
5. To construct CrtX_EGFP_pts or CrtX_EGFP_n, mix the two halves of PCR products together at an equimolar ratio and then do overlapping PCR under the same PCR conditions (see Note 10).
6. Repeat steps 3–11 in Subheading 3.1.

3.3. Pichia Transformation

1. Linearize 1–2 μg plasmid with 5 U of *Avr*II in 100 μL reaction mixture at 37°C for 5 h (see Note 11), heat-inactivate the reaction mixture at 65°C for 30 min, and then clean up with QIAquick Gel Extraction kit.
2. Transform 0.5–1 μg the linearized plasmid into 200 μL *P. pastoris* X-33 competent cells in 4-mm cuvettes that are precooled on ice by Gene Pulser electroporation system.
3. After transformation, allow the transformants to recover in 1 mL YPDS at 30°C for 2 h.
4. Plate 100–200 μL the cells on YPDS–zeocin prewarmed at 30°C.
5. Incubate the plates at 30°C until colonies are visible (see Note 12).

3.4. Culture Conditions

1. Select a single colony on YPDS–zeocin plate, streak it on fresh YPD–zeocin plate, and then incubate the plates at 30°C for 48–72 h.
2. Select a fresh single colony on YPD–zeocin plate and culture in 10 mL YPD–zeocin in a 125-mL flask at 30°C and 275 rpm for 28 h.
3. Harvest cells by centrifugation at 3,000 × *g* for 5 min, washed with 20 mL DDW, resuspended in 10 mL SCE buffer, and then used for further studies.

3.5. Fluorescence Microscopy

1. Drop 5 μL cells in SCE buffer on a glass plate and then carefully put a coverslip on the spot.
2. Check fluorescence of a single cell under a Leica fluorescence microscope with an AxioCamMR camera and MRGrab™ software version 1.0.0.4.
3. Take snapshots (see Note 13).

4. Notes

1. Invitrogen Corporation (http://www.invitrogen.com) sells several *E. coli–P. pastoris* shuttle vectors and *Pichia* strains that are right for your experiments. Thus, you can choose any vector or strain or combination for your experimental purposes.
2. "*Pichia* Protocols" in Methods in Molecular Biology, edited by D. R. Higgins and J. Cregg, will be helpful to be familiar to *Pichia* and its expression system. It collects key experimental procedures for use of the *P. pastoris* Expression System.
3. Zeocin™ is very sensitive to light, high ionic strength, and pH. Reduction of NaCl concentration to 5 g/L in LB and adjustment of the pH to 7.0–7.5 are recommended.
4. Plasmids carrying the genes encoding carotenogenic enzymes can be obtained from researchers working on carotenoid pathway engineering. Otherwise, these genes can be obtained by PCR amplification using specific primers.
5. We use the yeast lytic enzyme from ICN, but other yeast cell wall-degrading enzymes are available commercially.
6. If there are nonspecific PCR products, a QIAquick Gel Extraction kit (Qiagen) can be used instead of a PCR purification kit.
7. DDW is recommended if the purified PCR product is immediately used for restriction enzyme digestion. If not, 0.1 mM TE (pH 8.0) should be used.
8. Molar ratio of a PCR product (insert DNA) to pGAPZB (vector DNA) needs to be optimized.
9. X in CrtX, pUC-crtX, CrtX-F, and fXEGFP-I3 is E, B, or I in this study.
10. PCR condition may need to be optimized, especially annealing temperature.
11. Whether genes of interest have the same restriction site should be checked. If the enzyme site is present, site-directed mutagenesis PCR will be needed to remove the restriction enzyme site of the gene.
12. It takes usually 48–72 h for cells to grow on YPDS–zeocin.
13. You can see fluorescence of recombinant *P. pastoris* cells as shown in Fig. 1.

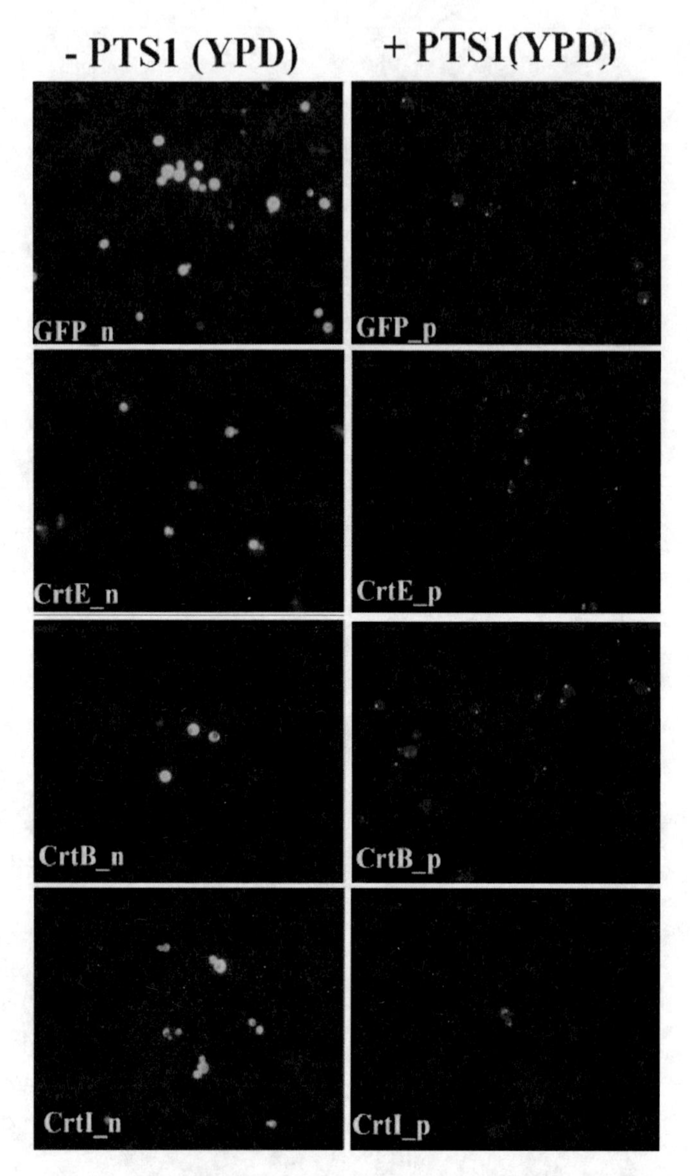

Fig. 1. Localization of control EGFP and the fused carotenogenic enzymes with (+PTS1) and without (−PTS1) PTS1 in *P. pastoris* grown on YPD. GFP_n, CrtE_n, CrtB_n, and CrtI_n represent the enzymes without PTS1 and GFP_p, CrtE_p, CrtB_p, and CrtI_p represent the enzymes with PTS1. Reproduced from ref. 8 with permission from Elsevier.

Acknowledgments

This work was supported by Mid-career Researcher Program through NRF funded by MEST (2011-0018057). Many parts of this manuscript was obtained from a reprint from Journal of Biotechnology, 140/3–4, Pyung Cheon Lee, Young geol Yoon,

References

1. Chae HS, Kim KH, Kim SC et al (2010) Strain-dependent carotenoid productions in metabolically engineered *Escherichia coli*. Appl Biochem Biotechnol 162:2333–2344
2. Kim JR, Kong MK, Lee SY et al (2010) Carbon sources-dependent carotenoid production in metabolically engineered *Escherichia coli*. World J Microbiol Biotechnol 26:2231–2239
3. Kim SH, Park YH, Schmidt-Dannert C et al (2010) Redesign, reconstruction, and directed extension of *Brevibacterium linens* carotenoid pathway in *Escherichia coli*. Appl Environ Microbiol 76:5199–5206
4. Lee PC, Momen AZR, Mijts BN et al (2003) Biosynthesis of structurally novel carotenoids in *Escherichia coli*. Chem Biol 10:453–462
5. Yoon YG, Haug CL, Koob MD (2007) Interspecies mitochondrial fusion between mouse and human mitochondria is rapid and efficient. Mitochondrion 7:223–229
6. Bhataya A, Schmidt-Dannert C, Lee PC (2009) Metabolic engineering of *Pichia pastoris* X-33 for lycopene production. Process Biochem 44:1095–1102
7. Poirier Y, Erard N, MacDonald-Comber Petétot J (2002) Synthesis of polyhydroxyalkanoate in the peroxisome of *Pichia pastoris*. FEMS Microbiol Lett 207:97–102
8. Lee PC, Yoon YG, Schmidt-Dannert C (2009) Investigation of cellular targeting of carotenoid pathway enzymes in *Pichia pastoris*. J Biotechnol 140:227–233
9. Britton G (1998) Overview of carotenoid biosynthesis. In: Britton G (ed) Carotenoids: biosynthesis and metabolism, 1st edn. Birkhäuser, Basel
10. Lee PC, Petri R, Mijts BN et al (2005) Directed evolution of *Escherichia coli* farnesyl diphosphate synthase (IspA) reveals novel structural determinants of chain length specificity. Metab Eng 7:18–26
11. Kovacs WJ, Olivier LM, Krisans SK (2002) Central role of peroxisomes in isoprenoid biosynthesis. Prog Lipid Res 41:369–391
12. McNew JA, Goodman JM (1994) An oligomeric protein is imported into peroxisomes in vivo. J Cell Biol 127:1245–1257
13. Lazarow PB, Fujiki Y (1985) Biogenesis of peroxisomes. Annu Rev Cell Biol 1:489–530
14. Gould SJ, McCollum D, Spong AP et al (1992) Development of the yeast *Pichia pastoris* as a model organism for a genetic and molecular analysis of peroxisome assembly. Yeast 8:613–628

Chapter 11

Production, Extraction, and Quantification of Astaxanthin by *Xanthophyllomyces dendrorhous* or *Haematococcus pluvialis*: Standardized Techniques

Alma Rosa Domínguez-Bocanegra

Abstract

For many years, benefits and disadvantages of pigments production either by microalgae or yeasts have been under analysis. In this contribution we shall deal with *Xanthophyllomyces dendrorhous* (formerly *Phaffia rhodozyma*) and *Haematococcus pluvialis*, which are known as major prominent microorganisms able to synthesize astaxanthin pigment. Then, the usual trend is to look for optimal conditions to conduct astaxanthin synthesis. From one side, pigment production by *H. pluvialis* is promoted under cellular stress conditions like nutrient deprivation, exposition to high light intensity, aeration. On the other side, *X. dendrorhous* is able to show significant increase in astaxanthin synthesis when grown in natural carbon sources like coconut milk, grape juice. The main aim of this chapter is to describe optimal environmental conditions for astaxanthin production by *X. dendrorhous* or *H. pluvialis*.

Key words: *Haematococcus pluvialis*, *Phaffia rhodozyma*, *Xanthophyllomyces dendrorhous*, Astaxanthin synthesis, Coconut milk

1. Introduction

Astaxanthin is an abundant carotenoid pigment that can be synthesized by few microorganisms, among which are *Brevibacterium*, *Mycobacterium lacticola* (1), *Agrobacterium auratim* (2), *Haematococcus pluvialis* (3), and the yeast *Xanthophyllomyces dendrorhous* (4). Several marine species get the color by ingestion of suitable microorganisms (5–7). Due to its antioxidant properties, there is a growing economic interest in aquaculture, chemical, pharmaceutical, and alimentary industries (8, 9). The renewed interest in natural sources of the pigment is twofold: from one side synthetic versions of astaxanthin are rather expensive ($2,500–3,000/kg) and on the other may induce human cancer (10).

José-Luis Barredo (ed.), *Microbial Carotenoids From Fungi: Methods and Protocols*, Methods in Molecular Biology, vol. 898,
DOI 10.1007/978-1-61779-918-1_11,

Microalgal astaxanthin purified from *H. pluvialis* is indeed a powerful antioxidant agent in vitro under both hydrophobic and hydrophilic conditions (11). Viability of large scale production for commercial purposes seems to be a realistic issue (12–14).

X. dendrorhous is able to ferment glucose and other sugars, and to produce carotenoids such as astaxanthin (15). Among the recommended media for cellular growth and pigment production, we can mention the following: cane molasses (16), sugar cane juice (17), corn wet-milling by-products (18), alfalfa residual juice (19), grape juice (20), hydrolyzed peat (21, 22), and coconut milk (23). Let's mention that carotenoid biosynthesis is regulated by singlet oxygen and peroxide radicals (7), and light (13, 14).

H. pluvialis presents extremely complex relationships between nutrients, growth, cell yield, cell type, and astaxanthin formation (24–27). For instance, high astaxanthin accumulation in *H. pluvialis* may induce changes in primary metabolism (28, 29). Moreover, as a secondary metabolite, astaxanthin production by *H. pluvialis* is enhanced by oxidative stress, claiming that could remove free radicals (30). But also, a high C/N ratio may enhance astaxanthin accumulation according to Kakizono et al. (31).

The aim of this chapter is to propose a systematic methodology to set experimental parameters which play a key role in astaxanthin synthesis by two prominent microorganisms, namely *X. dendrorhous* and *H. pluvialis*. When dealing with *X. dendrorhous*, the main issue is the selection of culture media. In the case of *H. pluvialis*, attention is paid to culture conditions like aeration, light intensity (continuous or photoperiod), systematic shaking of the culture, etc. Finally, standardization of the extraction, quantification, and production techniques for astaxanthin is proposed here.

2. Materials

Materials and methods for both *X. dendrorhous* and *H. pluvialis* are detailed here. Complementary descriptions can be found (23, 32).

2.1. Microorganisms

1. *X. dendrorhous* NRRL-10921 (ARS culture collection, United States Department of Agriculture USDA, Peoria, IL, USA) (Fig. 1).
2. *H. pluvialis* NIES-144 (National Institute for Environmental Studies (NIES), University of Tokyo, Japan) (Fig. 2).

2.2. Culture Media and Reagents

1. YM: 10 g/L of dextrose (glucose), 3 g/L of yeast extract, 5 g/L of bacto-peptide, and 3 g/L of malt extract.
2. YPG: 20 g/L of dextrose (glucose), 3 g/L of yeast extract, and 10 g/L of bacto-peptide.

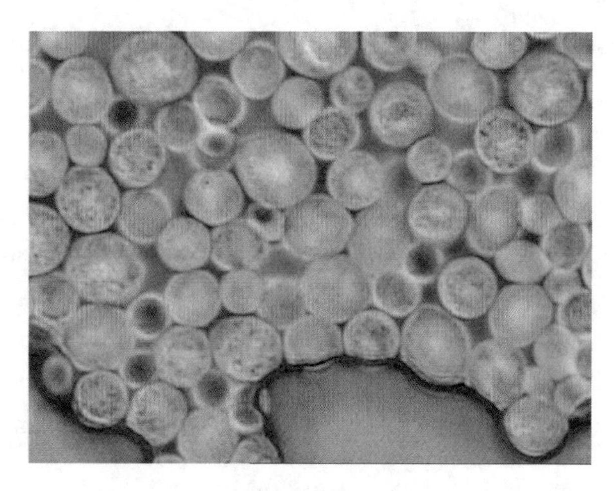

Fig. 1. *X. dendrorhous* view under microscope Zeiss model Axios II Plus (augmentation ×40).

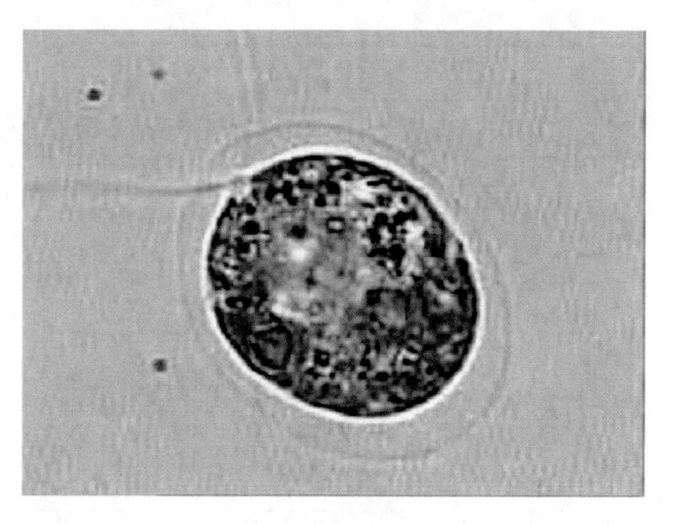

Fig. 2. *H. pluvialis* view under microscope Zeiss model Axios II Plus (augmentation ×40).

3. Coconut milk with this composition: 16 g/L of total carbohydrates, 0.25 g/L of sodium, 0.1 g/L of magnesium, 2.49 g/L of potassium, and 1 g/L of calcium (see Note 1).
4. BBM: 1.0 g/L $NaNO_3$, 0.025 g/L $CaCl_2·2H_2O$, 0.075 g/L $MgSO_4·7H_2O$, 0.075 g/L K_2HPO_4, 0.175 g/L KH_2PO_4, 0.025 g/L NaCl, 0.005 g/L $FeCl_3·6H_2O$, 4.2×10^{-5} g/L $NaMoO_4$, 4.4×10^{-4} g/L $ZnSO_4·7H_2O$, 3.6×10^{-3} g/L $MnCl_2·4H_2O$, 1.6×10^{-5} g/L $CuSO_4·5H_2O$, and 5.7×10^{-3} g/L H_3BO_4 (see Note 2).
5. BG-11: 1.5 g/L $NaNO_3$, 0.036 g/L $CaCl_2·2H_2O$, 0.075 g/L $MgSO_4·7H_2O$, 0.4 g/L K_2HPO_4, 0.001 g/L EDTA, 0.006 g/L citric acid, 0.2 g/L Na_2CO_3, 2.1×10^{-5} g/L $NaMoO_4$, 2.2×10^{-4} g/L $ZnSO_4·7H_2O$, 1.8×10^{-3} g/L $MnCl_2·4H_2O$, 0.8×10^{-5} g/L $CuSO_4·5H_2O$, and 2.8×10^{-3} g/L H_3BO_4 (see Note 2).

6. BAR: 0.25 g/L $NaNO_3$, 0.025 g/L $CaCl_2 \cdot 2H_2O$, 0.075 g/L $MgSO_4 \cdot 7H_2O$, 0.075 g/L K_2HPO_4, 0.175 g/L KH_2PO_4, 0.025 g/L NaCl, 0.025 g/L EDTA, 0.0025 g/L $FeCl_3 \cdot 6H_2O$, 3.5×10^{-4} g/L $NaMoO_4$, 4.4×10^{-3} g/L $ZnSO_4 \cdot 7H_2O$, 7.3×10^{-4} g/L $MnCl_2 \cdot 4H_2O$, 7.7×10^{-4} g/L $CuSO_4 \cdot 5H_2O$, 2.3×10^{-4} g/L $CoCl_3$, and 1.0 g/L CH_3COONa (see Note 2).
7. Dinitrosalicylic (DNS) acid reagent (34).
8. NaCl 20% w/v.
9. Methanol: water: hexane (95:4:1 v/v/v).
10. Solvent A: 100:70:70 of hexane: acetone: ethyl alcohol.
11. Astaxanthin standard (Hoffmann-La Roche, Basel, Switzerland).

2.3. Equipment

1. Centrifuge.
2. Orbital shaker.
3. Incubator.
4. Microscope.
5. Analytic balance.
6. 0.45-μm Filter (Millipore, Billerica, MA, USA).
7. Glass pearls (425–600 μm diameter).
8. Spectrophotometer.
9. Artificial lighting equipment.
10. Neubauer haemocytometer.
11. High-performance liquid chromatography (HPLC).
12. Vortex.

3. Methods

3.1. Inoculum Preparation of *X. dendrorhous*

1. From a storage tube, place aseptically the cell suspension in a 250-mL baffled Erlenmeyer flask containing 50 mL of YM medium.
2. Incubate the flask in an orbital shaker at controlled temperature of 22°C and stirring speed of 150 rpm, for 48 h.
3. Convey the obtained cellular growth in a second 250-mL baffled Erlenmeyer flask keeping a relation of 10% volume of seed culture.
4. Incubate at 22°C and 150 rpm for 48 h (see Note 3). The recommended inoculum appearance is shown in Fig. 3.

3.2. Inoculum Preparation of *H. pluvialis*

1. Within test tubes, inoculate 1 mL of *H. pluvialis* type strain in a total volume of 10 mL of BG-11.
2. Maintain the cultures with continuous light (177 μmol photon/m^2s) for 120 h at room temperature (28°C).

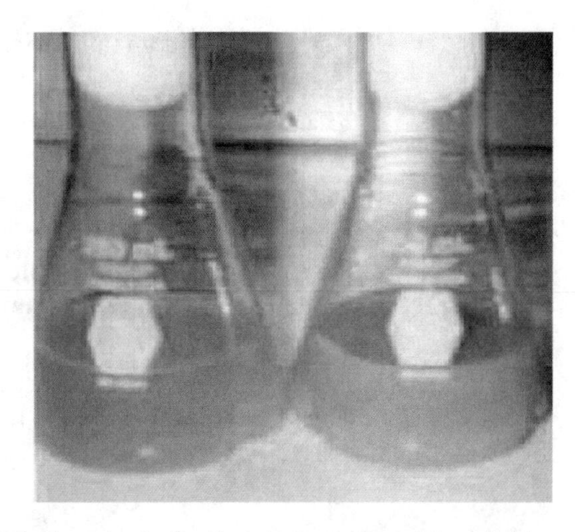

Fig. 3. *X. dendrorhous* standardized inoculum and astaxanthin synthesis.

Fig. 4. *H. pluvialis* standardized inoculum.

3. Convey the cultures to 250-mL Erlenmeyer flasks with 100 mL of BG-11 medium.
4. Keep the same conditions of continuous lighting (177 μmol photon/m^{2}s) and temperature (28°C) for 120 h, taking care of stirring by hand once a day for 10 s.
5. Inoculate again with the seed culture at 10% v/v in an Erlenmeyer flask of 250 mL with 100 mL of BG-11.
6. Incubate at 28°C, aeration of 1.5 vvm, and continuous illumination (177 mol photons/m^{2}s) for 120 h (see Note 4). The recommended inoculum appearance is shown in Fig. 4.

3.3. Dry Weight Analysis of *X. dendrorhous*

1. Centrifuge samples of 2 mL from the cultured medium for 5 min at 3,500 × *g*, room temperature (28°C).
2. Suspend again the cell pellet; wash it twice with 2 mL of distilled water and filter it through a 0.45-μm pre-weight filter.

3. Dry the filters containing the biomass at 60°C for 24 h.
4. Quantify and register the dry weight on a chosen analytic balance.

3.4. Dry Weight Analysis of H. pluvialis

1. Centrifuge samples of 5 mL from the culture medium for 5 min at 3,000 × *g* in a clinical centrifuge.
2. Suspend again the cell pellet; wash it with 3 mL of distilled water and centrifuge again under the same conditions. Perform previous operation two times more.
3. Filter the cellular package through a 0.45-μm pre-weight filter, leaving them to dry for 24 h at 60°C.
4. Quantify and register the dry weight on a chosen analytic balance.

3.5. Cellular Number of X. dendrorhous and H. pluvialis

With a Pasteur pipette, inoculate a sample of the culture (either *X. dendrorhous* or *H. pluvialis*) in a Neubauer haemocytometer. Then, perform the cellular counting every 24 h with a microscope, as long as the growth kinetics and pigment production processes are carried out, typically 120 h for *X. dendrorhous* and 168 h for *H. pluvialis*. Microorganisms' cellular shapes are shown in Figs. 1 and 2.

3.6. Optical Density of X. dendrorhous and H. pluvialis

1. Take a sample of 5 mL culture.
2. Measure optical density (OD) in a spectrophotometer: OD_{650} for *H. pluvialis* and OD_{470} for *X. dendrorhous.*
3. Perform this analysis every 24 h, as long as the experiment requires.

3.7. Chlorophyll Content Determination H. pluvialis

To determine chlorophyll by the APHA technique (33) proceed as follows:

1. Incubate triplicates of 5 mL samples from three different Erlenmeyer flasks and centrifuge the samples at 3,000 × *g* for 5 min.
2. Heat the cell package at 80% in a suspension, 90:10 (v/v) methanol:water, for 3 min in the dark and centrifuge once more at 3,000 × *g* for 5 min.
3. Determine OD_{665} in a spectrophotometer.

3.8. Sugar Content Evaluation X. dendrorhous

The DNS method (34) is used to determine residual reducing sugars in the culture media, as follows:

1. Incubate a 1 mL sample and centrifuged at 3,500 × *g* for 5 min.
2. Then, supply 1 mL of DNS reagent to the supernatant.
3. Cover the tubes and heat them up to boiling point for 5 min; place them immediately in an ice bath for rapid cooling, after which 8 mL of distillated water is added.

4. Shake the samples with a vortex for 5 min.
5. OD_{575} measurement. To calibrate the spectrophotometer use a standard reference curve.

3.9. Pigment Extraction from X. dendrorhous

1. Centrifuge at 3,500 × *g* triplicate samples of 1 mL aliquots of the saturated culture from three different Erlenmeyer flasks, for 5 min.
2. Then, wash the cell pellets twice with 1 mL of distilled water and 1 mL of acetone, until solvent evaporates.
3. Add 1 mL of glass pearls (425–600 μm diameter) together with 1 mL of dimethylsufoxide (DMSO) at 60°C.
4. In order to extract the colorant from its organic phase, break the cells by vigorous shaking using a vortex, for 5 min.
5. Add, subsequently, 1 mL of acetone, 1 mL of petroleum ether, and 1 mL of NaCl at 20% w/v.
6. Centrifuge the samples during 5 min at 3,500 × *g*, to separate the petroleum ether containing the colorant.
7. Check the presence of carotenoids by OD_{474} using an extinction coefficient of 1% = 2,100 (35).
8. Now, quantify astaxanthin with a commercial standard using the HPLC method (36). The eluting solvent is methanol: water: hexane (95: 4: 1 v/v/v) with a flow rate set at 0.5 mL/min. The recommended culture appearance is shown in Fig. 5.

3.10. Pigment Extraction from H. pluvialis

Here pigments are determined using a modification of the Britton procedure (37).

1. Inoculate triplicates with 5 mL samples from three different Erlenmeyer flasks and centrifuge the samples at 900 × *g* for 5 min.

Fig. 5. Different kinetic states of *X. dendrorhous* in astaxanthin production within coconut milk.

Fig. 6. *H. pluvialis* kinetics. From cellular growing in medium BBM to pigment cyst formation in BAR medium.

2. Wash the cell pellet twice with 5 mL of distilled water and add the same volume of glass pearls together with 5 mL of solvent A (100:70:70 of hexane: acetone: ethyl alcohol).
3. Break the cells by vigorous shaking using a vortex in order to extract the colorant from the organic phase.
4. Monitor the presence of carotenoids by OD_{474} using an extinction coefficient of 1% = 2,100 for the total carotenoids and 1% = 1,600 for astaxanthin (35).
5. Finally, quantify astaxanthin using a commercial standard by the HPLC method (36). The recommended culture appearance is shown in Fig. 6.

3.11. Growth Kinetics of *X. dendrorhous*

The optimal conditions for cellular production of *X. dendrorhous* are as follows:

1. Place samples of 5 mL from standardized inoculum in 250-mL baffled Erlenmeyer flasks containing 50 mL of coconut milk, YPG, or YM medium (see Subheading 3.1 and Note 3).
2. Ensure a homogeneous mixing using a shaker at 150 rpm, 22°C, and pH = 4.5 for 5 days. Parallel experiments must be run by triplicate.

3.12. Growth Kinetics of *H. pluvialis*

The optimal conditions for cellular production of *H. pluvialis* are as follows:

1. Place samples of 70 mL from standardized inoculum in 1,000-mL Erlenmeyer flasks containing 700 mL of BBM or BG-11 media (see Subheading 3.2 and Note 4).

2. Ensure regulated temperature at 28°C, daily manual agitation, aeration of 1.5% CO_2, continuous illumination or 12 h light/12 h dark cycle (177 μmol photon/m^{2}s). Usually, maximal cell growth is reached between 7 and 8 days.

3.13. Pigment Production of X. dendrorhous

The pigment production goes together with the growth of the yeast. The conditions are then exactly the same as described in Subheading 3.11. The key difference is the use of coconut milk as a carbon source without any additional nutrient (see Note 5). Usually, optimal pigment production is reached after 120 h. Figure 5 shows the desired culture pigmentation.

3.14. Pigment Production of H. pluvialis

1. Take a 1,000-mL Erlenmeyer flask, previously fitted with 700 mL of BAR medium, to inoculate 20% v/v of the seed culture optimally grown according to Subheading 3.12.
2. Maintain the following culture conditions: Non-aeration, continuous illumination at 345 μmol photon/m^{2}s, constant room temperature (28°C), and manual agitation every 24 h. Normally, the maximal pigment figures are obtained between 168 and 192 h.

It is worth noticing that conditions for cellular growth are completely different for pigment production, in terms of the media, aeration, and light intensity. In this sense, this second treatment is mandatory to get high pigment production (see Note 6). Figure 6 shows the desired culture pigmentation.

4. Notes

1. Raw residual coconut milk was obtained from a local candy industry located in Mexico City. It was sterilized by filtration and no nutrient was added.
2. The media reported here are widely used and may be found as follows: BBM (38), BG-11 (39), FAB is as BBM medium but without element trace, and BAR (40).
3. The transfer of *X. dendrorhous* in a second 250-mL baffled Erlenmeyer flask at 10% volume for 48 h is critical to ensure recommended exponential phase growth conditions. Figure 3 shows the standardized inoculum, suitable for astaxanthin production in different media (YM, YPG, or coconut milk). This step is mandatory.
4. The transfer of *H. pluvialis* seed in a second 250-mL baffled Erlenmeyer flask at 10% volume for 120 h is critical to ensure

recommended exponential phase growth conditions. Figure 4 shows the standardized inoculum. This step is mandatory as far as further microorganism growth is very sensitive initial conditions of inoculation.

5. Optimal cellular growth versus pigment production in *X. dendrorhous*: Natural carbon sources present in cane sugar molasses, grape juice, or coconut milk, promote both the cellular growth and the synthesis of astaxanthin in different *X. dendrorhous* yeasts. The very rich nutrient composition of coconut milk avoids the need of additional nutritional compounds added to the media. Further improvements of astaxanthin production may be obtained with the use of mutated version of the yeast (41).
6. Optimal cellular growth versus pigment production in *H. pluvialis*: Optimal factors in astaxanthin production are exposure to a high light intensity (345 μmol photon/m^2s), the absence of aeration, and the use of sodium acetate as a source of inorganic carbon in the BAR medium. Such factors corrupt the cell cycle and therefore limit its growth and photosynthetic process. Indeed, such stress growing conditions enhance the process of cyst formation, i.e., the formation of aplanospores with high contents of astaxanthin (42–48).

Acknowledgments

The author expresses her gratitude to TESE for the experimental facilities and to CINVESTAV for giving me a non-salary work permit along the development of this research work. I would also like to thank Dr. Jorge A. Torres-Muñoz for the fruitful discussions along the preparation of the manuscript.

References

1. Neils HJ, Leenheer AP (1991) Microbial sources of carotenoid pigments uses in foods and feeds. J Appl Bacteriol 70:181–191
2. Yokayama T, Miki W (1995) Composition and presumed biosynthetic pathway of carotenoids in the astaxanthin producing bacterium *Agrobacterium aurantiacum*. FEMS Microbiol Lett 28:139–144
3. Lotan T, Hirschberg J (1995) Cloning and expression in *Escherichia coli* of the gene encoding β-C-4-oxygenates, that converts β-carotene to the ketocarotenoid canthaxanthin in *Haematococcus pluvialis*. FEBS Lett 364:125–128
4. Andrews AG, Phaff HJ, Starr MP (1976) Carotenoids of *Phaffia rhodozyma* red pigmented fermenting yeast. Phytochemistry 15:10003–10007
5. Gu WL, An GH, Johnson EA (1997) Ethanol increases carotenoid productin in *Phaffia rhodozyma*. J Ind Microbiol Biotechnol 19:114–117
6. Schroeder WA, Johnson EA (1995) Carotenoids protect *Phaffia rhodozyma* against singlet oxygen damage. J Ind Microbiol 14:502–507
7. Johnson EA, Schroeder WA (1995) Microbial carotenoids. In: Fiechter A (ed) Advances biochemical engineering and biotechnology, vol 53. Springer, Berlin, pp 119–178

8. Fang TJ, Chiou TY (1996) Batch cultivation and astaxanthin production by to mutant of the net yeast, *Phaffia rhodozyma* NCHU-FS501. J Ind Microbiol 16:175–181
9. Yamane Y, Higashida K, Nishio N (1997) Influence of oxygen and glucose on primary metabolism and astaxanthin production by *Phaffia rhodozyma* in Fed-Bath cultures. Kinetic and stoichiometric analysis. Appl Environ Microbiol 63:4471–4478
10. Newsome RL (1986) Food colors. Food Technol 40:49–56
11. Kobayashi M, Sakamoto Y (1999) Singlet oxygen quenching ability of astaxanthin esters from the green alga *Haematococcus pluvialis.* Biotechnol Lett 21:265–269
12. Olaizola M (2000) Commercial production of astaxanthin from *Haematococcus pluvialis* using 25,000-liter outdoor photobioreactors. J Appl Phycol 12:499–506
13. de la Fuente JL, Rodríguez-Sáiz M, Schleissner C, Díez B, Peiro E, Barredo JL (2010) High-titer production of astaxanthin by the semi-industrial fermentation of *Xanthophyllomyces dendrorhous.* J Biotechnol 148:144–146
14. Rodriguez-Sáiz M, de la Fuente JL, Barredo JL (2010) *Xanthophyllomyces dendrorhous* for the industrial production of astaxanthin. Appl Microbiol Biotechnol 88:645–658
15. Johnson EA, An GH (1991) Astaxanthin from microbial sources. Crit Rev Biotechnol 11:297–326
16. Haard N (1988) Astaxanthin formation by the yeast *Phaffia rhodozyma* on molasses. Biotechnol Lett 10:609–614
17. Fontana JD, Guimaraes MF, Martins NT, Fontana CA, Baron M (1996) Culture of the astaxanthin ogenic yeast *Phaffia rhodozyma* in low-cost media. Appl Biochem Biotechnol 57–58:413–422
18. Hayman TG, Mannarelli BN, Leathers TD (1995) Production of carotenoids *by Phaffia rhodozyma* grown on media composed of corn wet-milling co-products. J Ind Microbiol 115:173–183
19. Okagbue RN, Lewis MW (1984) Autolysis of the red yeast *Phaffia rhodozyma*: a potential tool to facilitate extraction of astaxanthin. Biotechnol Lett 6:247
20. Meyer PS, du Preez JC (1994) Astaxanthin production by *Phaffia rhodozyma* mutant on grape juice. World J Microbiol Biotechnol 10:178–183
21. Acheampong E, Martin A (1995) Kinetic studies on the yeast *Phaffia rhodozyma.* J Basic Microbiol 35:147–155
22. Martin AM, Acheampong E, Patel TR (1993) Production of astaxanthin by *Phaffia rhodozyma* using peat hydrolysates as substrate. J Chem Tech Biotechnol 58:223–230
23. Domínguez-Bocanegra AR, Torres-Muñoz JA (2004) Astaxanthin hyperproduction by *Phaffia rhodozyma* (now *Xanthophyllomyces dendrorhous*) with coconut milk as sole source of energy. Appl Microbiol Biotechnol 66: 249–252
24. Droop MR (1954) Conditions governing haematochrome formation and loss in the alga *Haematococcus pluvialis.* Arch Microbiol 20:391–397
25. Droop MR (1955) Carotenogenesis in *Haematococcus.* Nature 175:42
26. Pringsheim EG (1966) Nutritional requirements of *Haematococus pluvialis* and related species. J Phycol 2:17
27. Cordero B, Otero A, Patiño M, Arredondo BO, Fábregas J (1996) Astaxanthin production from the green alga *Haematococcus pluvialis* with different stress conditions. Biotechnol Lett 18:213–218
28. Kobayashi M, Kakizono T, Yamaguchi K, Nishio N, Nagai S (1992) Growth and astaxanthin formation of *Haematococcus pluvialis* in heterotrophic and mixotrophic conditions. J Ferment Bioeng 74:17–20
29. Kobayashi M, Kakizono T, Nagai S (1992) Effects of light intensity, light quality and illumination cycle on astaxanthin formation in a green alga. J Ferment Bioeng 74:61–63
30. Kobayashi M, Kakizono T, Nagai S (1993) Enhanced carotenoid biosynthesis by oxidative stress in acetate induced cyst cells of to green alga *Haematococcus pluvialis.* Appl Environ Microbiol 59:867–873
31. Kakizono T, Kobayashi M, Nagai S (1992) Effect of carbon/nitrogen ratio on the encystment accompanied with Astaxanthin formation in to green alga, *Haematococcus pluvialis.* J Ferment Bioeng 74:403–405
32. Domínguez-Bocanegra AR, Guerrero-Legarreta I, Martínez-Jerónimo F, Tomassini-Campocosio A (2004) Influence of environmental and nutritional factors in the production of astaxanthin from *Haematococcus pluvialis.* Bioresour Technol 92:209–214
33. APHA, WCPF (1992) Standard methods for the examination of waters and wastewaters, 17th edn. APHA, WCPF, Washington, DC
34. Miller GL (1959) Use of dinitrosalicylic acid reagent for determination of reducing sugar. Anal Chem 31:426–428
35. Sedmak JJ, Weerasinghe DK, Jolly SO (1990) Extraction and quantification of astaxanthin by *Phaffia rhodozyma.* Biotechnol Tech 4:107–112
36. Jian-Ping Y, Xian-Di G, Feng C (1997) Separation and analysis of carotenoids and

chlorophylls in *Haematococcus lacustris* by high-performance liquid chromatography photodiode array detection. J Agric Food Chem 45:1952–1956
37. Britton G (1985) General carotenoids. Methods Enzymol 111:115–149
38. Nichols HW, Bold HC (1969) *Trichsarcina polyinorpha* gene. et sp. nov. J Phycol 1:34–38
39. Boussiba S, Vonshak A (1991) Astaxanthin accumulation in the green alga *Haematococcus pluvialis*. Plant Cell Physiol 7:1077–1082
40. Barbera E, Thomás X, Moya M, Ibañez T, Molins M (1993) Significance tests in the study of the specific growth rate of *Haematococcus lacustris*: influence of carbon source and light intensity. J Ferment Bioeng 5:403–405
41. Domínguez-Bocanegra AR, Ponce-Noyola T, Torres-Muñoz JA (2007) Astaxanthin production by *Phaffia rhodozyma* and *Haematococcus pluvialis*: a comparative study. Appl Microbiol Biotechnol 75:783–791
42. Zlotnik I, Sukenik A, Dubinsky Z (1993) Physiological and photosynthetic changes during the formation of red aplanospores in the chlorophyte *Haematococcus pluvialis*. J Phycol 29:463–469
43. Harker M, Tsavalos AJ, Young AJ (1996) Factors responsible for astaxanthin formation in the chlorophyte *Haematococcus pluvialis*. Bioresour Technol 55:207–214
44. Harker M, Tsavalos AJ, Young AJ (1996) Autotrophic growth and carotenoid production of *Haematococcus pluvialis* in to 30 liter air-lift photobioreactor. J Ferment Bioeng 2:113–118
45. Gong XD, Chen F (1988) Influence of medium components on astaxanthin content and production of *Haematococcus pluvialis*. Process Biochem 33:385–391
46. Tripathi U, Sarada R, Ramachandra R, Ravishankar GA (1999) Production of astaxanthin in *Haematococcus pluvialis* cultured in various media. Bioresour Technol 68:197–199
47. Boussiba S, Bing W, Yuan JP, Zarka A, Chen F (1999) Changes in pigments profile in the green alga *Haematococcus pluvialis* exposed to environmental stresses. Biotechnol Lett 21:601–604
48. Hagen C, Grunewald K, Xylander M, Rothe E (2001) Effect of cultivation parameters on growth and pigment biosynthesis in flafellated cells of *Haematococcus pluvialis*. J Appl Phycol 13:79–87

Chapter 12

Isolation and Selection of New Astaxanthin Producing Strains of *Xanthophyllomyces dendrorhous*

Diego Libkind, Martín Moliné, and Celia Tognetti

Abstract

Astaxanthin is a xanthophyll pigment of high economic value for its use as a feeding component in aquaculture. *Xanthophyllomyces dendrorhous* is a basidiomycetous fungi able to synthesize astaxanthin as its major carotenoid, and the only known yeast species bearing the capability to produce this type of carotenoid. Recently, the habitat and intraspecific variability of this species have been found to be wider than previously expected, encouraging the search for new wild strains with potential biotechnological applications. Here we describe effective procedures for isolation of *X. dendrorhous* from environmental samples, accurate identification of the strains, analysis of their astaxanthin content, and proper conservation of the isolates.

Key words: Astaxanthin, Phaffia, Xanthophyllomyces, ITS, Selective culture media, Natural environments, Mycosporines, Sequencing

1. Introduction

The ability of the basidiomycetous yeast *Xanthophyllomyces dendrorhous* (*Phaffia rhodozyma*) to accumulate the antioxidant, astaxanthin, as its major carotenoid is responsible for the industrial use of this yeast as a microbial source of pigments for aquaculture (1). Astaxanthin is economically important because it is the most expensive aquaculture feed component (2); it is used for the artificial pigmentation of fish and crustaceans (3). Astaxanthin is, in fact, one of the most expensive components of salmon farming, accounting for about 15% of total production costs (4).

Original isolations of *X. dendrorhous*, which have served for most of the studies involving this species, were carried out in the 1960s by Phaff et al. (5). These isolates were obtained from slime exudates of various broad-leaved trees in different mountainous regions in the northern hemisphere, such as Japan and Canada.

José-Luis Barredo (ed.), *Microbial Carotenoids From Fungi: Methods and Protocols*, Methods in Molecular Biology, vol. 898,
DOI 10.1007/978-1-61779-918-1_12, © Springer Science+Business Media New York 2012

More recently, isolations of this species have been carried out in Italy (6), Germany (7), Argentina (8), and the USA (9).

The isolations carried out in Argentina (NW Patagonia) were the first to be reported in the southern hemisphere, and unlike all previous isolations, they were not from sap flow, but rather from ascostroma of *Cyttaria hariotii* (8) parasiting the tree *Nothofagus dombeyi*. A few isolates were also obtained from a mountain lake. Patagonian strains of *X. dendrorhous* bear genetic differences with collection *X. dendrorhous* strains. These differences have been hypothetically explained by geographic isolation and habitat specificity (8). Even more recently, a new astaxanthin producing yeast isolate was obtained from a leaf of *Eucalyptus globulus* in Chile (10). This strain is remarkably different from all previous strains regarding the rRNA genes containing the internal transcribed spacers (ITS) sequences. These differences probably imply that the Chilean strain represents a novel species in the genus *Xanthophyllomyces*.

Thus, it becomes clear that the natural geographic distribution of *Xanthophyllomyces* is wider than expected, and that its habitat is not restricted to slime fluxes, but rather, includes sugary rich fungal stromata, leaves, and freshwater. Furthermore, the genetic variability of this astaxanthin producing yeast is also more diverse than what was previously thought, and it is likely that there is more than one species in the genus. The biotechnological potential of the recently discovered Patagonian isolates has only been preliminary assessed (11).

In this chapter, we describe the strategies currently used in our laboratory for the effective isolation of *Xanthophyllomyces* sp. from environmental samples, the accurate identification of the strains, the analysis of their astaxanthin content, and the proper conservation of the isolates.

2. Materials

Prepare all solutions using distilled sterile water and analytical grade reagents. Prepare and store all reagents at room temperature (unless otherwise indicated). Follow all waste disposal regulations when disposing waste materials.

2.1. Isolation of X. dendrorhous from Environmental Samples

1. YM agar: 3 g/L yeast extract, 3 g/L malt extract, 5 g/L peptone, 10 g/L dextrose, and 20 g/L agar. Adjust pH 4.5–5.0 and supplement with 200 mg/L chloramphenicol.
2. YPD agar: 10 g/L yeast extract, 20 g/L peptone, 20 g/L glucose, and 20 g/L agar. Adjust pH 4.5–5.0 and supplement with 200 mg/L chloramphenicol.

3. 0.9% NaCl.
4. Filter membranes (0.45 μm) (Millipore Corporation, Bedford, MA, USA).
5. Sterile filtering apparatus.

2.2. Identification of Wild X. dendrorhous Isolates Using Phenotypic Characteristics

1. YM agar (see Subheading 2.1).
2. Lugol's iodine: 0.33% w/v iodine (I_2) and 0.66% w/v potassium iodide (KI) in distilled water, to make a brown solution with a total iodine concentration of 150 mg/mL.
3. Fermentation base medium: 4.5 g/L yeast extract, 7.5 g/L peptone, 20 g/L glucose, and enough Bromothymol blue to obtain a sufficiently dark green color.
4. Durham tubes.
5. Illumination apparatus emitting visible light (photosynthetically active radiation) with or without UVA (see Note 1).
6. 25% v/v Methanol solution.
7. UV–visible spectrophotometer, with quartz cuvettes.
8. Thermostatic bath.
9. Incubation chamber with controlled temperature (20°C).
10. Ribitol agar: 5 g/L ribitol and 20 g/L agar.

2.3. Molecular Identification of Wild X. dendrorhous Isolates

1. YM agar (see Subheading 2.1).
2. ITS5 forward primer: 5′-GGA AGT AAA AGT CGT AAC AAG G-3′.
3. LR6 reverse primer: 5′-CGC CAG TTC TGC TTA CC-3′.
4. NL1 forward primer 5′-GCA TAT CAA TAA GCG GAG GAA AAG-3′.
5. NL4 reverse primer 5′-GGT CCG TGT TTC AAG ACG G-3′.
6. ITS1 forward primer 5′-TCC GTA GGT GAA CCT GCG G-3′.
7. ITS4 reverse primer 5′-TCC TCC GCT TAT TGA TAT GC-3′.
8. PCR mix kit.
9. Thermocycler.
10. Genomic DNA extraction kit.
11. Lysing buffer: 50 mmol Tris-HCL, 250 mmol NaCl, 50 mmol EDTA, and 0.3% w/v SDS, pH 8.
12. PCR product purification kit.

2.4. Extraction, Detection, and Quantification of Astaxanthin

1. C18 HPLC column (5 µm, 250 × 4.6 mm).
2. Acetonitrile/methanol/isopropanol: 40 mL acetonitrile, 50 mL methanol, and 10 mL isopropanol.
3. Astaxanthin standard (Sigma, St Louis, MO, USA).
4. Carotenoid extract from a reference *X. dendrorhous* collection strain (e.g., CBS 7918^{T}) (12).

2.5. Preservation of X. dendrorhous Wild Strains

1. YM agar (see Subheading 2.1).
2. Ultrafreezer (−80°C).
3. 20% Glycerol sterile solution.
4. Sterile 425–600 µm Ø glass beads.

3. Methods

Carry out all procedures at room temperature unless otherwise specified.

3.1. Isolation of X. dendrorhous from Natural Environments

To date, *X. dendrorhous* has been found in association to different types of natural substrates, including slime exudates, fungal stromata, leaves and, less frequently, in water samples. Slime exudates are typically found during the spring on birch or beech stumps, and when colonized by *X. dendrorhous*, have an orange-characteristic color. Photos of fluxes colonized by *X. dendorhous* can be seen in Weber et al. (7). Such fluxes have so far only been found in the northern hemisphere in tree species of the genera *Betula*, *Cornus*, *Carpinus*, and *Fagus*. In contrast, there are no reports of the presence of this yeast in exudates of trees from the south hemisphere. However, *X. dendrorhous* has been found in the stromata of the *Nothofagus* spp. parasitic fungus *Cyttaria*, and less frequently in leaves (*Nothofagus* and *Eucalyptus*) and freshwater samples (8, 10). In this subheading, we provide hints for an effective isolation of this yeast species from the different substrates, each of which require a specific procedure to free yeast cells.

3.1.1. Isolation of X. dendrorhous from Exudate Samples

1. Collect exudate samples aseptically and streak out directly onto agar plates.
2. Alternatively, plate out 0.1 mL aliquots of a dilution series obtained by shaking 1 g (fresh weight) of exudate in 10 mL 0.9% sterile aqueous NaCl or distilled water, and stepwise dilution of the resulting suspension in 0.9% NaCl or distilled water.

3.1.2. Isolation of X. dendrorhous from Stromata Samples

1. Collect aseptically stromata of *C. hariotii*, an ascomycetous fungus parasite of *N. dombeyi* ("Coihue").
2. Place them in sterile plastic bags with a known volume of distilled sterile water (5–50 mL).
3. Crush manually *Cyttaria* fruit bodies through the plastic bag and agitate in an orbital shaker for 30 min at 300 rpm.
4. Dilute the extract conveniently and spread aliquots on agar plates.

3.1.3. Isolation of X. dendrorhous from Leaves Samples

1. Collect aseptically whole leaves (as fresh as possible).
2. Place leaves (1–15 depending on size) in a 10- to 15-mL Falcon tube with 8–10 mL distilled water (just enough to cover leaves).
3. Vortex vertically at maximum speed for 5 min.
4. Shake samples overnight at 150 rpm in an orbital shaker.
5. Dilute conveniently the extract.
6. Spread aliquots on agar plates.

3.1.4. Isolation of X. dendrorhous from Water Samples

1. Collect water samples in sterile flasks.
2. Filter them preferably in situ (<15 in. of mercury) through filter membranes (0.45 μm) with a sterile filtering apparatus.
3. Place the filters directly on agar plates.

3.1.5. Isolation of X. dendrorhous by Enrichment

This is an additional isolation method for yeasts that may be applied to any of the previous substrates.

1. Incubate small samples in shaken flasks (100 mL liquid medium with antibiotics) in an orbital shaker (120 rpm) for 48 h.
2. Filter through a sterile glass wool.
3. Do dilution series in fresh medium.
4. Plate out on agar plates.

Recommended temperature range for incubation of plates is 15–20°C, with a 17°C optimum. Normally, 5 days of incubation are enough for colony emergence (see Note 2). Transfer orange colonies to YM or YPD fresh medium and purify.

3.2. Identification of Wild X. dendrorhous Using Phenotypic Characteristics

X. dendrorhous possess such unique morphological and physiological characteristics that makes identification based on phenotypic traits a viable alternative. Even if not used as only mean of identification, the tests described in this subheading are useful for reducing the number of putative *X. dendrorhous* isolates to the minimum. The distinctive combination of phenotypic traits includes: ability to produce amyloid compounds, glucose fermentation capabilities (13), mycosporine production (Fig. 1) (14),

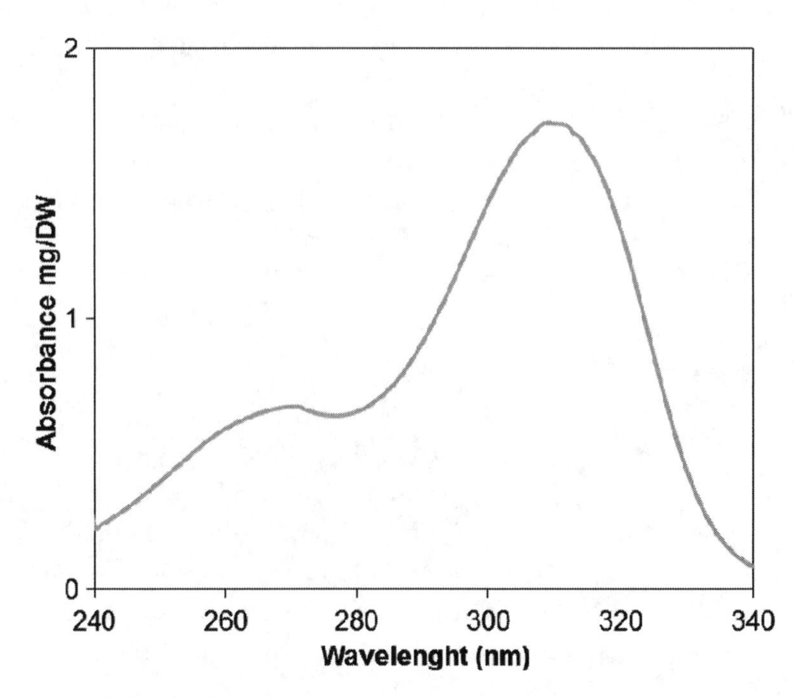

Fig. 1. UV–visible spectrum of mycosporine-glutaminol-glucoside in X. dendrorhous extract. DW dry weight.

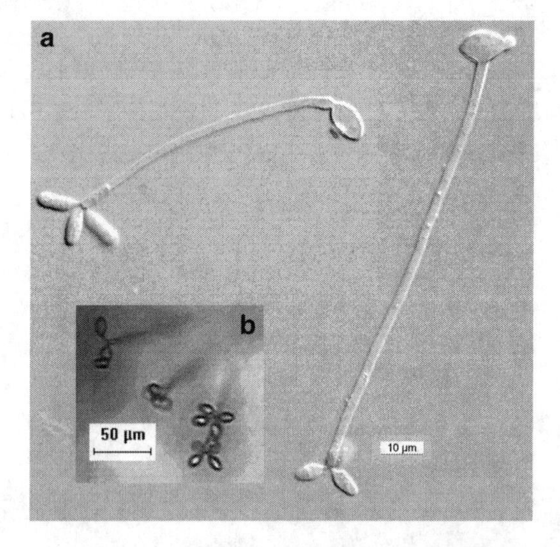

Fig. 2. Sexual structures of *X. dendrorhous*. (**a**) Daughter cell–mother cell conjugation (pedogamy) and holobasidium with terminal basidiospores. (**b**) Terminal basidiospores.

development of unique sexual structures (Fig. 2) (12), and synthesis of astaxanthin as the main carotenoid pigment (15). Any one of these features can be tested to eliminate non-*X. dendrorhous* isolates and the combination of all of these features will provide an accurate identification.

3.2.1. Production of Starch Compounds

1. Grow wild isolates on YM agar for 1 week at 20°C.
2. Flood the plates with five times diluted Lugol's iodine (see Note 3).

3.2.2. Glucose Fermentation

1. Dispense 5–6 mL of base fermentation medium into 150× 12 mm plugged test tubes containing a small inverted tube (Durham tube), approximately 50 mm high and 6 mm Ø.
2. Inoculate with 100 μL of a dense cell suspension of the yeast isolate.
3. Incubate for 1 week at 20°C (see Note 4).

3.2.3. Mycosporine Production

1. Grow yeast isolates on YM agar plates under illumination conditions for 3 days (see Note 1).
2. Harvest biomass and transfer to Eppendorf tube containing 1 mL of 25% (v/v) methanol.
3. Incubate in a water bath at 45°C for 2 h.
4. Clear extracts by centrifugation at 16,000 × g for 10 min.
5. Obtain full spectrum within the range 260–400 nm in an UV–visible spectrophotometer using quartz cuvettes.
6. Compare with reference strain or with Fig. 1.

3.2.4. Sexual Structure Formation

1. Inoculate young cells (48 h in YM agar) in Ribitol agar.
2. Incubate plates for 1 week at 17°C.
3. Compare with reference strain or with Fig. 2.

3.2.5. Astaxanthin Biosynthesis

Extraction, detection, and quantification of astaxanthin are described below (see Subheading 3.4).

3.3. Molecular Identification of Wild X. dendrorhous Isolates

As described above, *X. dendrorhous* has a unique combination of phenotypic characteristics that in some cases may be sufficient for an appropriate identification. However, nowadays, accurate identification requires the use of molecular techniques. Molecular techniques, such as DNA sequencing, allow phylogenetic analysis using additional sequences from public gene databases (e.g., GenBank). Useful regions for *X. dendrorhous* identification through DNA sequencing are the gene-encoding D1/D2 domains of the large ribosomal subunit (26S) and the region comprising the ITS1 and 2, and the 5.8S ribosomal subunit (hereinafter D1/D2 and ITS regions, respectively). The D1/D2 region is highly conserved in *X. dendrorhous*, and it is a good choice for accurate identification. The ITS region not only allows correct species assignation but also, given to a high intraspecific variability in *X. dendrorhous*, may provide information for distinguishing natural populations (8).

1. Prepare a fresh culture of the *X. dendrorhous* strain on YM agar.
2. Extract the genomic DNA using appropriate commercial kits or alternative procedures (see Note 5).
3. Amplify by polymerase chain reaction (PCR) a specific region of rRNA genes or using primers ITS5 and LR6. Use the following PCR program: (1) Initial denaturing step at 95°C for 5 min. (2) 40 cycles of 45 s at 93°C, 60 s at 50°C, and 60 s at 72°C. (3) A final extension step of 6 min at 72°C. Expect a fragment size of approximately 1,400 bp.
4. Purify the corresponding PCR product using an appropriate extraction kit, and sequence it using the pair of primers NL1/NL4 for D1/D2 region (ca. 660 bp) and/or ITS1/ITS4 (ca. 715 bp) for ITS region (see Note 6).

3.4. Extraction, Detection, and Quantification of Astaxanthin

X. dendrorhous strains typically produce astaxanthin as their major carotenoid pigment (approximately 80% of total carotenoids); consequently, this yeast possesses significant biotechnological relevance as a natural source of astaxanthin pigment for aquaculture feed. Other minor carotenoids produced by *X. dendrorhous* are β-carotene, echinenone, 3-hydroxyechinenone, phoenicoxanthin, and canthaxanthin (15). This pool of pigments substantially differs from that of the most pigmented yeasts, such as those of the genus *Rhodotorula*. Astaxanthin is a xanthophyll and is highly polar; hence, in HPLC chromatograms it normally appears first (Fig. 3a). *X. dendrorhous* extracts UV–Vis spectra has a typical broad spectrum band with no evident absorption peaks and a maximum at 477 nm (Fig. 3b).

1. Harvest the cells of 20 mL of culture by centrifugation at 16,000 × *g*.
2. Wash twice with distilled water.
3. Suspend the pellet in 4.5 mL of distilled water and divide in four aliquots: two tubes with 1 mL each for carotenoid extraction, and two previously tared tubes with 1 mL each for dry weight.
4. For carotenoid extraction, add 1.5 mL of glass beads (425–600 μm Ø) and incubate on ice for 2 min.
5. Mix in vortex for 2 min.
6. Add 2 mL of acetone and incubate on ice for 2 min.
7. Mix in vortex for 2 more minutes.
8. Centrifuge at 1,500 × *g* and collect the supernatant in clean tubes. Store at –20°C.
9. Repeat the process until a white pellet is obtained.

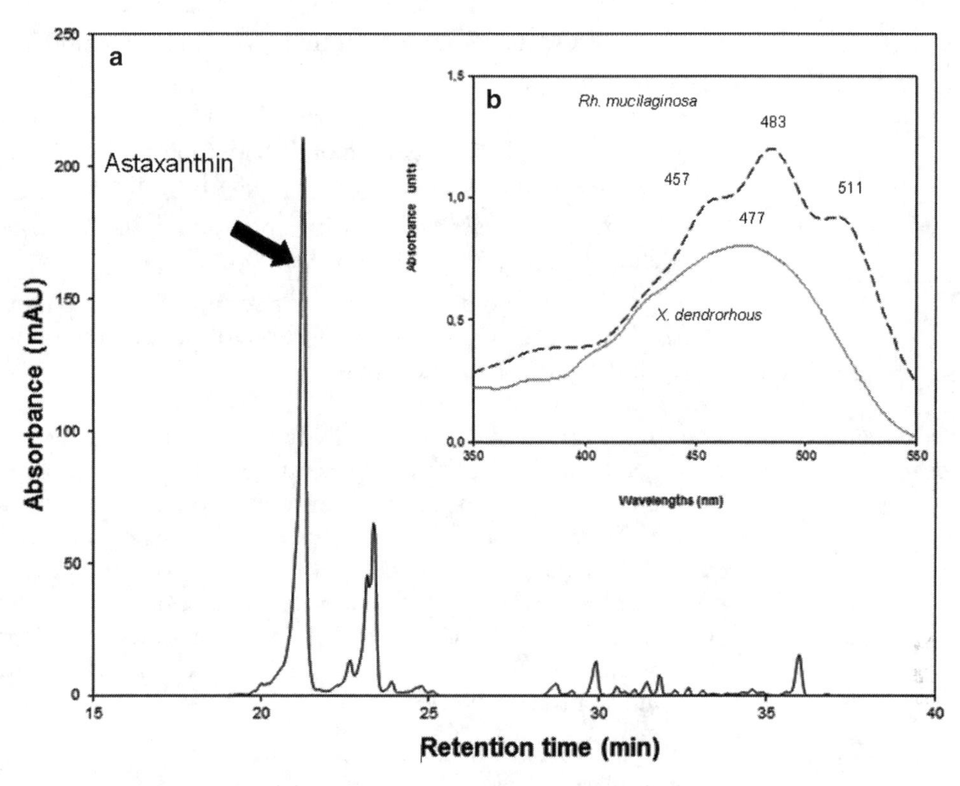

Fig. 3. Spectrophotometric and HPLC analysis of *Xanthophyllomyces dendrorhous* extracts. (**a**) Typical HPLC chromatogram of *X. dendrorhous*. (**b**) Comparison of carotenoid extracts UV–visible spectra with *Rhodotorula mucilaginosa* (ubiquitous pigmented yeast). *AU* absorbance units.

10. Add 2 mL of petroleum ether (35–60°C) at 5°C to the tube containing the supernatants and mix in vortex for 2 min.
11. Centrifuge the tubes at 10,000 × *g* for 10 min at 5°C and collect the supernatant, which contains the pigments, in a new tube.
12. Evaporate the petroleum ether under a continuous nitrogen flux.
13. Add 1 mL of *n*-hexane or petroleum ether for spectrophotometric quantification.
14. Measure the absorbance of the sample at 474 nm.
15. Estimate total carotenoid concentration using the following simplified formula.

$$\text{Carotenoids}_t(\mu g/g) = \frac{A_{474} \times 1 \times 10^4 (\mu g/mL)}{2{,}100 \times dw(g/mL)}.$$

A_{474}: average absorbance (474 nm) of the two tubes (see step 11), dw: average dry weight of the two tubes (see step 2), and 2,100: extinction coefficient of astaxanthin.

16. For HPLC detection of astaxanthin, suspend extracts in 100 μL acetone.
17. Analyze the carotenoid content using a C18 column, an isocratic mobile phase consisting of 85% acetonitrile/10% methanol/5% isopropanol, and a flow rate of 1 mL/min.
18. Monitor the eluted fractions with a photodiode array detector, scanning from 270 to 600 nm every 2 s.
19. Identify carotenoids by their retention times and by comparison of the spectral features with those of pure compounds or with those of a reference strain.

3.5. Preservation of *X. dendrorhous* Wild Strains

Long-term preservation of fungal strains is essential for their in-depth study; however, both the viability and the stability of living cells should be ensured during the preservation period. Wild isolates of *X. dendrorhous* normally do not survive long on regular culture media and need to be subjected to cryopreservation for prolonged preservation. Different methods have been reported for yeast preservation (16, 17); although, the method described in this subheading has proven to be effective for preserving *X. dendrorhous* wild strains (see Note 7).

1. Prepare a fresh culture of *X. dendrorhous* strain on a small YM agar plate.
2. Spread 1 mL of a sterile solution of 20% v/v glycerol over the colony.
3. Add 10–15 glass beads (2–3 mm diameter) onto the plate.
4. Use a sterile small spoon to mix beads, glycerol, and yeast biomass, and transfer yeast-soaked beads into a sterile cryotube.
5. Store tubes at −80°C (see Note 8).
6. To obtain a fresh culture, remove one of the beads using a sterile pair of tweezers; rub bead on YM agar plate and incubate at 17°C.

4. Notes

1. This species normally constitutively produces detectable levels of MYCs, though photostimulation is recommended. Any illuminating device with a significant light intensity will stimulate Mycosporine synthesis in *X. dendrorhous*. Temperature control is important when incubating under light due to the heat emitted by the lamps.
2. An additional 2–3 days at 5°C helps to intensify colony colors, and thus helps in the detection of *X. dendrorhous*-like colonies, which are orange or salmon-orange.

3. *X. dendrorhous* colonies turn to a characteristic intense blue to violet color when flooded with diluted lugol.
4. A positive result is obtained when gas is accumulated inside the inverted Durham tube.
5. An alternative DNA extraction procedure is described by Libkind et al. (18). Briefly, two loopfuls of YM agar grown cultures are suspended in Eppendorf tubes containing 500 μL of lysing buffer and a volume of 200 μL of 425–600 μm of glass beads. After vortexing for 3 min, tubes are incubated for 1 h at 65°C. After vortexing again for 3 min, the suspensions are centrifuged for 15 min at 4°C and 20,000 × *g*. Finally, the collected supernatant is diluted 1:750, and 5 μL are used directly for PCR studies.
6. Type strain of *X. dendrorhous* is CBS 7918 and the rDNA sequences for comparison are AF075496 (D1/D2) and AF139628 (ITS).
7. An alternative preservation method for *X. dendrorhous* mutants, based on dehydrated gelatin drops, was described by Baeza et al. (19).
8. In our experience, survival is improved if tubes are initially stored at −20°C (2–3 days) and then transferred to −80°C.

Acknowledgments

This work was supported by UNComahue-CRUB grant B143 and ANPCYT-FONCYT grant PICT 1745 to DL.

References

1. Rodríguez-Sáiz M, de la Fuente JL, Barredo JL (2010) *Xanthophyllomyces dendrorhous* for the industrial production of astaxanthin. Appl Microbiol Biotechnol 88:645–658
2. Johnson EA (2003) *Phaffia rhodozyma*: colorful odyssey. Int Microbiol 6:169–174
3. Johnson EA, Conklin DE, Lewis MJ (1977) The yeast *Phaffia rhodozyma* as a dietary pigment source for salmonids and crustaceans. J Fish Res Board Can 34:2417–2421
4. Mann V, Harker M, Pecker I, Hirschberg J (2000) Metabolic engineering of astaxanthin production in tobacco flowers. Nat Biotechnol 18:888–892
5. Phaff HJ, Miller MW, Yoneyama M, Soneda M (1972) A comparative study of the yeast florae associated with trees on the Japanese Islands and on the West Coasts of North America. In: Terui G (ed) Proceedings of the 4th IFS: fermentation technology today meeting. Society of Fermentation Technology, Osaka, pp 759–774
6. Weber RWS, Davoli P (2005) *Xanthophyllomyces* and other red yeasts in microbial consortia on spring sap-flow in the Modena province (Northern Italy). Atti Soc Nat Mat Modena 136:127–135
7. Weber RWS, Davoli P, Anke H (2006) A microbial consortium involving the astaxanthin producer *Xanthophyllomyces dendrorhous* on freshly cut birch stumps in Germany. Mycologist 20:57–61
8. Libkind D, Ruffini A, van Broock M, Alves L, Sampaio JP (2007) Biogeography, host specificity, and molecular phylogeny of the basidiomycetous yeast *Phaffia rhodozyma* and

its sexual form, *Xanthophyllomyces dendrorhous*. Appl Environ Microbiol 73:1120–1125

9. Fell JW, Scorzetti G, Statzell-Tallman A, Boundy-Mills K (2007) Molecular diversity and intragenomic variability in the yeast genus *Xanthophyllomyces*: the origin of *Phaffia rhodozyma*? FEMS Yeast Res 7:1399–1408
10. Weber RWS, Becerra J, Silva MJ, Davoli P (2008) An unusual *Xanthophyllomyces* strain from leaves of *Eucalyptus globulus* in Chile. Mycol Res 112:861–867
11. Libkind D, Moliné M, de García V, Fontenla S, van Broock M (2008) Characterization of a novel South American population of the astaxanthin producing yeast *Xanthophyllomyces dendrorhous* (*Phaffia rhodozyma*). J Ind Microbiol Biotechnol 35:151–158
12. Golubev WI (1995) Perfect state of *Rhodomyces dendrorhous* (*Phaffia rhodozyma*). Yeast 11:101–110
13. Miller MW, Yoneyama M, Soneda M (1976) *Phaffia*, a new yeast genus in the Deuteromycotina (Blastomycetes). Int J Syst Bacteriol 26:286–291
14. Libkind D, Moliné M, van Broock MR (2010) Production of the UVB absorbing compound mycosporine-glutaminol-glucoside by *Xanthophyllomyces dendrorhous* (*Phaffia rhodozyma*). FEMS Yeast Res. doi:10.1111/ j.1567-1364.2010.00688.x
15. Andrewes AG, Phaff HJ, Starr MP (1976) Carotenoids of *Phaffia rhodozyma*, a red-pigmented fermenting yeast. Phytochemistry 15:1003–1007
16. Malik KA, Hoffmann P (1993) Long-term preservation of yeast cultures by liquid-drying. World J Microbiol Biotechnol 9:372–376
17. Bond CJ (2007) Cryopreservation of yeast cultures. In: Day JG, Pennington MW (eds) Cryopreservation and freeze-drying protocols, vol 38, Methods in molecular biology. Humana Press, Totowa, pp 39–47
18. Libkind D, Brizzio S, Ruffini A, Gadanho M, van Broock MR, Sampaio JP (2003) Molecular characterization of carotenogenic yeasts from aquatic environments in Patagonia, Argentina. Antonie Van Leeuwenhoek 84:313–322
19. Baeza M, Retamales P, Sepúlveda D, Lodato P, Jiménez A, Cifuentes V (2009) Isolation, characterization and long term preservation of mutant strains of *Xanthophyllomyces dendrorhous*. J Basic Microbiol 49:135–141

Chapter 13

Isolation and Characterization of Extrachromosomal Double-Stranded RNA Elements in *Xanthophyllomyces dendrorhous*

Marcelo Baeza, María Fernández-Lobato, and Víctor Cifuentes

Abstract

Double-stranded RNA (dsRNA) molecules are widely found in yeasts and filamentous fungi. It has been suggested that may play important roles in the evolution of eukaryote genomes and may be a valuable tool in yeast typing. The characterization of these extrachromosomal genetic elements is usually a laborious process, especially when trying to analyze a large number of samples. In this chapter, we describe a simple method to isolate dsRNA elements from yeasts using low amounts of starting material, and their application to different *Xanthophyllomyces dendrorhous* strains. Furthermore, the methodologies for enzymatic and hybridization characterizations, and quantification of relative dsRNA abundance are detailed.

Key words: dsRNA, *Xanthophyllomyces dendrorhous*, Mycovirus, Astaxanthin

1. Introduction

Cytoplasmic inherited double-stranded RNA (dsRNA) molecules are extrachromosomal genetic elements (EGEs) encapsidated in virus-like particles (VLPs) commonly found in yeast and filamentous fungi. Their function is not well understood, and only in a few cases they have been unmistakably associated with a detectable phenotype in the host, such as mycocinogeny and hypovirulence (1). In strains of the astaxanthin-producing yeast *Xanthophyllomyces dendrorhous*, large variations in the size and number of dsRNAs have been described (2, 3). In addition, dsRNA patterns have been shown to be specific to the region of isolation of different strains. It has been proposed that, together with other molecular markers, the dsRNA pattern may be a valuable tool for yeast typing (4, 5). The isolation of dsRNA from yeast may be difficult and time consuming depending on the type of yeast and the number of samples.

José-Luis Barredo (ed.), *Microbial Carotenoids From Fungi: Methods and Protocols*, Methods in Molecular Biology, vol. 898, DOI 10.1007/978-1-61779-918-1_13, © Springer Science+Business Media New York 2012

In this chapter, the characterization of dsRNAs in *X. dendrorhous* will be used to exemplify a simple method for the isolation and determination of their chemical nature and relative abundance. Furthermore, the analysis of nucleotide similarity among dsRNAs using dot-blot hybridization is detailed.

2. Materials

2.1. Strains and Culture Media

1. *X. dendrorhous* UCD 67-385 (Phaff Yeast Culture Collection, CA, USA).
2. *X. dendrorhous* VKM Y-2266 (The All-Russian Collection of Microorganisms, Moscow, Russia).
3. *X. dendrorhous* VKM Y-2059 (The All-Russian Collection of Microorganisms, Moscow, Russia).
4. *X. dendrorhous* VKM Y-2786 (The All-Russian Collection of Microorganisms, Moscow, Russia).
5. *Saccharomyces cerevisiae* V18 (6).
6. Complete yeast malt medium (YM): 0.3% yeast extract, 0.3% malt extract, and 0.5% peptone.
7. Trace element solution: To 90 mL distilled water at room temperature, add salts in the following order: 5 g of citric acid·$1H_2O$, 5 g of $ZnSO_4 \cdot 7H_2O$, 1 g of $Fe(NH_4)_2(SO_4)_2 \cdot 6H_2O$, 0.25 g of $CuSO_4 \cdot 5H_2O$, 0.05 g of $MnSO_4 \cdot 1H_2O$, 0.05 g of H_3BO_3, and 0.05 g of $Na_2MoO_4 \cdot 2H_2O$. Adjust the volume to 100 mL, add 1 mL of chloroform, and store at room temperature in the dark.
8. Biotin stock solution: Dissolve 5 mg of biotin in 50 mL of water or 50% of ethanol. Store aliquots of 2.5 mL at –20°C.
9. Vogel-Bonner medium E (50× stock solution): To 755 mL distilled warm water (45°C) in a 2-L flask placed on a magnetic stirring hot plate, add salts in the following order: 125 g of Na_3 citrate·$2H_2O$, 250 g of KH_2PO_4, 100 g of NH_4NO_3, 10 g of $MgSO_4 \cdot 7H_2O$, 5 g of $CaCl_2 \cdot 2H_2O$, 5 mL of trace element solution, and 2.5 mL of biotin stock solution. It is important to allow each salt to dissolve completely before adding the next. Adjust the volume to 1,000 mL, add 5 mL of chloroform, and store at room temperature in the dark. Dilute to 1× for use.

2.2. Component for Isolation and Characterization of dsRNAs

1. Mini-Beadbeater (Biospec, Bartlesville, OK, USA).
2. Ethidium bromide (0.5 μg/mL).
3. 6 M KI.
4. Dimethylsulfoxide (DMSO).
5. Glass beads 0.5 mm diameter.

6. Chloroform:isoamyl alcohol (24:1).
7. TE buffer: 10 mM Tris–HCl, and 1 mM EDTA, pH 8.0.
8. TAE buffer: 2 M Tris base, ~1.56 M glacial acetic acid, and 0.05 M EDTA, pH 8.0.
9. Loading buffer (10×): 0.1% xylene cyanol, 0.1% bromophenol blue, 0.5% SDS, 0.1 M EDTA, and 50% glycerol.
10. 0.01× SSC buffer: 0.003 M NaCl and 0.0003 M sodium citrate, pH 7.0 (7, 8).
11. 2× SSC buffer: 0.6 M NaCl and 0.06 M sodium citrate, pH 7.0 (7, 8).
12. Wash buffer GC: 50 mM NaCl, 10 mM Tris–HCl, pH 7.5, 2.5 mM EDTA, and 50% v/v ethanol.
13. TBE buffer: 89 mM Tris base, 89 mM boric acid, and 2 mM EDTA, pH 8.0.
14. PBS buffer: 137 mM NaCl, 2.7 mM KCl, 10 mM Na_2HPO_4, and 2 mM KH_2PO_4, pH 7.4.
15. DNase I.
16. Nuclease S1.
17. RNase H.
18. RNase A.
19. DNase I buffer: 50 mM sodium acetate, 10 mM $MgCl_2$, and 2 mM $CaCl_2$, pH 6.5.
20. Nuclease S1 buffer: 30 mM potassium acetate, 0.3 M NaCl, 1 mM $ZnSO_4$, and 5% glycerol, pH 4.5.
21. RNase H buffer: 20 mM Tris–HCl, 0.1 M KCl, 10 mM $MgCl_2$, 0,1 mM dithiothreitol (DTT), and 5% sucrose, pH 7.5.
22. Saturated acid phenol: Add 500 g of phenol to an amber flask and incubate at 60°C until the phenol melts completely. Add 300 mL of distilled water, mix well by shaking, and incubate at room temperature for 12 h. Remove the water and add 200 mL of 50 mM sodium acetate, pH 4.0. Mix well by shaking, and incubate at room temperature for 12 h. Store at room temperature.
23. Glassmilk: Dissolve 1 g of silica in 10 mL of PBS buffer, allow the silica to settle for 2 h, remove the supernatant (containing fine particulate matter), and add 10 mL of PBS buffer. Repeat the procedure twice. Centrifuge at 2,000×*g* for 2 min and resuspend the silica in 10 mL of 3 M NaI. Store at 4°C in the dark.
24. λ *Hind*III molecular weight marker.
25. 1 kb DNA ladder.
26. 1D Image Analysis Software version 2.0.1 (Kodak Scientific Image System, NY, USA).

2.3. Hybridization

1. Nucleotide mix: 10 mM each dTAP, dGTP, dCTP, and dTTP.
2. Random hexamers.
3. ^{32}P-dCTP.
4. Sephadex G-50 (GE Healthcare, Fairfield, CA, USA).
5. Nylon membrane.
6. Autoradiographic film.
7. M-MuLV reverse transcriptase.
8. Hybridization solution: 0.5 M Na_2HPO_4 and 7% SDS, pH 7.2.
9. Wash solution H1: 40 mM Na_2HPO_4 and 5% SDS, pH 7.2.
10. Wash solution H2: 40 mM Na_2HPO_4 and 1% SDS, pH 7.2.
11. ImageJ software (9).

3. Methods

The methods described below are used for the characterization of dsRNA elements of *X. dendrorhous.* Cultures of this yeast are grown at 22°C in complete YM medium or in minimal saline Vogel-Bonner medium E (10, 11) supplemented with glucose or succinate (2% w/v), which are used as fermentable and nonfermentable carbon sources, respectively. The method used for the isolation of RNA has been optimized for the processing of a high number of samples, obtaining a large enough quantity for their enzymatic characterization and molecular studies. The entire process is performed in Eppendorf tubes at room temperature using ~0.1 g of yeast pellet (wet weight) obtained from culture aliquots using centrifugation at 7,000 × *g* for 5 min (see Note 1).

3.1. Isolation of Total RNA

1. Wash the yeast pellet twice with 1 mL of TE buffer and suspend in 0.4 mL of TE buffer.
2. Add 0.4 mL of 0.5-mm diameter glass beads and 0.4 mL of acid phenol and shake the tubes in a cell disruptor for 3–4 min (see Note 2). The results of RNA isolations obtained at various times of cell disruption are shown in Fig. 1a.
3. Centrifuge the samples at 10,000 × *g* for 2 min and transfer the aqueous phase to a tube containing 0.4 mL of saturated acid phenol (pH 4.0).
4. Shake in vortex for 2 min and centrifuge at 10,000 × *g* for 1 min (see Note 3).
5. Transfer the aqueous phase to a tube containing 0.4 mL of chloroform:isoamyl alcohol (24:1), vortex for 2 min, and centrifuge at 10,000 × *g* for 1 min.

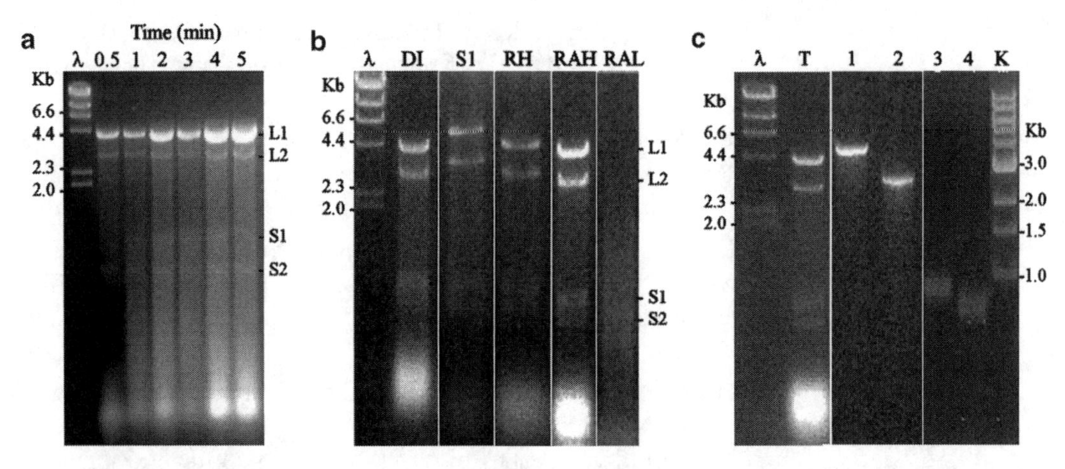

Fig. 1. Isolation and characterization of dsRNAs in *Xanthophyllomyces dendrorhous* UCD 67-385. (**a**) Total RNA obtained after different times of mechanical yeast rupture. (**b**) Determination of chemical nature using enzymatic digestions: DI, DNase I; S1, nuclease S1; RH, RNase H; RAH, RNase A at high ionic strength; RAL, RNase A at low ionic strength. (**c**) Purification of particular dsRNA from agarose gels: T, total RNA sample; 1, purified L1 dsRNA; 2, purified L2 dsRNA; 3, purified S1 dsRNA; 4, purified S2 dsRNA. Samples were resolved in 1% agarose gels and stained with ethidium bromide. Molecular weight markers: λ *Hind*III (λ) and 1 kb DNA ladder (K).

6. Perform this last step twice.
7. Mix the aqueous phase with 1 volume of isopropanol and incubate at −20°C for 2 h.
8. Centrifuge at 14,000 × *g* for 15 min, discard the supernatant, dry the pellet, and suspend it in 10–20 μL of nuclease-free water.
9. Add 1–2 μL of loading buffer (10×) to RNA samples and run using 1% agarose gel electrophoresis (AGE).
10. After staining with a solution of 0.5 μg/mL ethidium bromide for 5–10 min, visualize acid bands under UV light. An example of the isolation of total RNA from *X. dendrorhous* can be seen in Fig. 1a, c.

This method has been used successfully for the isolation of total RNA from yeast belonging to other genera, including *Rhodotorula*, *Cryptococcus*, *Bullera*, *Candida*, *Kluyveromyces*, and *Pichia*, with good results.

3.2. Determination of Chemical Nature

The chemical nature of EGEs can be assessed enzymatically using a set of four enzymes: DNase I for degrading the deoxyribonucleic acids, nuclease S1 for degrading single-stranded nucleic acids, RNase H for degrading RNA from a DNA/RNA hybrid, RNase A at low ionic strength (0.01× SSC buffer) for degrading all ribonucleic acids, and RNase A at high ionic strength (2× SSC buffer) for degrading only single-stranded RNAs.

1. Digestion with DNase I: Mix 0.5 μg of total RNA, 0.5 U of DNase I, and 2 μL of DNase I buffer; add nuclease-free water to reach 20 μL final volume and incubate for 1 h at 37°C.

2. Digestion with nuclease S1: Mix 0.5 μg of total RNA, 0.5 U of nuclease S1, and 2 μL of nuclease S1 buffer; add nuclease-free water to reach 20 μL final volume and incubate for 1 h at 37°C.
3. Digestion with RNase H: Mix 0.5 μg of total RNA, 0.5 U of RNase H, and 2 μL of RNase H buffer; add nuclease-free water to reach 20 μL final volume and incubate for 1 h at 37°C.
4. Digestion with RNase A at low ionic strength: Mix 0.5 μg of total RNA, 0.5 μg of RNase A, and 2 μL 0.01× SSC buffer; add nuclease-free water to reach 20 μL final volume and incubate for 1 h at 37°C.
5. Digestion with RNase A at high ionic strength: Mix 0.5 μg of total RNA, 0.5 μg of RNase A, and 2 μL 2× SSC buffer; add nuclease-free water to reach 20 μL final volume and incubate for 1 h at 37°C.
6. Analyze the samples by AGE, as previously indicated.

As shown in Fig. 1b, the EGEs of *X. dendrorhous* are resistant to DNase I, nuclease S1, RNase H, and RNase A at high ionic strength, but sensitive to RNase A at low ionic strength. With these results, it can be concluded that these four EGEs correspond to dsRNAs.

3.3. Purification of dsRNAs

An inexpensive method for the isolation of a specific dsRNA element with a high degree of purity was previously described (12).

1. Add loading buffer (10×) to samples of total RNA and subject them to 1% AGE in TAE buffer (see Note 4).
2. Stain with ethidium bromide and visualize under UV light (see Fig. 1c).
3. Excise the bands corresponding to each dsRNA molecule from the gel (see Note 5), place in an Eppendorf tube, and add 3 volumes of 6 M KI (see Note 6).
4. Incubate tubes at 55°C until the gel fragments completely dissolve.
5. Cool to room temperature and mix with 10 μL of glass milk.
6. Incubate on ice for 10 min and centrifuge at 10,000 × *g* for 30 s.
7. Discard the supernatant, and wash the pellet three times with 500 μL of wash buffer GC, centrifuging at 10,000 × *g* for 30 s between washing steps.
8. Remove the wash buffer and resuspend the pellet in 15–20 μL of nuclease-free water.
9. Incubate at 55°C for 10 min.
10. Centrifuge at 10,000 × *g* for 20 s and transfer the supernatant containing the nucleic acids to a new tube.

The results of purification of each type of dsRNA from *X. dendrorhous* UCD 67-385 are shown in Fig. 1c (see Note 7). Samples of dsRNAs purified from gels were used for the estimation of molecular size (see Note 8) and hybridization experiments.

3.4. Estimation of the Length of dsRNAs

1. Resolve the dsRNA samples in 1 and 2% agarose gels in TAE buffer.
2. Stain with ethidium bromide and photograph under UV light.
3. Estimate the molecular size values by comparison against λ *Hind*III and 1 kb DNA markers using 1D Image Analysis Software version 2.0.1.
4. Correct the values obtained by the difference in electrophoretic mobility between dsDNA and dsRNA (13).

Using this method, the following lengths were obtained for dsRNAs: L1, 5.0 kb; L2, 3.7 kb; S1, 0.9 kb and S2, 0.8 kb (see Fig. 1c).

3.5. Fixation of dsRNAs to Membrane

1. Incubate dsRNA samples at 95°C for 10 min in the presence of 18% DMSO.
2. Immediately cool in an ice/water bath.
3. Dot 10 μL (~200 ng) onto a nylon membrane
4. Incubate at 80°C for 15 min.

3.6. Synthesis and Cleaning of cDNA Probes

1. Incubate samples of 400 ng of dsRNA at 95°C for 15 min, and quickly cool in an ice/water bath.
2. Add 200 U of M-MuLV reverse transcriptase, 1 μL of 25 mM random hexamers, 1 μL of 10 mM nucleotide mix, and 3 μCi of ^{32}P-dCTP.
3. Incubate at 42°C for 90 min and stop the reaction by incubation at 70°C for 10 min.
4. Prepare a Sephadex G-50 column: Weigh 10 g of Sephadex G-50 powder, transfer to 200-mL Erlenmeyer flask, and add 100 mL of TE buffer. Incubate for 1 h at room temperature. Fill a 1-mL syringe barrel, plugged at the bottom with glass wool, with the equilibrated Sephadex G50, place into 15-mL centrifuge tube, and spin at 1,000 × *g* for 2 min. Then refill with equilibrated Sephadex G50 and spin at 1,000 × *g* for 2 min. Repeat the last step until the 90% volume of the syringe is filled. For the collection of the eluate place the bottom on the syringe into a microfuge tube, and both into a 15-mL centrifuge tube. In this way, after centrifugation step, the eluate is collected in the microfuge tube.
5. Load the sample onto Sephadex G-50 columns and centrifuge at 1,000 × *g* for 2 min.
6. Collect the eluate, incubate at 95°C for 5 min, and quickly cool in an ice/water bath (denatured cDNA probe).

3.7. Hybridization

1. Incubate membranes at 65°C for 10 min in 10 mL of hybridization buffer.
2. Add the denatured cDNA probe and incubate at 65°C for 16 h (see Note 9).
3. Wash the membranes twice at 65°C for 30 min with wash buffer H1.
4. Wash the membranes twice at 65°C for 30 min with wash buffer H2.
5. Dry membranes and expose to autoradiography films at –80°C (see Note 10).
6. Develop the films (see Fig. 2).

3.8. Determination of the Relative Abundance of dsRNAs

The relative abundance between two dsRNAs of *X. dendrorhous* was analyzed at different times of growth in complete or minimal medium supplemented with glucose or succinate.

1. Inoculate 1,000 mL of medium with 10 mL of a late exponential phase culture of *X. dendrorhous.*
2. Incubated at 22°C with constant agitation and harvest 10 mL of samples at several time points.
3. Make tenfold serial dilutions from 1 mL of culture sample.
4. Seed onto YM agar plates and incubate at 22°C for 3–5 days.

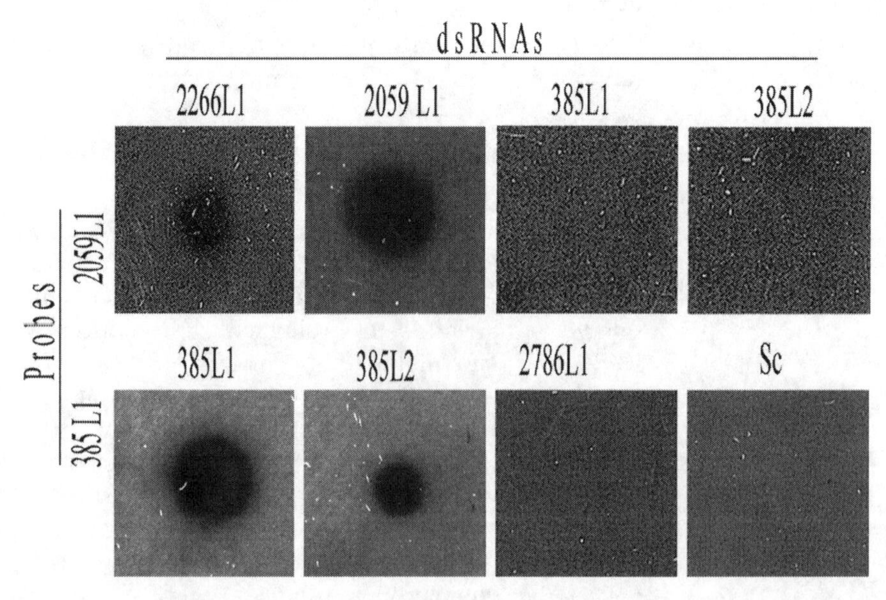

Fig. 2. Dot-blot hybridization. dsRNAs were purified from *Xanthophyllomyces dendrorhous* strains VKM Y-2266 (2266L1), VKM Y-2059 (2059L1), VKM Y-2786 (2786L1), and UCD 67-385 (385L1 and 385L2) and from *S. cerevisiae* strain V18 (Sc) (14). dsRNA samples were heat denatured and fixed onto nylon membranes. The radioactive cDNA probes were synthesized from L1dsRNA of *X. dendrorhous* UCD 67-385 (p385L1) and *X. dendrorhous* VKM Y2059 (p2059L1) strains.

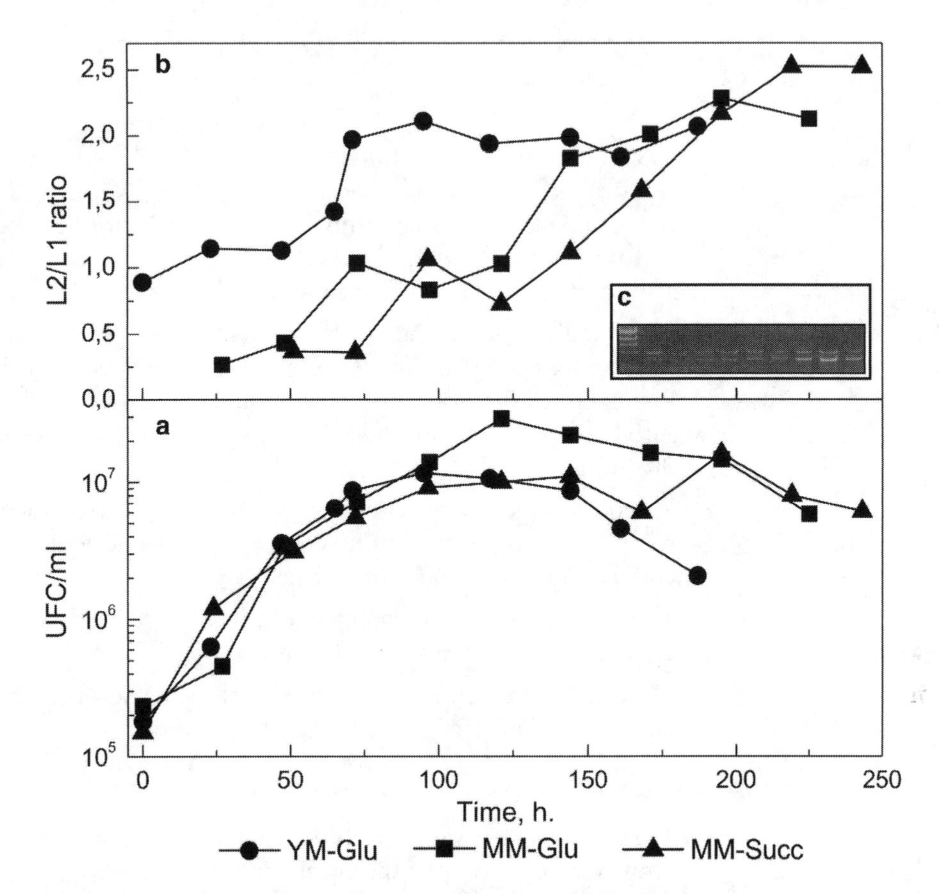

Fig. 3. The relative abundance of dsRNA elements in *Xanthophyllomyces dendrorhous* during different growth times. (**a**) *X. dendrorhous* UCD 67-385 was grown in YM medium supplemented with glucose (YM-Glu) and in Vogel-Bonner medium E supplemented with glucose (MM-Glu) or succinate (MM-Succ). Total RNA was extracted from culture samples collected at each point, and the L2/L1 dsRNA ratios (**b**) were calculated from the band intensities of each dsRNA elements separated in agarose gels, as shown in (**c**) for MM-Glu.

5. Calculate the colony-forming units (CFUs) and the CFU/mL for each dilution.
6. Isolate the total RNA from each culture sample, analyze by 1% AGE, stain with ethidium bromide, and photograph under UV light.
7. Perform the analysis in triplicate using different volumes (2–10 μL) of each RNA sample to obtain more information.
8. Compare intensities of dsRNA bands separated in the same gel using ImageJ software (9).

The relative abundance of L2 vs. L1 dsRNAs during different time points of *X. dendrorhous* growth in different media is shown in Fig. 3.

4. Notes

1. The volume of yeast culture to be centrifuged is variable and will depend on the growth phase of harvesting. Typically, 0.1 g of moist pellet is obtained from 7 to 10 mL of stationary-phase culture. Alternatively, yeast can be obtained from cultures on semisolid media.
2. The results shown here were obtained using the Mini-Beadbeater cell disruptor. The time of rupture must be defined for other cell disruptor equipment. An inexpensive alternative is to perform cell rupture by vigorously shaking in a vortex for 30–40 min.
3. When extraction with acid phenol is used, the DNA molecules solubilize in the organic phase, producing an aqueous phase enriched mainly in RNA molecules.
4. The purification of fragments from TBE-based gels is more complicated than from TAE-based gels.
5. Only the section of gel that contains the nucleic acids should be excised. This facilitates the dissolving of the gel and allows the use of several gel pieces in a single tube, increasing the amount of dsRNA recovered.
6. The reagent NaI was used in the original published protocol. However, we have used the cheaper reagent KI, obtaining similar results.
7. Several pieces of agarose gel were put in a single tube to obtain a high concentration of dsRNA.
8. The electrophoretic mobility of the dsRNA molecules in agarose gels is changed by the presence of small molecules, which generates error in the calculations of molecular sizes. These molecules include other dsRNAs or ribosomal RNA, as can be seen in Fig. 1b (compare lines DI and S1).
9. The incubation temperature must be adapted to the degree of stringency desired.
10. The time of exposure varies due to a wide range of possible identities among the cDNA probes and dsRNA targets. Because of this variance, at least three autoradiography films are placed in each cassette and exposed at different times. In this way, overexposure of the strong signals is avoided.

Acknowledgments

This work was supported by grants Fondecyt 11060157 and Innova CORFO 07CN13PZT-17.

References

1. Tavantzis SM (2002) dsRNA genetic elements. Concepts and applications in agriculture, forestry, and medicine. CRC PRESS, Boca Raton
2. Baeza M, Sanhueza M, Flores O, Oviedo V, Libkind D, Cifuentes V (2009) Polymorphism of viral dsRNA in *Xanthophyllomyces dendrorhous* strains isolated from different geographic areas. Virol J 6:160–166
3. Pfeiffer I, Kucsera J, Varga J, Parducz A, Ferenczy L (1996) Variability and inheritance of double-stranded RNA viruses in *Phaffia rhodozyma*. Curr Genet 30:294–297
4. Maqueda M, Zamora E, Rodriguez-Cousino N, Ramirez M (2010) Wine yeast molecular typing using a simplified method for simultaneously extracting mtDNA, nuclear DNA and virus dsRNA. Food Microbiol 27:205–209
5. Cubeta MA, Vilgalys R (1997) Population biology of the *Rhizoctonia solani* complex. Phytopathology 87:480–484
6. Baeza ME, Sanhueza MA, Cifuentes VH (2008) Occurrence of killer yeast strains in industrial and clinical yeast isolates. Biol Res 41:173–182
7. Pryor A, Boelen M (1987) A double-stranded RNA mycovirus from the maize rust *Puccinia sorghi*. Can J Bot 65:2380–2383
8. Castillo A, Cifuentes V (1994) Presence of double-stranded RNA and virus-like particles in *Phaffia rhodozyma*. Curr Genet 26:364–368
9. Rasband W (2007) National Institute of Health U: ImageJ. http://rsb.info.nih.gov/ij/ In: Book ImageJ. http://rsb.info.nih.gov/ij/
10. Retamales P, Hermosilla G, Leon R, Martinez C, Jimenez A, Cifuentes V (2002) Development of the sexual reproductive cycle of *Xanthophyllomyces dendrorhous.* J Microbiol Methods 48:87–93
11. Vogel HJ (1956) A convenient growth medium for *Neurospora* (medium N). Microbiol Genet Bull 13:42–43
12. Boyle JS, Lew AM (1995) An inexpensive alternative to glassmilk for DNA purification. Trends Genet 11:8
13. Livshits MA, Amosova OA, Lyubchenko Y (1990) Flexibility difference between double-stranded RNA and DNA as revealed by gel electrophoresis. J Biomol Struct Dyn 7: 1237–1249
14. Reyes E, Barahona S, Fischman O, Niklitschek M, Baeza M, Cifuentes V (2004) Genetic polymorphism of clinical and environmental strains of *Pichia anomala*. Biol Res 37:747–757

Chapter 14

Isolation of Carotenoid Hyperproducing Mutants of *Xanthophyllomyces dendrorhous (Phaffia rhodozyma)* by Flow Cytometry and Cell Sorting

Byron F. Brehm-Stecher and Eric A. Johnson

Abstract

Approaches for improving astaxanthin yields in *Xanthophyllomyces dendrorhous* include optimization of fermentation conditions and generation of hyperproducing mutants through random mutagenesis using chemical or physical means. A key limitation of classical mutagenesis is the labor-intensive nature of the screening processes required to find relatively rare mutants having increased carotenoid content, as these are present against a high background of low-interest cells. Here, flow cytometry is described as a high-throughput, single-cell method for primary enrichment of mutagenized cells expressing high levels of astaxanthin. This approach improves the speed and productivity of classical strain selection, enhancing the chances for isolating the carotenoid hyperproducing mutants (CHMs) needed to enable high-titer, economical production of natural astaxanthin.

Key words: *Xanthophyllomyces dendrorhous*, *Phaffia rhodozyma*, Astaxanthin, Mutant, Flow cytometry, Cell sorting

1. Introduction

Carotenoid biosynthesis is a fundamental process in plants, algae, photosynthetic bacteria, and certain fungi. Carotenoids play important roles as photoprotectants or precursors to hormones, and they may have as of yet undiscovered functions in producer organisms. In addition to their biological roles, carotenoids also have high economic value and are used in agriculture as vitamin precursors and colorants in feeds, particularly in aquaculture (1). Since many carotenoids are excellent antioxidants and quench singlet oxygen and other reactive oxygen species, there is also great interest in their use in nutraceutical or cosmeceutical applications to prevent various human diseases and to slow the aging process.

José-Luis Barredo (ed.), *Microbial Carotenoids From Fungi: Methods and Protocols*, Methods in Molecular Biology, vol. 898,
DOI 10.1007/978-1-61779-918-1_14,

An example of a high-value carotenoid at the nexus of several industrial markets (agriculture, food, nutraceutics, cosmetics, etc.) is astaxanthin (3,3′-dihydroxy-β,β-carotene-4,4′-dione). Although it is a natural product, much of the astaxanthin used today in applications such as aquaculture is produced synthetically from petrochemical feedstock (1, 2). The red yeast *Xanthophyllomyces dendrorhous* (the teleomorphic state of *Phaffia rhodozyma*) produces astaxanthin as its principal carotenoid. Along with the microalgae *Haematoccoccus pluvialis*, *X. dendrorhous* is a promising natural source of astaxanthin. However, in order to become competitive with synthetic production, the astaxanthin titer from microbial fermentations must be increased substantially (e.g., ~10–50-fold) (1, 3).

The current market value of synthetic astaxanthin is estimated to be $2,500 US per kg, but naturally produced astaxanthin can command a higher market price (1). This and additional factors, such as consumer preference for natural products and "greener" production methods, are fueling a resurgence of interest in methods for improved microbial production of this carotenoid (1). Approaches for improving astaxanthin yields in *X. dendrorhous* have included optimization of fermentation conditions, generation of hyperproducing mutants through random mutagenesis using chemical or physical means, and, more recently, metabolic engineering of microbial carotenoid pathways (1, 3). Despite the potential drawbacks of classical mutagenesis, such as spontaneous reversion or introduction of secondary mutations that may reduce strain fitness, the simplicity of these methods is attractive, especially when used in conjunction with optimized fermentation conditions (1, 3). However, a key limitation of classical mutagenesis is the labor-intensive nature of the screening processes required to find the "needle(s) in the haystack"—the relatively rare cells having increased carotenoid content that may be present against a high background of low-interest cells. High-throughput, single-cell methods for primary enrichment of mutagenized cells expressing high levels of astaxanthin would improve the speed and productivity of classical approaches and would enhance the chances for isolating the carotenoid hyperproducing mutants (CHMs) needed to enable high-titer, economical production of natural astaxanthin.

This chapter outlines a general approach for using flow cytometry and cell sorting (FCCS) to isolate CHMs of the red yeast *P. rhodozyma* (*X. dendrorhous*) after chemical mutation. The first description of this technique was published approximately 20 years ago (4). Since then, there have been many advances in instrumentation, including more sensitive detectors, more capable and user-friendly analysis software, and enhanced (approximately tenfold) sorting capacity. In 2008, Ukibe et al. revisited this original approach and made some improvements to the technique (5). Although the underlying concept remains the same, the method has been refined, with particular reference to the wavelength of light chosen to monitor astaxanthin fluorescence for the isolation of astaxanthin hyperproducing mutants. These refinements are reflected here.

2. Materials

2.1. Yeasts

1. *X. dendrorhous* ATCC 24230 (American Type Culture Collection, Manassas, Virginia, USA) (4) (see Note 1).
2. *X. dendrorhous* ATCC 96815 (American Type Culture Collection, Manassas, Virginia, USA) (see Note 1).
3. *X. dendrorhous* ATCC 74218 (American Type Culture Collection, Manassas, Virginia, USA) (see Note 1).

2.2. Culture Media and Mutagenesis

Complex media routinely used for the growth of yeasts are used.

1. YM: 3 g/L yeast extract, 3 g/L malt extract, 5 g/L peptone, and 10 g/L dextrose.
2. YM solid: YM and 20 g/L agar.
3. Ethyl methanesulfonate (EMS).
4. Phosphate buffer 0.05 M, pH 7.0.
5. Phosphate buffer 0.1 M, pH 7.0.

2.3. Carotenoid Extraction and Analysis

1. Carotenoid standards (Sigma-Aldrich Co., St. Louis, MO, USA).
2. Phosphate-buffered saline: 0.1 M potassium phosphate, 0.1 M NaCl, pH 7.4.
3. AMINCO French press (American Instrument Company, Silver Spring, MD, USA).
4. Acetone and hexane 1:2 (vol/vol).
5. N_2 gas.
6. HPLC (Rainin Instrument Co., Emeryville, CA, USA).
7. C18 reverse-phase column (250×4.6 mm, 5 μm particle size) (Alltech Econosphere, Grace, Deerfield, IL, USA).
8. Solvent A: 85% acetonitrile and 15% methanol.
9. Solvent B: 100% dichloromethane.
10. Dynamax UV-1 UV/vis detector (Rainin Instrument Co., Emeryville, CA, USA).

2.4. Flow Cytometry and Cell Sorting

1. Sorting instrument.
2. Potassium phosphate buffer 10 mM, pH 7.4.
3. Polystyrene tube (5-mL tube fitted with a cell-strainer cap) (Becton Dickinson, Franklin Lakes, NJ, USA).
4. Polystyrene tube (5-mL tube with a CellTrics® in-tube filter) (Partec GmbH, Görlitz, Germany).

3. Methods

Best practices for carrying out basic research should be followed, including use of analytical-grade reagents, ultrapure (18 MΩ) deionized water, and proper handling and disposal of biological and chemical wastes in accordance with existing regulations or institutionally approved practices.

3.1. Growth and Harvest of Yeast Cells

1. Seed flasks (100–300 mL) containing YM broth (20–30 mL) with isolated colonies from YM plates.
2. Incubate the flasks at 18–20°C and 200 rpm until the desired growth phase is reached:
   - 2.1. For mutagenesis, grow the cells for 2 days, harvest them by centrifugation and treat them as described below (see Subheading 3.2).
   - 2.2. For traditional isolation of CHMs, inoculate mutagenized cells onto YM plates, grow for 7–10 days and screen visually for the desired pigmentation. Then, grow isolated colonies for 5–7 days in flasks prior to carotenoid extraction and analysis via HPLC.
   - 2.3. For screening of CHMs via FCCS, inoculate mutagenized cells into fresh YM broth and grow for as little as 2 days after treatment with EMS, followed by cytometric screening (see Subheading 3.4).

3.2. Yeast Mutagenesis

As described elsewhere (3, 4), various chemical or physical methods for inducing mutations may be used for generation of carotenoid mutants in addition to the EMS method described here. These include UV light, *N*-methyl-*N*′-nitro-*N*-nitrosoguanidine (NTG), ethidium bromide, and acriflavine.

1. For EMS mutagenesis, harvest 2-day-old cultures of flask-grown cells via centrifugation (800 × *g*, 2 min) and wash them with potassium phosphate buffer (0.1–0.05 M, pH 7.0).
2. Adjust cell suspensions to an optical density at 660 nm of 0.3–0.4.
3. Resuspend in the same buffer containing 3–4% EMS (see Note 2).
4. Incubate treated suspensions with shaking at yeast growth temperatures (18–20°C) or simply allowed to stand for a period after treatment.
5. Add sodium thiosulfate to inactivate EMS (see Note 3).

3.3. Carotenoid Analysis

There are various approaches for recovery and analysis of cellular carotenoids (5, 6). The relatively straightforward approach reported by Echavarri-Erasun and Johnson (7) is described below. Astaxanthin and other carotenoid standards are commercially available.

1. Grow cells for 5 days in 30 mL YM broth (see Subheading 3.1), harvest via centrifugation and wash with distilled water.
2. Resuspend cells in phosphate-buffered saline and chill on ice for at least 10 min prior to lysis.
3. Lyse chilled cell suspensions at 1,500 psi pressure in a French press. Monitor microscopically the completeness of cell lysis.
4. Add 1:2 (vol/vol) mixture of acetone and hexane to each sample and mix samples vigorously for 2 min.
5. Separate aqueous and organic phases via centrifugation and remove the organic supernatant to a clean 10-mL glass tube (wrapped in aluminum foil to exclude light).
6. Repeat extractions until the cellular debris pellet is colorless, indicating complete extraction of carotenoids.
7. Dry the organic phase under N_2 and store the extract at –20°C until analyzed.
8. Carry out the quantitative carotenoid analysis via HPLC using a C18 reverse-phase column. Use a flow rate of 1 mL/min for each 18-min run, with the solvent gradient as follows: (a) 100% solvent A, 1 min, (b) linear gradient over 1.5 min to 32% solvent B; isocratic conditions with 32% solvent B for 11 min, (c) decreasing linear gradient of solvent B to 100% solvent A for 30 s, and (d) system reequilibration with 100% solvent A for 4 min.
9. Monitor astaxanthin absorption at 474 nm using an UV-1 UV/vis detector. Collect astaxanthin peaks eluting at 3.85 min, dry and store as described above.

3.4. Flow Cytometry and Cell Sorting

Flow cytometry technology is continually evolving and several sorting instruments are produced commercially. Because different cytometers will undoubtedly be available to individual researchers, no recommendation for specific instrumentation is given here, other than that the system used must be capable of high-throughput analysis and physical sorting of target cells based on their physical or physicochemical characteristics, specifically high astaxanthin content. Available sorting instruments include those from BD Biosciences, Beckman Coulter, iCyt, and Partec, with some instruments delivering sort rates of 100,000–200,000 cells/s. Other instruments, while no longer commercially manufactured, may still be available in academic labs. As noted above, FCCS-based screening can be carried out on yeast populations after as little as 2 days growth following mutagenesis.

1. Wash cells with and resuspend in 10 mM phosphate buffer pH 7.4.
2. Filter cell suspensions through a 40-μm mesh and collect into a 5-mL polystyrene tube. For example, a 5-mL tube fitted with

a cell-strainer cap (Becton Dickinson) or with a CellTrics® in-tube filter (Partec) may be used (see Note 4).

3. Adjust the concentration of the cell suspension to be analyzed to achieve quality sorting of CHMs (see Note 5).
4. Use a sorting flow cytometer equipped with a 488-nm excitation laser. Commercial sheath fluid or 0.1 M sodium phosphate, pH 7.0 may be used as sheath fluid (see Note 6). Calibrate flow cytometers daily using commercially available fluorescent bead kits and associated software (see Note 7).
5. Analyze cells in log/log mode and collect multiple parameters, including forward scatter (FSC), side scatter (SSC), and fluorescence. Use factory default band-pass filters designed to transmit light centered at specific wavelengths (see Note 8).
6. Use defined control cultures to facilitate proper definition of sort regions and to demonstrate experimentally that the approach is capable of detecting and isolating CHMs having high astaxanthin content (see Note 9).

3.5. Isolation of Carotenoid Hyperproducing Mutants via FCCS

1. Open an acquisition plot of FL4 (675 nm) versus FL1 (525 nm).
2. Run samples of wild-type and astaxanthin hyperproducer control strains, examining where each population falls on this plot.
3. Draw four contiguous rectangular gates as shown in Fig. 1, so that when a 1:100 mixture of CHM to wild-type strains is examined, the greatest number of events is centered in gate A and those events having the highest fluorescence fall into gate C.
4. Apply the sort regions established above to the examination of mutagenized cell populations grown for ~2 days, as described above.
5. Sort those events falling into region C (or region D, if applicable) and collect these into glass tubes containing sterile YM broth. With accurate sorting, this process should yield an approximately tenfold physical enrichment for *X. dendrorhous* cells having high astaxanthin content (see Note 10).
6. Collect cells having increased fluorescence at 675 nm via FCCS and deposit into YM broth (a primary screen).
7. Plate cells onto YM agar and screen as in the traditional approach. This step represents a secondary screen carried out on the FCCS-enriched population.
8. Grow cells for 7–10 days and examine visually for highly pigmented colonies.
9. Grow colonies of interest for 5–7 days in flasks prior to carotenoid extraction and analysis via HPLC.

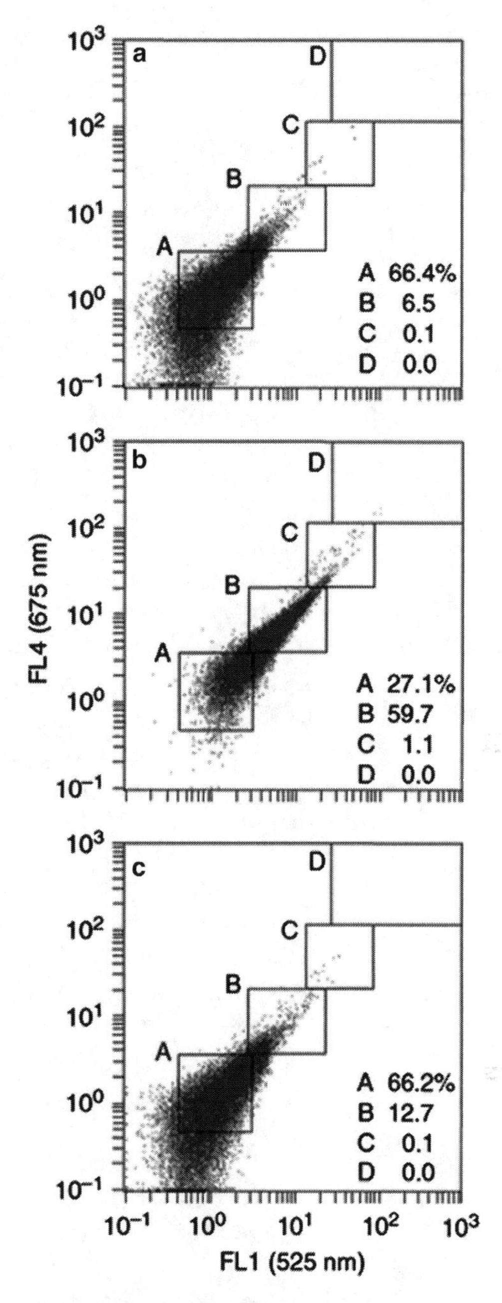

Fig. 1. Gating solution for sorting of *Xanthophyllomyces dendrorhous* based on cellular astaxanthin content. This figure shows an experimental validation of a gating solution for sorting of *X. dendrorhous* cells with increased astaxanthin content. Log/log plots of FL4 (675 nm) and FL1 (525 nm) fluorescence were used to analyze three different samples: (**a**) A wild-type strain (Y989), (**b**) a CHM strain (Y2342), and (**c**) a 1:100 mixture of CHM and wild-type strains. Four contiguous rectangular gates were drawn based on the FL4/FL1 responses of the wild-type and CHM strains when assessed individually. For the 1:100 cell mixture, gate *A* contained the bulk of the events and gate *C* contained the most highly fluorescent events. Cells from gates *A–D* were sorted and plated onto YM agar. Plates inoculated with cells sorted from gate *C* were enriched ninefold for the CHM strain (Y2342). The same approach was applied to EMS-mutagenized cells, enabling the authors to select for mutants producing between 1.5 and 3.8-fold more astaxanthin than parent strains. The overall efficiency of the FCCS-based method was five times greater than for the traditional plating method, allowing isolation of 39 CHM strains versus only seven for plating. Figure reproduced from Ukibe et al. (5), with permission from John Wiley & Sons.

4. Notes

1. Wild-type strains of *X. dendrorhous* used in the lab for basic research or for process development may be used as parent strains for mutagenesis and selection of CHM's. These strains are typically well characterized and possess known and desirable attributes, such as sufficient growth rates, cell yield, good basal level of expression for the carotenoid of interest, etc. Examples of yeasts used as the basis for generation and FCS-based screening for CHMs include natural isolates such as *X. dendrorhous* ATCC 24230 (4) or industrial strains (6). Defined carotenoid mutants (albino strains such as ATCC 96815 and CHMs isolated via traditional methods, such as ATCC 74218 and related strains) are used as negative and positive controls when setting up and validating FCCS gating strategies (4, 6) (see Subheading 3.5).
2. EMS is a hazardous compound, is an inhalational and contact hazard, and is toxic by ingestion. Exposure to EMS must be avoided through proper containment procedures and use of appropriate personal protective equipment. Refer to the manufacturer's Material Safety Data Sheet (MSDS) for additional information. As with EMS, alternative mutagens should be considered hazardous and treated with the same care as described here for use of EMS.
3. Although EMS may be inactivated with the addition of sodium thiosulfate, this treatment may further reduce yeast viability. Alternatively, EMS may be removed using successive water washes. EMS exposure parameters should be chosen to yield a 60–70% survival rate after treatment. Variables expected to affect yeast survival rate include the strain used, EMS dosage, cell density, and incubation conditions. Initial experiments should be performed to achieve the dosage conditions producing the desired survival rate.
4. This step is important because the assay is intended to screen for and physically isolate individual cells having higher astaxanthin content. Therefore, the presence of cell aggregates could be a potentially confounding factor for this single-cell technique. Filtered cell suspensions are diluted further in the same buffer and diluted to an optimal concentration for sorting.
5. An et al. (3) analyzed a suspension at ~10^7 CFU/mL and Ukibe et al. (5) used a suspension at ~2×10^6 CFU/mL. However, optimal sorting concentrations may differ among various instruments. Therefore, initial experiments should be performed to optimize this parameter. As an example, with the BD FACSAria III, the best sort rates are typically achieved by

balancing the cell concentration and the sample pressure settings, with the highest possible cell concentration and lowest possible sample pressure yielding optimal results. Appropriate control populations should be used to determine and validate the optimal sorting parameters.

6. Although An et al. (3) used a laser operating at 200 mW, other instruments equipped with lower-power lasers may be used, such as the 20-mW laser used in the Becton Dickinson FACSAria III instrument or the 15-mW laser used in the Beckman Coulter EPICS Elite ESP instrument (5). A 60- to 70-μm nozzle is used, depending on the instrument used. Commercial sheath fluid is formulated with sodium azide and surfactants. No information on the impact of azide on *X. dendrorhous* recovery after sorting is available. If desired, azide-free 0.1 M phosphate buffer, pH 7.0 may be used.
7. Because flow cytometers are inherently variable, absolute scatter or fluorescence values may differ from instrument to instrument, or even from day to day on the same instrument. Therefore, instruments are calibrated daily using commercially available fluorescent bead kits and associated software, according to manufacturer's instructions. Fluorescent beads are also used to set the sort delay prior to sorting of cells.
8. For example, the ranges reported by Ukibe et al. (5) for their instrument were FL1 (490–550 nm), FL2 (570–580 nm), FL3 (600–620 nm), and FL4 (655–685 nm). Although "factory presets," or default filter sets are useful for selection of *X. dendrorhous* CHMs, custom filters are available for use, if deemed warranted. These are available from custom vendors such as Chroma Technology Corporation (Bellows Falls, VT, USA). Threshold settings for FSC and SSC, detector voltages and fluorescence compensation settings will differ among instrument makes and among individual instruments. These are therefore determined locally for each experiment. Typically, microorganisms such as bacteria and yeasts are thresholded on SSC. SSC threshold settings are obtained by examining both unstained yeasts and buffer without yeasts. The setting that provides clear SSC signals from yeasts while minimizing background noise (checked using buffer control) is used.
9. Flow cytometry is a valuable tool that investigators can use to study complex populations according to specific cell traits. It is expected that differences in fluorescence staining among cell populations can be linked to underlying biological phenomena. In order to demonstrate such linkage, sound experimental design is required, including use of appropriate controls. These controls are critical to data collection and interpretation. Both albino and CHM strains are available through the American Type Culture Collection (ATCC). Albino strains available

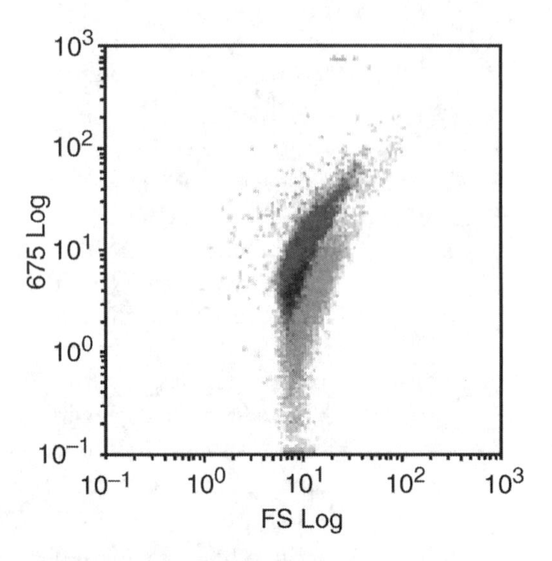

Fig. 2. Astaxanthin-related differences in fluorescence of *Xanthophyllomyces dendrorhous* cells can be detected by flow cytometry. An astaxanthin-overproducing strain of *X. dendrorhous* (Y2342) was grown in YM with (a) the addition of 50 μM diphenylamine, an inhibitor of astaxanthin biosynthesis (*light gray* population) or (b) without added diphenylamine (no inhibition of astaxanthin, *darker* population). Plotting of fluorescence in FL4 (675 nm) against FSC allows clear differentiation between astaxanthin-producing cell populations and those in which astaxanthin synthesis was chemically inhibited. These results suggest that similar differentiation of CHM and wild-type cells, based on astaxanthin content, can be carried out via cytometric analysis of chemically mutagenized *X. dendrorhous* populations. Figure reproduced from Ukibe et al. (5), with permission from John Wiley & Sons.

include ATCC 96815. CHM strains include ATCC 74218, ATCC 74219, ATCC 74220, and ATCC 74221. When screening for *X. dendrorhous* CHMs, appropriate controls include *X. dendrorhous* strains representative of the lower-, middle-, and upper-tier responses for astaxanthin-related fluorescence. Positive and negative control cultures (e.g., astaxanthin CHMs and albino strains), as well as wild-type strains, are used. These strains should be analyzed alone and in defined mixtures. In the absence of defined mutant strains, chemical inhibition of carotenoid synthesis may also be used to generate cells having different carotenoid content. Figure 2 illustrates differences seen between an astaxanthin-overproducing strain of *X. dendrorhous* (Y2342) grown with 50 μM diphenylamine (an inhibitor of astaxanthin biosynthesis, gray population) or with 0 μM diphenylamine (no inhibition of astaxanthin, black population). When these strains were examined as a function of fluorescence in FL4 (675 nm), the astaxanthin-overproducing strain was notably shifted toward increased fluorescence in FL4 (5). For reporting purposes, experimental details should meet agreed upon standards for the minimal information required of flow cytometry experiments (MIFlowCyt) (8).

10. Other options for sorting, according to the instrument used, include sorting into 96-well plates or onto solid agar media.

Acknowledgments

The methods described here are based on the work of Gil-Hwan An, whose doctoral work in the Johnson laboratory was focused on "Improved astaxanthin production from the red yeast *P. rhodozyma*" (4), and on the work of Ukibe et al. (5) who recently revisited and revised our original protocol for FCCS-based selection of CHMs.

References

1. Rodríguez-Sáiz M, de la Fuente JL, Barredo JL (2010) *Xanthophyllomyces dendrorhous* for the industrial production of astaxanthin. Appl Microbiol Biotechnol 8:645–658
2. Johnson EA (2003) *Phaffia rhodozyma*: colorful odyssey. Int Microbiol 6:169–174
3. An GH, Bielich J, Auerbach R et al (1989) Isolation and characterization of carotenoid hyperproducing mutants of yeast by flow cytometry and cell sorting. Biotechnology (N Y) 9:70–73
4. An GH (1991) Improved astaxanthin production from the red yeast *Phaffia rhodozyma*. PhD thesis, University of Wisconsin, Madison
5. Ukibe K, Katsuragi T, Yoshiki T et al (2008) Efficient screening for astaxanthin-overproducing mutants of the yeast *Xanthophyllomyces dendrorhous* by flow cytometry. FEMS Microbiol Lett 286:241–248
6. An GH, Schuman D, Johnson E (1989) Isolation of *Phaffia rhodozyma* mutants with increased astaxanthin content. Appl Environ Microbiol 56:2944–2945
7. Echavarri-Erasun C, Johnson EA (2004) Stimulation of astaxanthin formation in the yeast *Xanthophyllomyces dendrorhous* by the fungus *Epicoccum nigrum*. FEMS Yeast Res 4:511–519
8. Lee JA, Spidlen J, Boyce K et al (2008) MIFlowCyte: the minimum information about a flow cytometry experiment. Cytometry A 73A:926–930

Chapter 15

Generation of Astaxanthin Mutants in *Xanthophyllomyces dendrorhous* Using a Double Recombination Method Based on Hygromycin Resistance

Mauricio Niklitschek, Marcelo Baeza, María Fernández-Lobato, and Víctor Cifuentes

Abstract

Generally two selection markers are required to obtain homozygous mutations in a diploid background, one for each gene copy that is interrupted. In this chapter is described a method that allows the double gene deletions of the two copies of a gene from a diploid organism, a wild-type strain of the *Xanthophyllomyces dendrorhous* yeast, using hygromycin B resistance as the only selection marker. To accomplish this, in a first step, a heterozygous hygromycin B-resistant strain is obtained by a single process of transformation (carrying the inserted *hph* gene). Following, the heterozygous mutant is grown in media with increasing concentrations of the antibiotic. In this way, the strains that became homozygous (by mitotic recombination) for the antibiotic marker would able to growth at higher concentration of the antibiotic than the heterozygous. The method can be potentially applied for obtaining double mutants of other diploid organisms.

Key words: Xanthophyllomyces dendrorhous, Astaxanthin, Recombination, Hygromycin B

1. Introduction

To obtain a homozygous mutant of the *Xanthophyllomyces dendrorhous* yeast strain UCD 67-385, a two-step transformation of the diploid wild-type strain is necessary. However, the molecular tools for transforming *X. dendrorhous* and quickly and reliably selecting mutant strains are insufficient, which has led to the development of the *D*ouble *R*ecombinant *M*ethod (DRM) (1). The DRM is based on the naturally high mitotic recombination frequency of this yeast (as compared to other yeasts, such as *Saccharomyces cerevisiae*), which allows the segregation of some characteristics when the yeast is grown under normal culture conditions. Basically

José-Luis Barredo (ed.), *Microbial Carotenoids From Fungi: Methods and Protocols*, Methods in Molecular Biology, vol. 898,
DOI 10.1007/978-1-61779-918-1_15, © Springer Science+Business Media New York 2012

the entire process had two steps: In a first step, a heterozygous mutant is generated by insertion of the resistance cassette in one copy of the target gene. In a second step, the heterozygous yeast strain is subcultivated in liquid media supplemented with increasing concentrations of the selective antibiotic. We used hygromycin B because the gene of resistance to this antibiotic shows a doses dependence to confer resistance at different concentrations of the antibiotic. The heterozygous yeast strain is first grown using a low concentration of antibiotic, and after it reaches a high OD is used as the inoculum for the next medium supplemented with higher concentration of the antibiotic. A yeast strain carrying two or more copies of a resistance gene will grow at higher antibiotic concentration that the heterozygous strain carrying only one copy of the selective marker gene. After this procedure is repeated several times, a culture that is enriched in strains homozygous for the selective marker gene is obtained.

Using this method, we obtain mutants of *X. dendrorhous* that are heterozygous and homozygous for the genes *crt*YB, *crt*I, or *crt*S. These mutants were analyzed at molecular level and in relation to production of pigments. As expected, the homozygous mutants are able to grow at higher concentrations of hygromycin B than the heterozygous mutants. The homozygous mutants to *crt*YB and *crt*I genes show an albino phenotype, and homozygous mutants for *crt*S gene develop yellow colonies.

2. Materials

2.1. Strains, Plasmids, and Culture Medium

1. *X. dendrorhous* UCD 67-385 (ATCC, Manassas, VA, USA).
2. *X. dendrorhous* T5 ($crtI^{+}/crtI^{-}\Delta RV$) (1).
3. *Escherichia coli* DH5α competent cells (Invitrogen, Carlsbad, CA, USA).
4. pBS4 (2) (see Note 1).
5. pL22 (see Note 2).
6. pMN-hph (see Note 3).
7. pXD-*crt*YB::*hph* (see Note 4).
8. pXD-*crt*I::*hph* (see Note 5).
9. pXD-*crt*S::*hph* (see Note 6).
10. pBluescript SK– (Stratagene, La Jolla, CA, USA).
11. Yeast-malt medium (YM): 0.3% yeast extract, 0.3% malt extract, 0.5% peptone, and 1% glucose (3).
12. LB: 10 g/L of tryptone, 5 g/L of yeast extract, and 10 g/L of NaCl, adjust pH to 7.0 with 5 N NaOH and sterilize by autoclaving (121°C for 15 min).

13. LB solid: LB and 15 g/L of agar.
14. LB–ampicillin: Add 1 mL of 100 g/L of ampicillin to 1 L LB before use. Store at 4°C.
15. X-gal (5-bromo-4-chloro-3-indolyl-β-D-galactopiranoside).
16. Hygromycin B.

2.2. PCR-Mediated DNA Amplification and DNA Manipulation

1. 0.8 M KCl.
2. Lysing enzymes (Sigma, St Louis, MO, USA).
3. 10 mM Tris–HCl pH 8.0 and EDTA 25 mM.
4. SDS 10%.
5. Ethidium bromide (0.5 μg/mL).
6. Proteinase K.
7. RNAse A.
8. Saturated phenol pH 8.0.
9. Phenol:chloroform:isoamyl alcohol (25:24:1).
10. Chloroform:isoamyl alcohol (24:1).
11. TE buffer: 10 mM Tris–HCl and 1 mM EDTA pH 8.0.
12. TAE buffer: 2 M Tris base, 0.05 M EDTA pH 8.0, and ~1.56 M glacial acetic acid.
13. Loading buffer (6×): 0.1% xylene cyanol, 0.1% bromophenol blue, 0.5% SDS, 0.1 M EDTA, and 50% glycerol.
14. 6 M KI.
15. Dimethylsulfoxide (DMSO).
16. Glass beads (0.5 mm diameter).
17. Wash buffer GC: 50 mM NaCl, 10 mM Tris–HCl pH 7.5, 2.5 mM EDTA, and 50% v/v ethanol.
18. TBE buffer: 89 mM Tris base, 89 mM boric acid, and 2 mM EDTA, pH 8.0.
19. Saturated basic phenol (pH 8.0).
20. Silica S-5631 (Sigma, St Louis, MO, USA).
21. PBS buffer: 137 mM NaCl, 2.7 mM KCl, 10 mM Na_2HPO_4, and 2 mM KH_2PO_4, pH 7.4.
22. BD solution: 50 mM K_2HPO_4 pH 7.0, and 25 mM dithiothreitol (DTT).
23. STM solution: 270 mM sucrose, 10 mM Tris–HCl, pH 7.5, and 1 mM $MgCl_2$.
24. Glassmilk (see Note 7).
25. λ*Hin*dIII molecular weight marker.
26. 1 kb DNA ladder.
27. *Pfu* polymerase.

Table 1
Primers used in this work

Primer	Sequence 5′→3′	Description	Position on the gene
*gpd*T-F	AGGCCTACGGTTCTCTCCAA	Forward *gpd* terminator	Y08366 2.431–2.452→
*gpd*T-R	ATGAGAGATGACGGAGATG	Reverse *gpd* terminator	Y08366 ←2.798–2.816
PEF-F-EV	GATATCGGCTCATCAGCCGAC	Forward promoter *TEF*-1α with *EcoRV* site	953–971→
PEF-R-Hyg	GCTTTTTCATGGTGAAGC TGT TCGAGATAG	Promoter *TEF*-1α reverse with *hph* gene adapter	←1,328–1,347
Hyg-F-PEF	CAGCTTCACCATGAAAAAGCCT GAACTCACC	Forward *hph* gene with Promoter *TEF*-1α adapter	1.174–1.189→
Hyg-R-*gpd*T	CCGTAGGCCTCTATTCCTTTGC CCTCGGAC	Reverse *hph* gene with adapter of terminator *gpd*	←2.183–2.202
*gpd*T-F-Hyg	AAAGGAATAGAGGCCTACGGTT CTCTCCAA	Forward terminador*gpd* with adapter gen *hph*	1.511–1,540→
Hyg-F	ATGAAAAAGCCTGAACTCACC	Forward *hph* gene	1.183–1.189→
Hyg-R	CTATTCCTTTGCCCTCGGAC	Reverse *hph* gene	←2.183–2.193
*gpd*T-R-EV	GATATCATGAGAGATGAC GGAGATG	Reverse terminator *gpd* with *EcoRV* site	←1.894–1.918
EF-aD	GGATCCTTCAAGTACGC	*TEF*-1α	1.501–1.517→
EF-aR	ACGTTCTTGACGTTGAA	*TEF*-1α	←2.531–2,547
C13D5	CGAGGCTTACCTTGTCTCTC	Forward *crt*I gene	889–918→
Pha4-8R	GAAAGCAAGAACACCAACGG	Reverse *crt*I gene	←5.024–5.043

28. M-MulV reverse transcriptase.
29. Restriction endonucleases.
30. Primers (see Table 1).
31. Geneclean Kit (MP Biomedicals, Solon, OH, USA).
32. 1D Image Analysis Software version 2.0.1 (Kodak Scientific Image System, Rochester, NY, USA).

2.3. HPLC Analyses

1. High performance liquid chromatography (HPLC) with diode array detector.
2. Column Lichrospher RP18 125-4 (Merck, Darmstadt, Germany).

3. Mobile phase: acetonitrile:methanol:isopropanol (85:10:5).
4. Carotenoids standards (Sigma, St Louis, MO, USA) (see Note 8).

3. Methods

The methods described below are used for the selection, cloning, and construction of the hygromycin B resistance cassette.

3.1. Genomic DNA purification

The protocol described here is optimized for obtaining a great yield of genomic DNA with the minimum degradation as possible.

1. Grow the yeast in 500–1,000 mL of YM medium at 22°C for 4–5 days.
2. Centrifuge at 7,000 × *g* for 10 min, discard the supernatant, and wash the cell pellet with 200 mL of 0.8 M KCl. Then centrifuge at 7,000 × *g* for 10 min, discard the supernatant, and resuspend cell pellet in 16–20 mL of 0.8 M KCl.
3. Divide the cells in four flasks (4–5 mL each) and add 1 mL of 20 mg/mL of lysing enzymes. Incubate at 22°C with gentle agitation for 12–16 h.
4. Centrifuge at 4,000 × *g* for 5 min, and resuspend in 6 mL of Tris–HCl 10 mM pH 8.0 and EDTA 25 mM.
5. Add SDS 10% to reach a final concentration of 1% and incubate at room temperature for 5 min.
6. Add 50 μL of 20 mg/mL of proteinase K and incubate at 55°C for 1 h.
7. Add 10 μL of 10 mg/mL of RNAse A and incubate at room temperature for 10 min.
8. Add 1 volume of saturated phenol pH 8.0 and mix gently.
9. Centrifuge at 4,000 × *g* for 5 min and transfer the aqueous phase to a new tube.
10. Add 2 mL of 10 mM Tris–HCl and 25 mM EDTA to the remnant phenolic phase and mix gently.
11. Centrifuge at 4,000 × *g* for 4 min and transfer the aqueous phase to the tube containing the previously collected aqueous phase.
12. Wash twice with saturated basic phenol.
13. Wash with 1 volume of phenol:chloroform:isoamyl alcohol (25:24:1).
14. Centrifuge at 4,000 × *g* for 4 min and transfer the aqueous phase to a new tube.
15. Wash with 1 volume of chloroform:isoamyl alcohol (24:1).

16. Centrifuge at 4,000 × *g* for 4 min and transfer the aqueous phase to a new tube.
17. Add 2.5 volumes of ethanol and mix gently to form a "DNA ball," get it out with a pipette and put it in a new tube.
18. Wash the "DNA ball" with ethanol 70%.
19. Let the "DNA ball" dry and resuspend in TE buffer.
20. Store the DNA at –20°C until it is used.

3.2. Purification of DNA from Agarose Gel

1. Add loading buffer (6×) to samples and subject them to 1% agarose gel electrophoresis in TAE buffer (see Note 9).
2. Stain with ethidium bromide and visualize under UV light.
3. Excise the band of interest from the gel (see Note 10); place in an Eppendorf tube and add 3 volumes of 6 M KI.
4. Incubate tubes at 55°C until the gel fragments completely dissolve.
5. Cool to room temperature and mix with 10 μL of glassmilk.
6. Incubate on ice for 10 min and centrifuge at 10,000 × *g* for 30 s.
7. Discard the supernatant, and wash the pellet three times with 500 μL of wash buffer GC, centrifuging at 10,000 × *g* for 30 s between washing steps.
8. Remove the wash buffer and resuspend the pellet in 15–20 μL of nuclease-free water.
9. Incubate at 55°C for 10 min.
10. Centrifuge at 10,000 × *g* for 20 s, and transfer the supernatant containing the nucleic acids to a new tube.

3.3. PCR Amplification of the TEF-1α Promoter from X. dendrorhous

Based on studies in filamentous fungi and yeasts (4), we decided to use the promoter of the gene encoding the translation elongation factor 1-α (*TEF*-1α) of *X. dendrorhous*.

1. To design the primers for the cloning of the *TEF*-1α promoter, compare all available coding sequences for fungal *TEF*-1α, and identify the most conserved regions by multiple alignments.
2. Use the two highly conserved regions to design the primers. Primers designed are shown in Table 1.
3. Amplify by PCR using the primers EF-aD and EF-aR and genomic DNA from *X. dendrorhous*, and this standard PCR reactions: initial denaturation at 95°C for 3 min, 35 cycles of denaturation at 94°C for 30 s, annealing at 55°C for 30 s, synthesis at 72°C for 3 min, and a final extension step at 72°C for 10 min.
4. Check the amplicons by agarose gel electrophoresis. In our model a unique 1 kb amplicon was obtained.

3.4. Cloning the TEF-1α Promoter from X. dendrorhous from a DNA Library

1. Use the primers EF-aD and EF-aR (see Table 1) to screen for the *TEF-1α* gene in an *X. dendrorhous* gDNA library constructed with *Bam*HI fragments (1).
2. Prepare a PCR reaction solution as follow: In a PCR tube mix 18.8 μL of nuclease-free water, 2.5 μL of *Taq* buffer (10×), 0.5 μL of dNTPs mix 10 mM, 1 μL of $MgCl_2$ 50 mM, 1 μL of each primer 25 mM, and 0.2 μL of *Taq* DNA polimerase (5 U/μL).
3. Using a toothpick take out the colony to analyze and suspend in PCR reaction solution.
4. Proceed to the amplification with the following program: Initial denaturation at 95°C for 3 min; 35 cycles of 94°C for 30 s, 55°C for 30 s, and 72°C × 3 min; and final elongation at 72°C for 10 min.
5. Check the PCR products by agarose gel electrophoresis.
6. Grow the positive clones in LB medium at 37°C for 16 h.

Purify the plasmid DNA by a kit, and sequence the DNA fragment by two strands and resolve any ambiguities by resequencing.

The contig of two positive clones (pBR60 and pTEF2-2) obtained for *TEF-1α* gene is shown in Fig. 1. The sequence analysis reveals that the *TEF-1α* gene is in a 14,107 bp DNA fragment containing *Bam*HI cleavage site. The ATG start codon, promoter region, and putative regulating boxes are defined (see Fig. 2).

3.5. Cloning of Transcription Terminator

To construct the hygromycin B resistance cassette, the transcription terminator of the glyceraldehyde-3 phosphate dehydrogenase-encoding gene, *gpd*, is used because it has been successfully used in other basidiomycetes yeast expression systems. It is necessary to obtain the sequence for the *X. dendrorhous gpd* gene before its translation stop codon-containing downstream region (5).

1. Use the primers gpdT-F and gpdT-R (see Table 1) to amplify the *gpd* terminator using genomic DNA of *X. dendrorhous.*
2. Separate the PCR products by agarose gel electrophoresis and purify the amplicon (~390 pb) from gel.
3. Mix the purified amplicon with plasmid pBluescript SK previously digested with *Eco*RV.
4. Add 5 U of T4 DNA ligase and incubate for 16 h at 14°C.
5. Desalt the reaction mix by dialysis.
6. Mix the desalted DNA with electrocompetent *E. coli* DH5α, apply electroshock (25 μF, 200 Ω, 2.5 kV), and suspend in 1 mL of LB medium.
7. Seed 100 μL of cell suspension onto LB agar plate supplemented with 100 μg/mL of ampicillin and 80 μg/mL of X-gal.

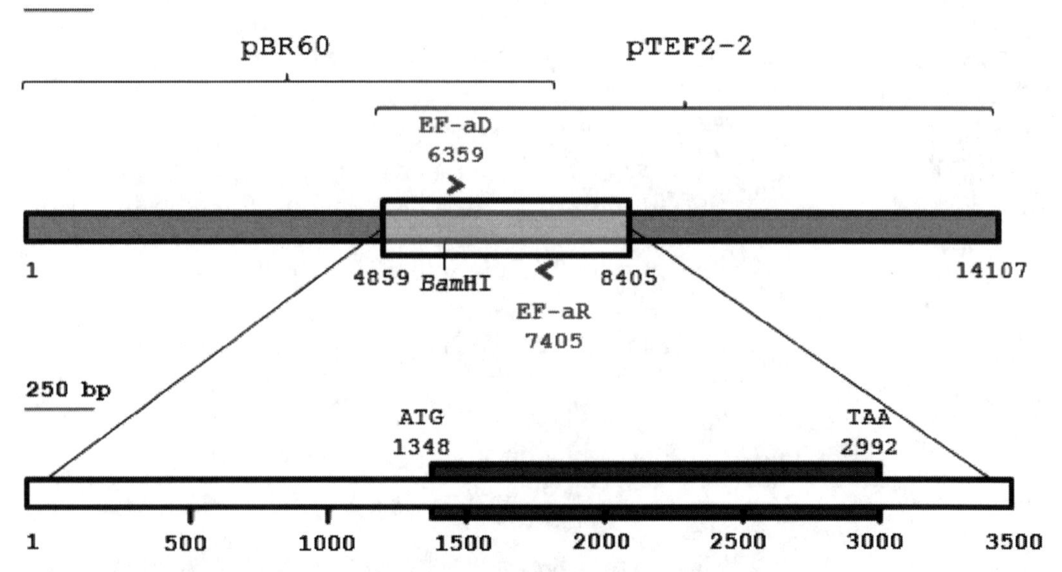

Fig. 1. Schematic representation of the sequences of the pBR60 and pTEF2-2 clones. The EF-aD and EF-aR primer binding sites are shown. A 3,547 bp DNA sequence that included 1,500 bp upstream of the EF-aD primer binding site, a 1,047-bp fragment amplified by the EF-aD and EF-aR primers, and 1,000 bp downstream of the EF-aR primer binding site was compared against GeneBank databases.

TCGCCAAGGTCGGTGGTTGGGTCTCTGAAAAAATTCCGCGAGAGAACGCATCTCGCGCAGGGAAGG
CGATCCTCTCTCGGATCGCGTGGGAGGAAGATGCTGACGTGAAACGGCTACGAGCTGAGCATTTCA
GCTTCTGGCAGAGCTGGGTGATCAAAAAAAGGTTAGGCTTGTTGGCAATCCTTTTTGGATTAGCAA
CAGCAGCAGGGGCCTGGAAAGGCAAAAGATACAAGCAAGCGATCCTTTCTCGATTGCGTGGAGGAA
GATGTTACGTGAACGGGTACGAGTGAGCATTTCAGCTTCTGGCAGAGTGGGTGATCAAAAAAAGGT
TAGGCTTGTTGGCAATCCTTTTTGGATTAGCAACACAGCAGGGGCCTGGAAATGGCAAGTATTTAG
GTGACGGCCTTAGTTACAATTATCAAAATGTTAAACTAGAAAGTATTCTATGTATAATTAATTTCA
TCCTCCTTTCTCTCCTTCTGCGTAGCTGATGATGGGACCTGCCGTTTTGTGTCAGCATGAAGGCAC
ATTTCGTACAATGAGCTGTGGTATTTAGGTGACGGCCTTAGTTACAATTATCAAAATGTTAAACTA
GAAAGTATTCTATGTATAATTAATTTCATCCTCCTTTCTCTCCTTCTGCGTAGCTGATGATGGGAC
CTGCCGTTTTGTGTCAGCATGAAGGCACATTTCGTACAATGAGCTGTGTTGAACGAGGAAGACAAA
ATTCAAGCAATATACATAGTAGGGTGTAGAGAGAAAAAGCCGACTTTCGACTGAAACTAGCTGTTT
GGCCAAGACTAACAGCACAAATGCTTCTGAACATATCTCGATCATGCGTAACTGGGGGCTCATCA
GCCGACAGTTCATCCGACACAAGCTCTTTGCCTAGATCGTCAAACGATCGACATCGACACGAGGTT
TGACATATCTCGATCATGGGTAACTGGGG*GCTCATCAGCCGAC*AGTTCATCCGACACAAGCTCTTT
GCCTAGATCGTCAAACGATCGACATCG**ACACGAAAACAAAT**CCCGATCAGTCAATCAGTAG**AACTG**
CCGAAGAATGAAACTGAAGACTGCTGTGACACGTGAC**TATAGAA**GCGGTGTCATCTGACTTGCGAA
T**T**TGCTTGTACAAAAGTCAGGTTGGCTGATTCGCTCGCGTTGTCGACAAGAATTAGGAAA**CCTCAT**
TTCCAGCCTTCCTTTTTTTCCCCTCAGAACACCTCACATCTTCTCACAAAAAAGTAAGATATTTTC
ACTTGGAGTTGGTGGATTTGGCGACTGAATGGAGAGACAGAAGCTTGATACTGATAAGACACTCTT
TTGTAATCTATCTCGAACAGCTTCAAA**ATG**

Fig. 2. Sequence analysis of the 3.5 kb region containing the *TEF*-1α gene. The sequence analysis identified the ATG translation start codon at position 1,348. The putative TATA box is located at position 1,094, the transcription start site is at position 1,124, the putative RPG box is located between positions 1,018 and 1,031, the HOMOL1 box is located between positions 1,052 and 1,064, and a CT-rich region is located between positions 1,183 and 1,213. The binding sites of the *TEF*-1α gene primers, EF-aD and EF-aR, are also shown.

8. Incubate the plates at 37°C for 16 h.
9. Select the white colonies (harboring the recombinant plasmid).

In this way we obtained the plasmid pBgpdT harboring the terminator of *gpd* gene, which was confirmed by sequencing.

The use of antibiotics has great advantages for the selection of transformants, a critical step in obtaining genetically modified organisms. Currently, *X. dendrorhous* transformant strains are selected using the antibiotic geneticin (G-418). However, the diploid nature of the *X. dendrorhous* UCD 67-385 strain requires the use of other selection markers. The antibiotic hygromycin B is compatible with geneticin, making this mix a good choice for selection (6). We used the *E. coli hph* gene encoding the hygromycin B phosphotransferase enzyme as a resistance marker (7). For this reason, we cloned the *hph* gene from the plasmid pBS4 for our selection system.

3.6. Construction of the Hygromycin B Resistance Cassette

1. Mix equimolar amounts (100–300 ng) of DNA fragment corresponding to the *TEF-1*α promoter and the *hph* resistance gene in a final volume of 20.8 μL. Add 2.5 μL of *Taq* buffer (10×), 0.5 μL of dNTPs mix 10 mM, 1 μL of $MgCl_2$ 50 mM, and 0.2 μL of *Taq* DNA polymerase (5 U/μL). Perform an elongation reaction with the following program: Initial denaturation at 95°C for 1 min, 10 cycles of denaturation at 94°C for 30 s, annealing at 50°C for 45 s, synthesis at 72°C for 1 min, and a final extension step at 72°C for 10 min.
2. Use 10 μL from the elongation reaction as a template to amplify the full fragment by PCR using the primers PEF-F-EV and Hyg-R-*gpd*T (see Table 1).
3. Separate PCR products by agarose gel electrophoresis and purify the amplicon (1.5 kb) from gel.
4. Mix the amplicon obtained with *gpd* gene terminator region and perform an elongation reaction.
5. Use 10 μL from the elongation reaction as a template to amplify the full fragment by PCR using the primers PEF-F-EV and gpdT-R-EV (see Table 1).
6. Separate by agarose gel electrophoresis and purify the amplicon from it.
7. Ligate into the *Eco*RV site of pBluescript SK–, desalt, transform into *E. coli* DH-5α, and select recombinant clones as described above.

A schematic representation of the construction of the resistance cassette is shown in Fig. 3. A plasmid named pMN-Hyg containing the desired construction is obtained.

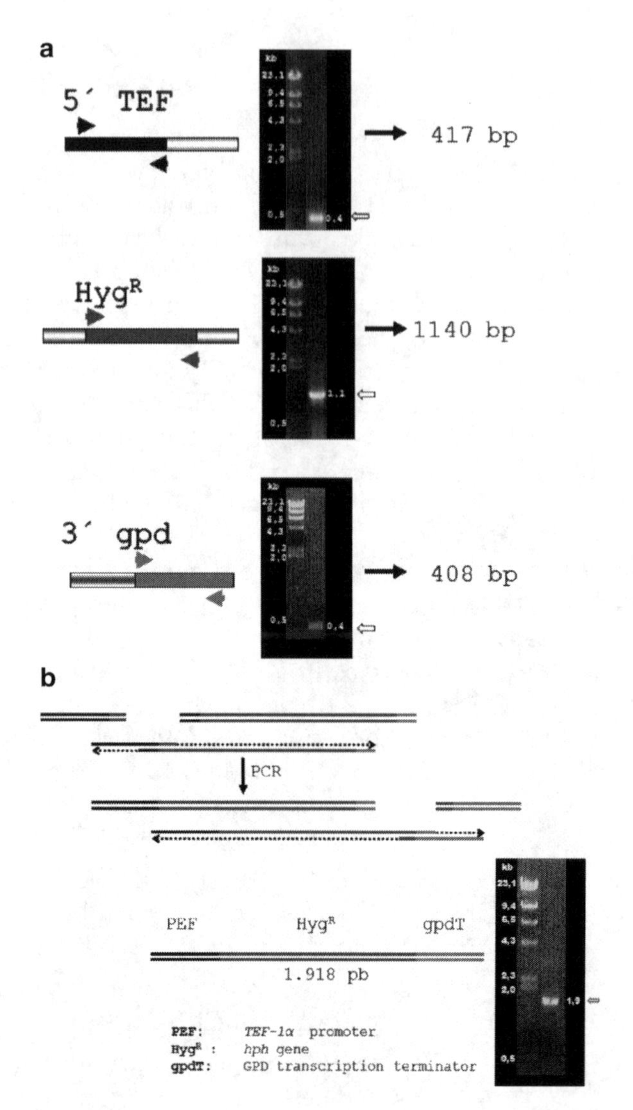

Fig. 3. Construction of *X. dendrorhous* hygromycin B resistance cassette. (**a**) Amplification of the *TEF*-1α gene promoter region with the primers PEF-F-EV and PEF-R-Hyg, amplification of the *hph* gene with the primers Hyg-F-PEF and Hyg-R-gpdT, and amplification of the *gpd* gene transcription termination region with the primers gpdT-F-Hyg and gpdT-R-EV. (**b**) Hygromycin B resistance cassette construction through fragment union by elongation and a subsequent PCR amplification. Molecular weight markers λ*Hin*dIII (λ) are indicated. The size of the amplified products are indicated by *arrows.*

3.7. Insertion of the Hygromycin B Resistance Cassette into the crtI Locus by Transformation of X. dendrorhous

To check the functionality of the hygromycin B resistance cassette in *X. dendrorhous*, it was used to the replacement of *crt*I, a gene involved in the biosynthesis of astaxanthin. In this way, the transformants obtained will be resistant to hygromycin B and will display different coloration respect to parental strain.

1. Grow *X. dendrorhous* in 500–1,000 mL of YM medium at 22°C, until an OD_{600} of 1.2 is reached.

2. Centrifuge at 4,000 × *g* for 5 min, suspend in 25 mL of BD solution, and incubate at 22°C for 15 min.
3. Wash the cells twice with 25 mL of cold STM solution and suspend in 1 mL of STM solution (see Note 12).
4. Mix 60 μL of electrocompetent cells with 10–20 μg of "transforming DNA" (see Note 13), deposit in the electroporation cuvette, and apply the electric pulse (125 μF, 600 Ω, and 0.45 kV).
5. Add to the cuvette 1 mL of YM medium, resuspend the cells well, and transfer to a sterile tube.
6. Incubate at 22°C for 5 h, seed aliquots of 100 μL onto YM plates supplemented with 10 μg/mL of hygromycin B, and incubate at 22°C until development of colonies.

In a first step, we used the pIH plasmid (see Note 11) to transform a wild-type (*crt*I$^+$/*crt*I$^+$) *X. dendrorhous* strain. A hygromycin B-resistant transformant was obtained and selected because it has a level of pigmentation lower than the parental strain (pale phenotype). Genomic DNA was extracted from clones, and used in PCR reactions with: (1) PEF-F-RV and gpdT-R-RV primers that amplify the whole cassette (1,900 pb); (2) Hyg-F and Hyg-R primers that amplify the *hph* gene (1,100 pb); (3) C13D5 and Hyg-R primers that would amplify a 5′ fusion region of *hph* and *crt*I genes (2,283 bp); and (4) pha4(8R) and Hyg-F primers that would amplify a 3′ fusion region of *hph* and *crt*I genes (2,990 bp). Results obtained from PCR reactions indicate that the pale transformants have the hygromycin B cassette integrated in one copy of *crt*I gene, therefore, are heterozygous mutants (*crt*I$^+$/*crt*I$^-$). These results strongly suggest a gene dose-dependency in the biosynthesis of astaxanthin, because the presence of only one allelic copy reduced the pigmentation Fig. 4.

3.8. Generation of Homozygous Mutants of *X. dendrorhous*

In a second step, the heterozygous mutant is transformed into homozygous mutant. For that the DRM is developed (1). The basis of the method is that cells which have a mitotic recombination event in the *crt*I locus will be homozygous for the mutation, increasing the copy of the *hph* gene and therefore its resistance to the antibiotic.

1. Grow the heterozygous strains at 22°C in YM medium supplemented with 50 μg/mL of hygromycin B, until late exponential phase (OD_{600} 8–10).
2. Use this culture to inoculate a new YM media containing 100 μg/mL of hygromycin B and incubate at 22°C until late exponential phase.
3. Repeat the procedure of subcultivation. In each step increase the concentration of antibiotic twice, increasing at twofold, until 800 μg/mL is reached.

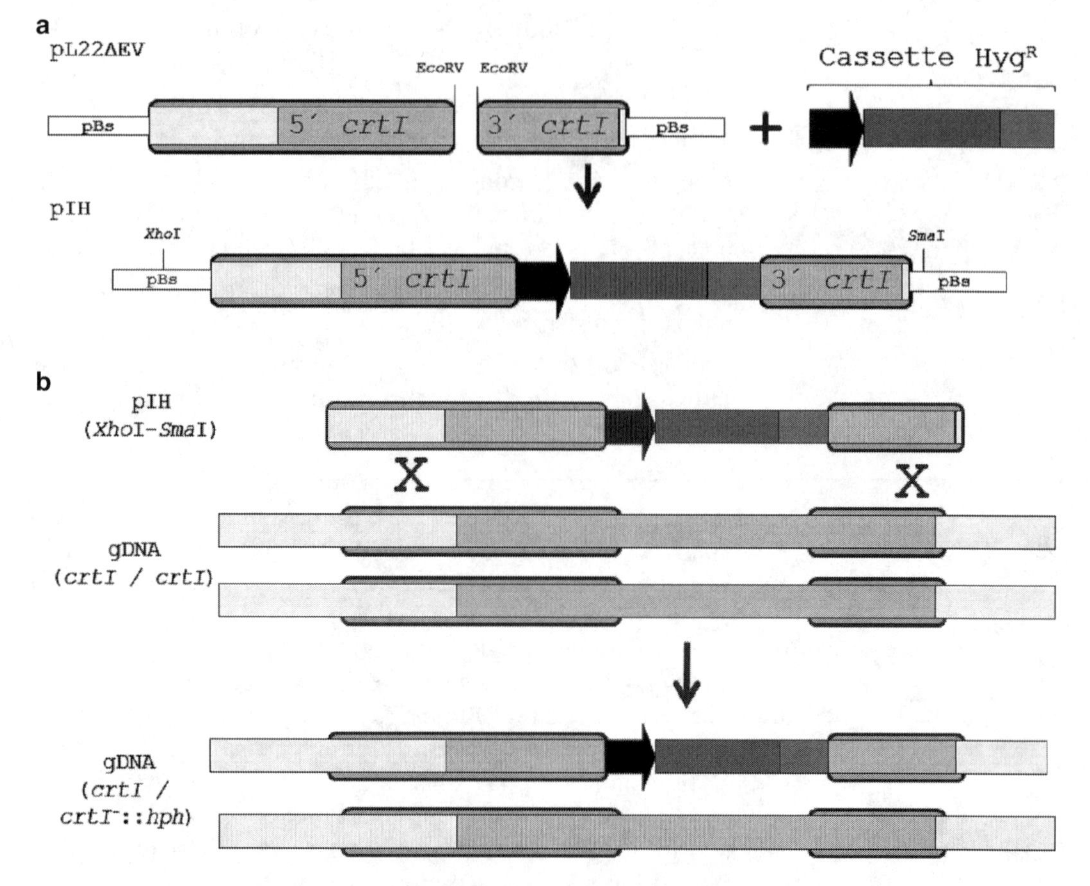

Fig. 4. Schematic representation of insertion of hygromycin B resistance cassette into *crt*I locus of *X. dendrorhous*. (**a**) The *X. dendrorhous* hygromycin B resistance cassette is inserted in the *Eco*RV site of the plasmid pL22ΔRV, resulting in the plasmid pIH that is used to transform *X. dendrorhous*. (**b**) The mechanism of the integration of the hygromycin B resistance cassette in the *crtI* locus by homologous recombination.

4. Collect aliquots of each culture, seed onto YM plates containing hygromycin B and incubate at 22°C until development of colonies.

The procedure and the results obtained are schematized in Fig. 5. As the subcultivations are made, the coloration of the colonies obtained varies from pale phenotype (heterozygous mutant) to albino phenotype (homozygous strains). Molecular analysis to confirm the homozygous mutants are made as described above. In this way the homozygous mutants *crt*YB$^-$::*hph*/*crt*YB$^-$::*hph*, *crt*I$^-$::*hph*/*crt*I$^-$::*hph* and *crt*S$^-$::*hph*/*crt*S$^-$::*hph* are obtained.

3.9. Pigment Extraction and Analysis

1. Centrifuge 50 mL of the yeast culture at 7,000×*g*.
2. Wash with 1 mL of water and suspend in 1 mL of water.
3. Add 0.5 mL of glass beads (0.5 mm diameter) and vortex for 2 min.

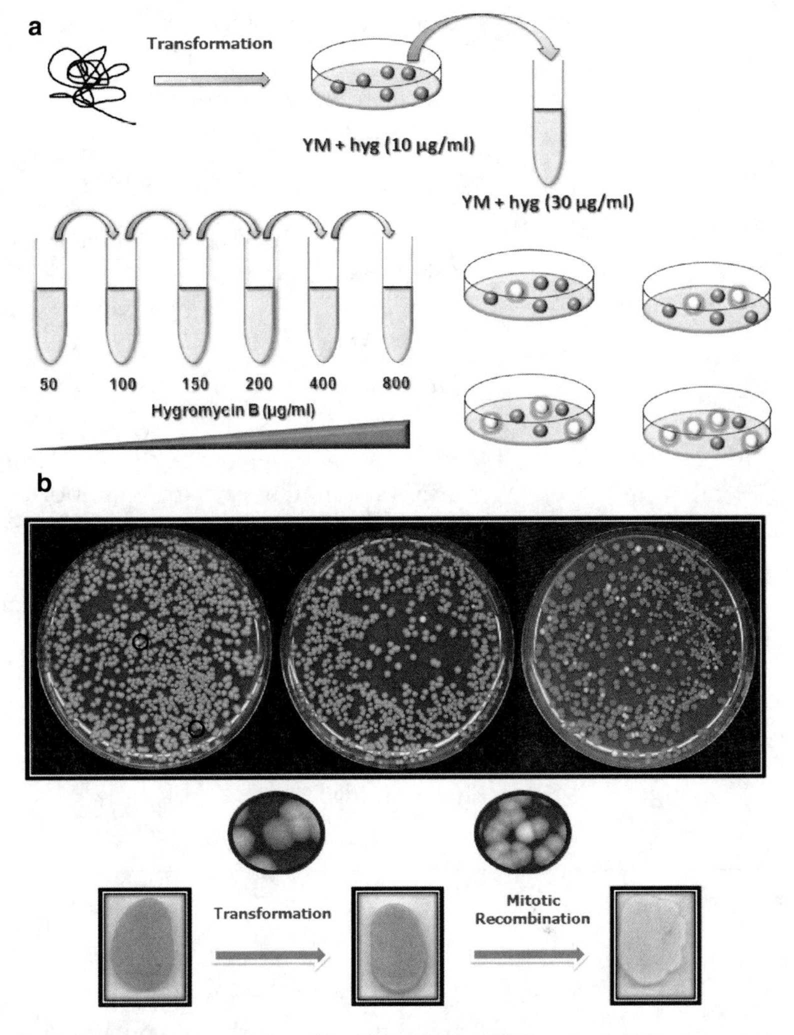

Fig. 5. Schematic representation of the *X. dendrorhous* DRM protocol. (**a**) The hygromycin B-resistant heterozygous strains are grown in liquid media with increasing concentrations of the antibiotic, and then plated on YM with 15 μg/mL of hygromycin B. (**b**) Petri dishes with colonies grown at 50, 100, and 150 μg/mL of hygromycin B.

4. Incubate on ice for 2 min and add 1 mL of acetone.
5. Vortex for 2 min and centrifuge at 12,000 × *g* at 4°C.
6. Transfer the organic phase to a new tube.
7. Add 2 mL of acetone to the cellular pellet and repeat the steps 5 and 6.
8. Repeat step 7 until the cellular pellet is completely discolored and join all organic phases collected in the same tube.
9. Add 1/5 volume of petroleum ether and vortex for 2 min.
10. Centrifuge at 12,000 × *g* for 3 min and transfer the petroleum ether phase to a new tube.

11. Determine the volume of petroleum ether obtained and determine OD_{474}.
12. Calculate the amount of total carotenoid pigments according to the following formula (3):

$$TCP = OD_{474} \times V_f \times 10^4 / 2,100 \times P$$

where TCP is the total carotenoids pigments (μg/g of sample), OD_{474} the absorbance at 474 nm, V_f the final volume of petroleum ether, P the grams of sample and the value 2,100 is the molar extinction coefficient of astaxanthin.

13. Dry the sample of petroleum ether under nitrogen and dissolve in 100 μL of acetone.
14. Separate the pigment by HPLC, using a RP-18. Lichrocart 125-4 column, in isocratic conditions using a solution of aceto nitrile:methanol:isopropanol (85:10:5) as mobile phase, at a flux of 1 mL/min.
15. Detect the pigments at 474 nm and determine the absorbance spectrum of the samples collected.
16. Identify the pigments according to retention time and absorbance spectrum.

The analysis of pigment of different mutants of *X. dendrorhous* is shown in Fig. 6.

4. Notes

1. pBS4 contains the *hph* gene.
2. pL22 is pBluescript SK– containing a 3,558-bp amplicon carrying the *crt*I gene of *X. dendrorhous.*
3. pMN-hph is pBluescript SK– containing, in the *Eco*RV site, a 1.8-kb cassette encoding the *Escherichia coli hph* hygromycin B resistance gene.
4. pXD-crtYB::*hph* is pBluescript containing a 5.9-kb *Eco*RV fragment carrying the *crt*YB gene with a deletion of a *Hpa*I 1,245 bp segment and an insertion of an *hph* cassette from pMN-*hph*.
5. pXD-*crt*I::*hph* is pL22 with a deletion of a 1,127 bp *Eco*RV fragment of the *crt*I gene and an insertion of an *hph* cassette from pMN-*hph*.
6. pXD-*crt*S::*hph* is pBluescript containing a 4 kb *Pst*I fragment carrying the *crt*S gene with a deletion of a *Stu*I-*Hpa*I 2,479 bp fragment and an insertion of an *hph* cassette from pMN-*hph*.

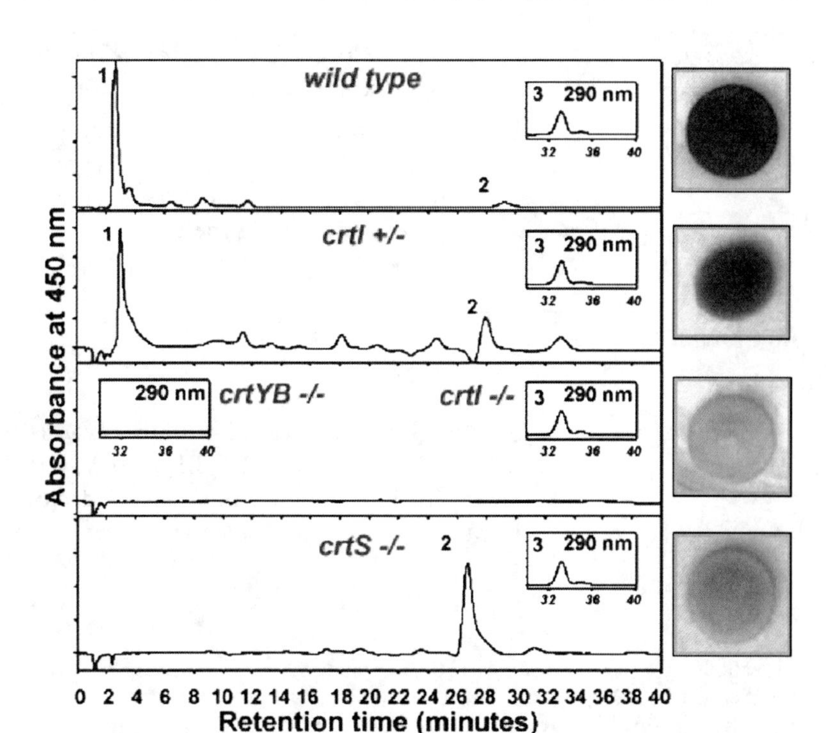

Fig. 6. RP-HPLC analysis at 450 nm of carotenoids from the parental wild type, heterozygous, and homozygous mutant strains. The numbers indicate the different carotenoids as follows: (1) astaxanthin, (2) beta-carotene, and (3) phytoene (290 nm). The strain genotype that corresponds to each chromatogram is indicated.

7. Glassmilk: Dissolve 1 g of silica in 10 mL of PBS buffer, allow the silica to settle for 2 h, remove the supernatant (containing fine particulate matter), and add 10 mL of PBS. Repeat the procedure twice. Centrifuge at 2,000×*g* for 2 min and resuspend the silica in 10 mL of 3 M NaI. Store at 4°C in the dark.
8. Each pigment is identified by comparison with specific standards based on retention time and absorption spectrum (8).
9. The purification of fragments from TBE-based gels is more complex than from TAE-based gels.
10. Only the section of gel that contains the nucleic acids should be excised. This facilitates the dissolving of the gel and allows the use of several gel pieces in a single tube, increasing the amount of DNA recovered.
11. Plasmid pMN-Hyg is digested with the *Eco*RV enzyme releasing the 1,912-bp *X. dendrorhous* hygromycin B resistance cassette. This fragment is purified by Geneclean and inserted into the *Eco*RV site of the pL22ΔRV plasmid, creating the pIH plasmid (see Fig. 4). The plasmid pL22ΔEV carries a *crt*I gene mutated by deletion of a fragment of 1,128 between *Eco*RV restriction sites.

12. Do not store the *X. dendrorhous* electrocompetent cells for long period of time. Use preferably just after production.
13. For high transformation efficiency, the DNA used for transformation must be lineal. In our case we use the pIH plasmid digested with *Xho*I and *Sma*I; these digestions release a DNA fragment containing the cassette.

Acknowledgments

This work was supported by Fondecyt 1100324, Innova CORFO 07CN13PZT-17, CICYT BIO2010-20508-C04-01/04, and by an institutional grant from the Fundación Ramón Areces to the Centro de Biología Molecular Severo Ochoa.

References

1. Niklitschek M, Alcaino J, Barahona S, Sepulveda D, Lozano C, Carmona M, Marcoleta A, Martínez C, Lodato P, Baeza M, Cifuentes V (2008) Genomic organization of the structural genes controlling the astaxanthin biosynthesis pathway of *Xanthophyllomyces dendrorhous*. Biol Res 41:93–108
2. Goldstein A, Mc Cusker J (1999) Three new dominants drug resistance cassettes for gene disruption in *Saccharomyces cerevisiae*. Yeast 15:1541–1553
3. An GH, Schhuman DB, Johnson EA (1989) Isolation of *Phaffia rhodozyma* mutants with increased astaxanthin content. Appl Environ Microbiol 55:116–124
4. Kitamoto K, Matsui J, Kawai Y, Kato A, Yoshino S, Ohmiya K, Tsukagoshi N (1998) Utilization of the *TEF*-1alpha gene (*TEF1*) promoter for expression of polygalacturonase genes, *pga*A and *pga*B, in *Aspergillus oryzae*. Appl Microbiol Biotechnol 50:85–92
5. Verdoes JC, Wery J, Boekhout T, Van Ooyen AJJ (1997) Molecular characterization of the glyceraldehyde-3-prosphate dehydrogenase gene of *Phaffia rhodozyma*. Yeast 13:1231–1242
6. Burland T, Pallotta D, Tardif M, Lemieux G, Dove W (1991) Fission yeast promoter-probe vectors based on hygromycin resistance. Gene 100:241–245
7. Gritz L, Davies J (1983) Plasmid-encoded hygromycin B resistance: the sequence of hygromycin B phosphotransferase gene and its expression in *Escherichia coli* and *Saccharomyces cerevisiae*. Gene 25:179–188
8. Britton G, Liaaen-Jensen S, Pfander H (eds) (2004) Carotenoids handbook. Birkhäuser Verlag, Basel

Chapter 16

Genetic Manipulation of *Xanthophyllomyces dendrorhous* and *Phaffia rhodozyma*

Guangyun Lin, Joanna Bultman, Eric A. Johnson, and Jack W. Fell

Abstract

The yeasts *Xanthophyllomyces dendrorhous* (teleomorph) and *Phaffia rhodozyma* (anamorph) are of basidiomycetous affinity and have the unique property among yeasts of producing the carotenoid pigment astaxanthin. Astaxanthin imparts the attractive coloration to salmonids, crustaceans, and several birds such as the flamingo, and it has considerable economic value. Microbiological and genetic techniques for manipulation are rudimentary in the yeast, while their utility would be valuable for strain development including hypermutants that overproduce astaxanthin. Here we describe methods for manipulation of the yeast, including induction of the sexual stage with basidiospore formation, methods for isolation of mutants (particularly mutants affected in carotenoid biosynthesis) as well as techniques for isolation and analysis of carotenoids. These methods are valuable for understanding the biology and enhancing the biotechnology value of the yeast.

Key words: *Xanthophyllomyces dendrorhous, Phaffia rhodozyma*, Basidiomycetous yeast, Basidiospores, Astaxanthin, Carotenoid biosynthesis, Knockout mutant, Chemical mutagenesis

1. Introduction

Xanthophyllomyces dendrorhous (anamorph *Phaffia rhodozyma*) is a basidiomycetous yeast discovered and extensively studied by Herman Jan Phaff and colleagues (1–4) that produces the high-value carotenoid astaxanthin (5–7). Astaxanthin is one of most common carotenoids in the biosphere, particularly in marine environments and is produced by a few species of bacteria, microalgae, thraustochytrids, and filamentous fungi (7). These organisms serve as the base of a food chain for various macro-creatures that leads to attractive pigmentation of birds, such as the flamingo and scarlet ibis, marine crustacea including shrimp and lobsters, and fish including farmed salmonids.

José-Luis Barredo (ed.), *Microbial Carotenoids From Fungi: Methods and Protocols*, Methods in Molecular Biology, vol. 898, DOI 10.1007/978-1-61779-918-1_16, © Springer Science+Business Media New York 2012

Wild strains of *X. dendrorhous* and *P. rhodozyma* strains produce low quantities of astaxanthin (200–500 μg/g dry yeast), while industrial strains and processes have been developed for *X. dendrorhous* strains that yield values of 6,000–15,000 μg/g. *X. dendrorhous* can also reach high cell densities of 100–130 g dry cell weight (DCW) per liter in submerged fermentations with sufficient mass transfer and oxygenation. Although astaxanthin, like many other fine chemicals produced by fungi have traditionally been produced by chemical synthesis, consumer demands for natural food and feed ingredients have led to renewed interest in natural sources of industrially important products.

For the past 30 years, our laboratories have studied numerous strains of *X. dendrorhous* and *P. rhodozyma* and developed and optimized several genetic techniques. These techniques are described in the following sections of this chapter.

1.1. Vegetative Growth and Perfect Activity in X. dendrorhous

Phaff, Miller, and colleagues attempted to show the perfect state of *P. rhodozyma* in the 1970s and 1980s, but it was not until 1995 that Golubev described the sexual state of certain *P. rhodozyma* strains and named the teleomorphic form *X. dendrorhous* (8). Subsequent studies have used similar methodologies to describe in further detail the homothallic life cycle of the yeast as well as the environmental parameters that stimulate sexual activity (9). However, due to the paucity of the number and variety of strains used in both studies, complete conclusions cannot be extrapolated to all known strains of *P. rhodozyma* and *X. dendrorhous*. The overall strategy and experimental protocols for various strains is described from experience in our laboratories.

1.2. Generation of Carotenoid Mutants by Chemical Mutagenesis

In one of the early studies of isolation of carotenoid mutants (10), plating of *X. dendrorhous* onto yeast malt agar containing 50 μM antimycin A gave rise to colonies of unusual morphology, characterized by a nonpigmented lower smooth surface that developed highly pigmented vertical papillae after 1–2 months. Isolation and purification of the pigmented papillae, followed by testing for pigment production in shake flasks, demonstrated that several antimycin isolates were increased two- to fivefold in astaxanthin content compared with the parental natural isolate (UCD-FST 67-385). One of the antimycin strains (ant-1) and a nitrosoguanidine derivative of ant-1 (ant-4) produced considerably more astaxanthin than the parent (ant-1 had 800–900 μg/g, ant-1-4 had 900–1,300 μg/g, and UCD 67-385 had 300–450 μg/g) (10, 11). Here, three methods are described to obtain the higher astaxanthin-producing mutants (Fig. 1).

1.3. Construction of Targeted Knockout Mutants

In this section, a method developed in E.A. Johnson's laboratory to knock out a specific gene in *X. dendrorhous* is described. To knock out a specific gene, a transfer vector which contains two

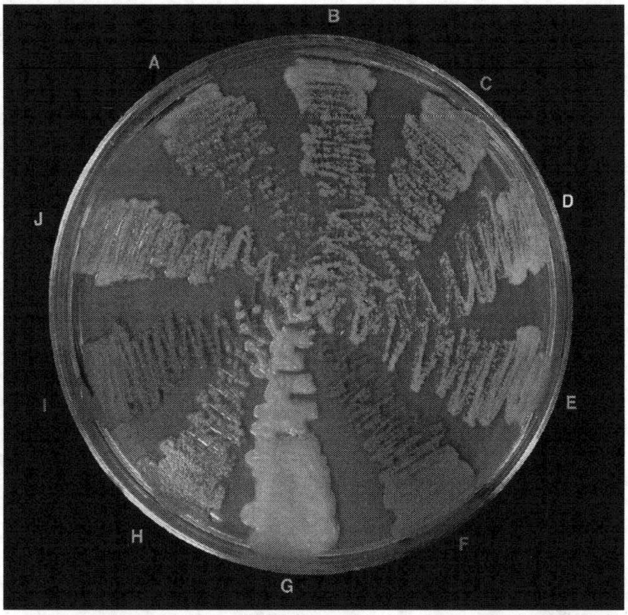

Fig. 1. Color mutants of *Xanthophyllomyces dendrorhous* isolated in EAJ laboratory. The parental strain was UCD 67-385, and mutants were mainly obtained by mutagenesis using *N*-methyl-*N'*-nitro-*N*-nitrosoguanidine (NTG) as described. (*A*) UCD 67-385 (*X. dendrorhous*), (*B*) CBS5909 (*Phaffia rhodozyma*), (*C*) Y174-30 (isolate of *X. dendrorhous* from Michigan; courtesy of Clete Kurtzman), (*D*) AF-1 (albino mutant of UCD 67-385), (*E*) YAN-1 (β-carotene mutant), (*F*) CAX (hyperproducer), (*G*) YAN-1-OP (lycopene mutant from UCD 67-385), (*H*) YAN-1-OO (β-carotene hyperproducer), (*I*) CAX-HP20 (astaxanthin hyperproducer), (*J*) YAN-1-HP (β-carotene hyperproducer). Photograph by C. Echavarri-Erasun from (6).

flanking regions of the gene needs to be first constructed. Flanking regions in this construct are necessary for double crossover homologous recombination. The size of the flanking region usually is approximately 250–1000 bp to promote homologous recombination. An antibiotic selection marker expression cassette is inserted between the flanking regions. The fragment containing the two flanking regions and the antibiotic selection gene expression cassette is transformed into competent yeast recipient cells to promote the double cross homologous recombination process. The specific targeted gene is then replaced by the antibiotic gene expression cassette after double crossover homologous recombination is achieved. In order to make sure that the gene knockout process does not alter genetic functions in the genome, a "repair" strain is generated to rescue the function of the replaced protein. In EAJ's laboratory, the *ast* gene was successfully knocked out in *X. dendrorhous*. This is used as an example to introduce the detailed methods of targeted mutation (Fig. 2) and verification (Fig. 3) of the knockout and repair mutants.

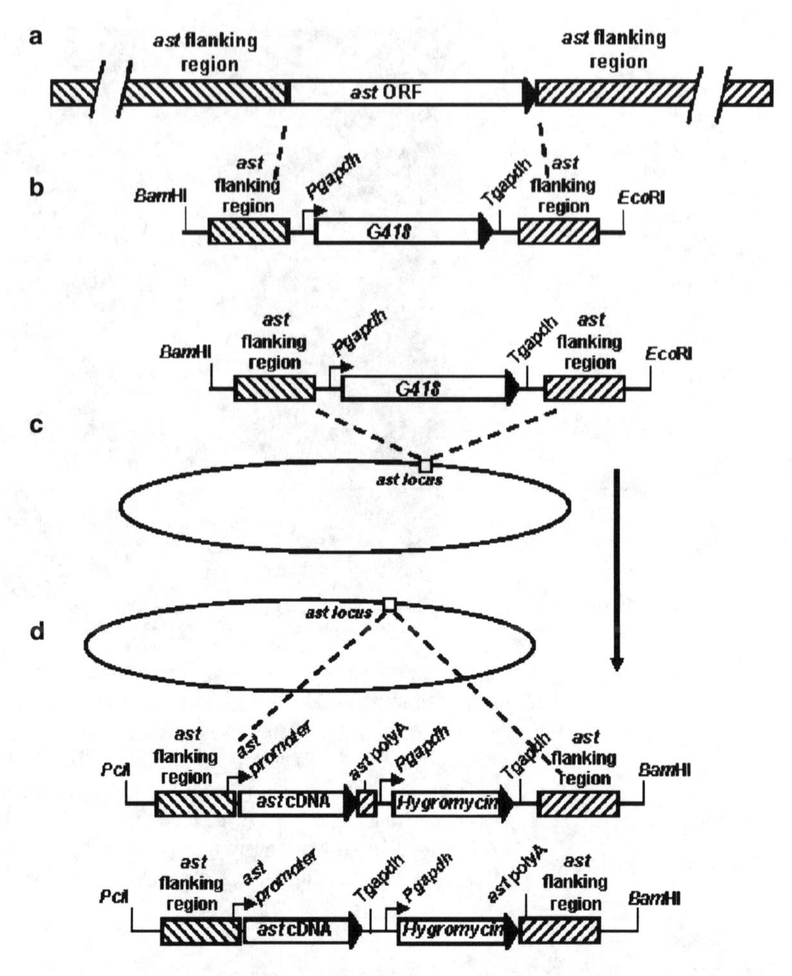

Fig. 2. Strategy for construction of the *ast*-null *X. dendrorhous* CBS 6938 by deletion of the CBS 6938 *ast* gene and its rescue by reinsertion of the wild-type CBS 6938 *ast* cDNA. (**a**) Relative locations and orientations of the *ast* locus in CBS 6938. (**b**) Organization of the transfer vector DNA used to generate the *ast* knockout in *X. dendrorhous* CBS 6938 by recombination. (**c**) The gene organization of the *ast*-null mutant (CBS 6938^{astKO}) is shown. CBS 6938^{astKO} contains the G418 cassette in the *ast* locus and the entire *ast* orf was removed. (**d**) The *ast*-null repair strain CBS 6938$^{astKO\text{-}REP}$ was derived from CBS 6938^{astKO} by insertion of the wild-type *ast* cDNA into the *ast* locus by homologous recombination. G418 expression cassette in the *ast*-null mutant was replaced by *ast* cDNA and the hygromycin gene cassette. The drawing is not to scale.

2. Materials

2.1. Generation of Carotenoid Mutants by Chemical Mutagenesis

1. Antimycin A (Sigma-Aldrich Co., St Louis, MO, USA).
2. Thenoyltrifluoroacetone (TTFA).
3. Sodium cyanide.
4. β-Carotene (Sigma-Aldrich Co., St Louis, MO, USA).

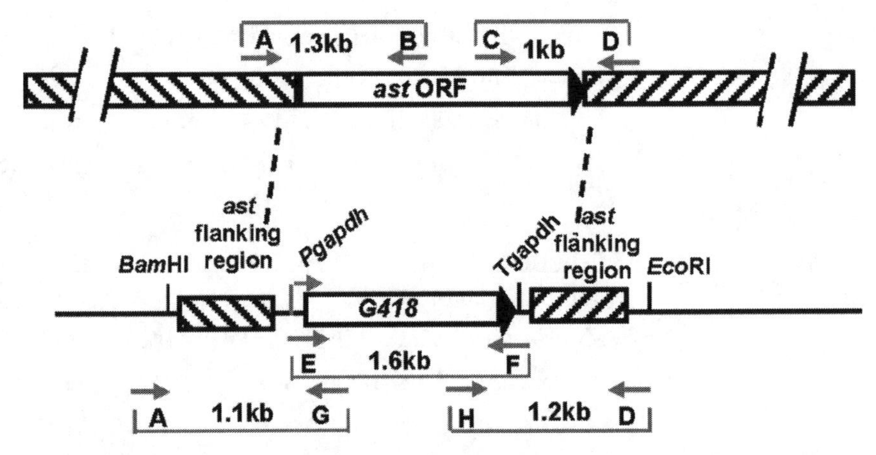

Fig. 3. Confirmation of the gene replacements in CBS 6938^{astKO}. The strategy for PCR analysis of the *ast* locus in CBS 6938^{astKO} is indicated by the positions of primer pairs (*arrows* and *brackets*). The *top* diagram shows the structure of the wild-type (CBS 6938) *ast* locus, and the *lower* diagram shows the structure of CBS 6938^{astKO}. To confirm the deletion of the *ast* gene and insertion of the G418 expression cassette in the *ast* locus CBS 6938^{astKO}, primer pairs A+B, C+D, A+G, and H+D were used to examine the recombination junctions by PCR analysis. The sizes of expected PCR amplification products are shown with each primer pair on the diagram. The drawing is not to scale.

5. *N*-methyl-*N′*-*N*-nitrosoguanidine (NTG) (Sigma-Aldrich Co., St Louis, MO, USA).
6. Ethyl methanesulfonate (EMS) (Sigma-Aldrich Co., St Louis, MO, USA).
7. 2-Deoxyglucose (Sigma-Aldrich Co., St Louis, MO, USA).
8. YM broth: 10 g/L glucose, 3 g/L yeast extract, 3 g/L malt extract, and 5 g/L Bacto peptone. Adjust to pH 5–6. Autoclave at 121°C for 20 min.
9. YM agar: YM and 2% agar.
10. UV light.
11. Hood.
12. Bench centrifuge.
13. Spectrophotometer.

2.2. Growth of *X. dendrorhous*

1. YM: see above.
2. 50 mL tubes.
3. PBS buffer: 0.1 M sodium phosphate buffer, and 0.1 M NaCl, pH 7.4.

2.3. Pigment Extraction and Analysis

1. French Press.
2. Solvent mixture: acetone and hexane (1:2).
3. 10-mL Glass tubes.
4. Aluminum foil.

5. Solvent A: 85% acetonitrile and 15% methanol.
6. Solvent B: 100% dichloromethane.
7. Reverse phase C18 column (250×4.6 mm, 5 μm particle size).
8. Astaxanthin, β-carotene, and lycopene standards (Sigma-Aldrich Co., St Louis, MO, USA).

2.4. Genomic DNA Extraction and Analysis

1. ESC buffer: 60 mM EDTA, 1.2 mM sorbitol, and 0.1 M trisodium citrate, pH 7.0.
2. TE: 50 mM Tris, and 20 mM EDTA, pH 7.5.
3. Glucanex Sigma-Aldrich Co., St Louis, MO, USA.
4. Sodium dodecyl sulfate.
5. Phenol–chloroform–isoamylalcohol (25:24:1).
6. YPD: 10 g/L Bacto yeast extract, 20 g/L Bacto peptone, and 20 g/L dextrose. Autoclave at 121°C for 20 min.

2.5. Construction of Targeted Knockout Mutants

1. *X. dendrorhous*, haploid strain CBS 6938 (ATCC 96594) (ATTC, Manassas, VA, USA).
2. 50% Glycerol (see Note 1).
3. Plasmid pFAST 72 (see Note 2).
4. Restriction enzymes (New England Biolabs, Boston, USA).
5. pUC19 (Invitrogen, Grand Island, NY, USA).
6. TOPO PCR cloning kits (Invitrogen, Grand Island, NY, USA).
7. TOPO Blunt PCR cloning kits (Invitrogen, Carlsbad, CA, USA).
8. TOPO XL PCR cloning kits (Invitrogen, Carlsbad, CA, USA).
9. TOP 10 *Escherichia coli* competent cells (Invitrogen, Carlsbad, CA, USA).
10. YPD medium: 10 g Bacto yeast extract, 20 g Bacto peptone, and 20 g dextrose.
11. STM buffer: 270 mM sucrose, 10 mM Tris–HCl, pH 7.5, and 1 mM $MgCl_2$.
12. Potassium phosphate buffer: 50 mM potassium phosphate buffer, pH 7.0.

3. Methods

3.1. Vegetative Growth

Vegetative cells reproduce by enteroblastic budding, are ellipsoidal, occur singly and in pairs, and occasionally in short chains (Fig. 4). Chlamydospores with refractile granules may be formed.

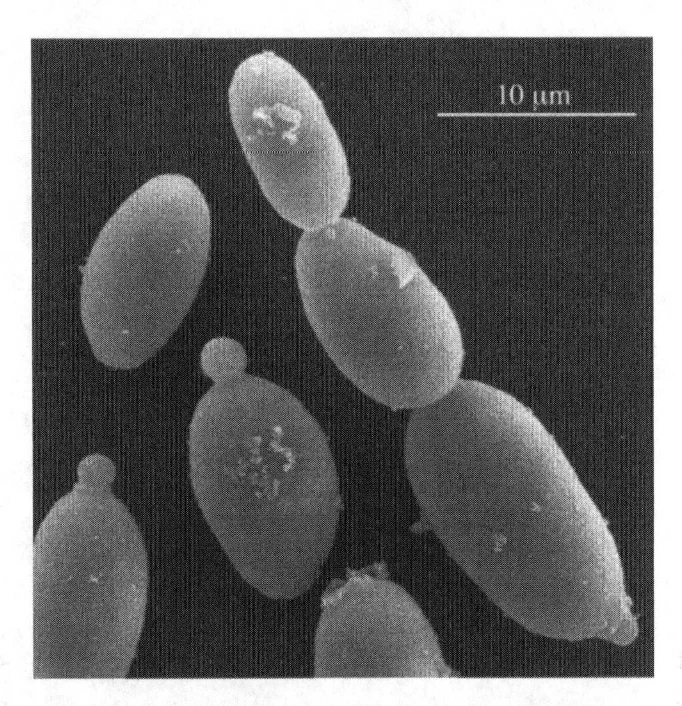

Fig. 4. *Xanthophyllomyces dendrorhous* strain UCD 67-385. Scanning electron micrograph of cells grown for 24 h on YM agar at 20°C. Photograph is by N. Pinel and Jack Fell.

1. Cultivate *X. dendrorhous* and *P. rhodozyma* using the carbon and nitrogen sources described previously (2–4).
2. Grow the cultures at maximum temperature of 25°C (see Note 3).
3. Store stock samples in 50% glycerol at −80°C.

3.2. Formation of Basidiospores

1. Incubate *X. dendrorhous* and *P. rhodozyma* at 18°C on agar media containing polyols (ribitol, D-glucitol, L-arabitol, or D-xylitol), pentoses (D-ribose, D-xylose, or D-arabinose), or D-xylose, except for D-mannitol and L-arabinose.
2. Observe basidiospores formation after 2–3 weeks (see Note 4). Usually 3–4 (up to 6) thin-walled oval or ellipsoidal spores 3–6 × 5–12 μm are produced on the terminal apex of the basidium (Fig. 5). The sessile basidiospores occur on short basidiophores and germinate by budding. Ballistospores are not produced.

3.3. Generation of Carotenoid Mutants by UV Light

1. Inoculate *P. rhodozyma* in 30 mL YM medium and grow the cells until an optical density 0.3–0.4 at 660 nm is obtained.
2. Transfer cells into a sterile Petri dish in a hood.
3. Expose the cells under UV light (220–280 nm) approximately 15–30 min to kill ≥95% of the cells.

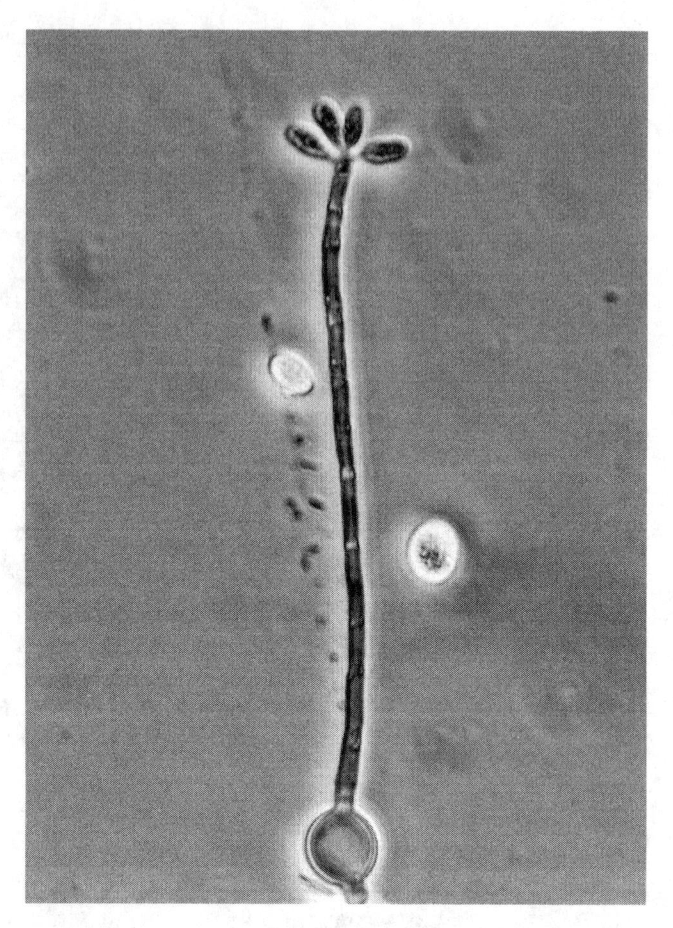

Fig. 5. CBS 9090. Basidium formed by cell/bud mating after 2 weeks at 18°C on polyol media. Terminal basidiospores occur on basidiophores (Photograph by Carlos Echavarri-Erasun).

4. Subculture UV-exposed cells for 18 h in the dark, then plate the cells on selective agar or YM agar for screening for alterations in pigmentation (see Subheading 3.6 and Note 5).

3.4. Generation of Carotenoid Mutants by EMS

1. Spin down 30 mL of *P. rhodozyma* cells to collect the cell pellet.
2. Resuspend the cell pellet with 0.96 mL phosphate buffer pH 7.0 in a 1.5-mL Eppendorf tube.
3. Add EMS to the cells to a final concentration of 4%, mix well by vortexing.
4. Incubate the cells for 10 min at room temperature.
5. Wash cells four times with phosphate buffer pH 7.0, then followed three more washes with water to eliminate EMS.
6. Grow the treated cells overnight in YM broth.
7. Plate the cells on YM plates or selective agar.

3.5. Generation of Carotenoid Mutants by NTG

1. Collect fresh cultured cells by centrifugation and wash cells twice with 5 mL of sodium citrate (0.1 M, pH 5.5).
2. Resuspend the cells to an optical density of 1.5–2 (660 nm) with sodium citrate (0.1 M, pH 5.5).
3. Aliquot 6.7 mL of the yeast suspension and add 0.28 mL of NTG (1 mg/mL in sodium citrate) to a final concentration of 40 μg/mL, mix well by vortexing. Incubate the cell suspension for approximately 20 min at room temperature to kill >95% cells.
4. Aliquot 1 mL treated cells into 1.5-mL sterile Eppendorf tubes. Collect cells by centrifugation. The supernatant containing the mutagen is treated with hydrochloric acid before being discarded (see Note 6).
5. Wash cells three times with 1 mL of 0.1 M potassium phosphate (pH 7.0).
6. Inoculate the treated cells into YM broth for overnight growth before plating on selective medium.

3.6. Selection Procedures for Astaxanthin Production

1. Plate the treated cells to 30–100 YM agar plates containing 50 μM antimycin, seal the Petri dish plates and incubate them at 20–25°C for 1.5–3 months.
2. Pick and inoculate the resistant colonies into YM broth medium, culture for 5 days for analysis of carotenoids.

3.7. *X. dendrorhous* Growth and Pigment Extraction

1. Grow cells in 50 mL YM media at 20°C in dark conditions with shaking for 96 h.
2. Collect cells in 50-mL propylene tubes by centrifugation at 4,000 × *g*, wash cell pellets twice with 15 mL distilled H_2O.
3. Resuspend the cell pellets in 5 mL PBS buffer and incubate the cell suspension in ice for at least 10 min.
4. Lyse the cells using a French Press or by vigorous shaking with glass beads.
5. Extract carotenoids by adding 10 mL of a solvent mixture consisting of acetone and hexane (1:2), followed with shaking the cell suspension at room temperature for 3 min.
6. Separate the organic phase from the aqueous phase by centrifugation at 4,000 × *g* for 6 min.
7. Transfer the upper organic phase to a 10-mL glass tube wrapped with aluminum foil to exclude light.
8. Repeat this process until the pellet is colorless, usually requiring three extractions.
9. Dry the pooled organic phase under nitrogen and stored at –20°C in the dark.

10. Place the same volume of cells in aluminum panel plates and dry them in an oven at 80°C for 24 h. The dry weight is used with the HPLC data to calculate the amount of pigments in μg/g DCW. All experiments are performed in at least duplicate (see Note 7).

3.8. Analysis of Carotenoids

1. During chromatography on reversed phase HPLC Monitor carotenoid peaks using a variable wavelength UV/visible absorbance detector set at 474 nm or at multiple wavelengths with suitable equipment.
2. Use a flow rate of 1 mL/min and run for approximately 20 min in duration. Run a solvent gradient as follows: (a) 100% solvent A for 1 min, (b) linear gradient over 1.5 min to 32% solvent B, (c) isocratic with 32% solvent B for 12 min, (d) decreasing linear gradient of solvent B to 100% solvent A for 30 s, and (e) a final system re-equilibration with 100% solvent A for 5 min.
3. Collect HPLC fractions eluting at ca. 6.8 min (astaxanthin) and ca. 14.8 min (β-carotene) and dry under nitrogen gas.
4. Pool the selected HPLC fractions from 3 to 4 runs to yield enough pigment for later spectroscopic analysis. Dry samples in nitrogen and rechromatographed for further purification (12).
5. Analyze the pigments by dissolution in hexane and spectrophotometric analysis in the 215–600 nm UV/visible range using a spectrophotometer.
6. Use astaxanthin, β-carotene, and lycopene standards as controls in the HPLC and UV/visible scanning methodologies (see Note 7).

3.9. Genomic DNA Extraction and Analysis

1. Inoculate a single *X. dendrorhous* colony in 20 mL of YPD media and grow as described above.
2. Collect cells by centrifuging for 5 min at 5,000 × *g* and wash cells twice with 20 mL of ESC buffer.
3. Resuspend the cell pellet in 3 mL of ESC buffer containing 19 mg of Glucanex per mL and incubate it at 30°C for 3 h to form spheroplasts.
4. Collect the cells by centrifugation at 2,000 × *g* for 5 min and resuspend the cell pellet in 10 mL of TE buffer containing 0.5% sodium dodecyl sulfate.
5. Add 10 mL phenol–chloroform–isoamylalcohol (25:24:1) into the cell suspension to extract DNA. Mix well and centrifuge 5 min at 14,000 × *g* to separate the DNA extraction phase from phenol–chloroform–isoamylalcohol phase. Protein will remain in the interphase.
6. Carefully transfer the upper phase to another clean tube.
7. Repeat the extraction procedures twice.

8. Add 2 volumes of 100% ethanol into the DNA extraction solution and incubate at –20°C for at least 2 h to precipitate the DNA.
9. Collect the DNA by centrifugation and wash the DNA pellet twice with 70% ethanol, then air dry the DNA.
10. Dissolve the DNA with 0.5 mL sterile TE buffer or distill water (see Note 8).
11. Check the DNA quality on agarose gel and DNA concentration using spectrophotometer.

3.10. Electro-transformation of X. dendrorhous

1. Inoculate a colony into 30 mL of YPD medium and incubate the culture at 21°C with shaking for 48 h.
2. Transfer the precultured cells into 100 mL of YPD medium and shake for 18–20 h at 21°C until reaching an OD_{600} of approximately 1.2.
3. Collect cells from 100 mL of culture by centrifugation for 5 min at 5,000 × *g* at room temperature.
4. Resuspend the cell pellet in 25 mL of potassium phosphate buffer containing 25 mM dithiothreitol (DTT) and incubate for 15 min at room temperature.
5. Collect the cell pellets by centrifugation for 5 min at 5,000 × *g* at 4°C and wash the cell pellets twice with 25 mL of ice-cold STM.
6. Resuspend the washed cell pellet in 1 mL of STM buffer and store the cells on ice and used the same day.
7. Mix 1 μg DNA with 100 μl competent cells and then transfer them to an electroporation cuvette.
8. Electroshock the DNA into competence cells using an appropriate electroporation system. EAJ's lab has used the BioRAD GenePulser Xcell Electroporation System.
9. Add 500 μL YPD in the cuvette to dilute the cells (see Note 9) and plate 100 μL/plate cells on YPD agar plates containing 40 mL/mL G418 sulfate.
10. Incubate the plates at 21°C for 4 days.
11. Select colonies growing on the antibiotic plates based on their color.

3.11. Generation of the ast Knockout Transfer Vector

See Table 1 for the primer sequences used in this procedure.

1. PCR amplify, an approximately 1.5 kb G418 cassette which contains the G418 gene, a *gapdh* promoter and terminator from the plasmid pFAST 725, and *Kpn*I and *Sac*I restriction sites are added to the 5′ and 3′ ends, respectively. The primers for the amplification of the G418 cassette are the"G418 cassette 5′ forward primer" and the "G418 cassette 3′ reverse primer."

Table 1
Primers used in the generation of the transfer vectors and verification of the mutant

Primers names	Primers sequences
G418 cassette 5′ forward primer	5′-GGTACCTGGTGGGTGCATGTATGTACGTG-3′ *Kpn*I
G418 cassette 3′ reverse primer	5′-GAGCTCCTTGATCAGATAAAGATAGAGAT-3′ *Sac*I
5′ Flanking region forward primer	5′-GGATCCGACCACGTTGTGTAACCAGTTCGATTG-3′ *Bam*HI
5′ Flanking region reverse primer	5′-GGTACCATGGAGAAAGTAGGTGGCAAGAATCGG-3′ *Kpn*I
3′ Flanking region forward primer	5′-GAGCTCGTTGATTCTTCATATGTTAAGAGAAG-3′ *Sac*I
3′ Flanking region reverse primer	5′-GAATTCCTGCAGTCGACAAACATGAAGATCAC-3′ *Eco*RI
5′ Location checking primer	5′-CGAACAATCCTGGCGCGCCTGGAGGAGC-3′
G418 inside 3′ primer	5′-GGAACACGGCGGCATCAGAGCAGCCG-3′
3′ flanking region checking primer	5′-GAAGATAAAGTTGTCTGGATGGAGTTCC-3′
G418 inside 5′ primer	5′-GGTCTTGTCGATCAGGATGATCTGG-3′
Partial *ast* reverse primer	5′-CTTCGATGTCGCTTATGAGCTTCACC-3′
Partial *ast* forward primer	5′-TCTGATAGACGCGGACCAATTCAGCG-3′
G418 forward primer	5′-ATGATTGAACAAGATGGATTGCACGCAGGTTC-3′
G418 reverse primer	5′-CAGAAGAACTCGTCAAGAAGGCGATAGAAGGCG-3′

2. Clone the G418 expression cassette fragment into the TOPO XL PCR cloning vector.
3. Excise the G418 cassette fragment by *Kpn*I and *Sac*I double digestion.
4. Clone the fragment into the plasmid pUC19 which has been previously digested with *Kpn*I and *Sac*I to generate the plasmid *pUC-G418*.
5. PCR amplify, an approximately 300 bp *ast* 5′ flanking region from the plasmid pFAST725, and *Bam*HI and *Kpn*I restriction sites are added to the 5′ and 3′ ends, respectively. The primers for the amplification of the *ast* 5′ flanking region are the "5′ flanking region forward primer" and the "5′ flanking region reverse primer."
6. Clone the PCR product (*ast* 5′ flanking region) into the TOPO Blunt PCR plasmid.

7. Excise the *ast* 5′ flanking region fragment by *Bam*HI and *Kpn*I double digestion.
8. Ligate the *ast* 5′ flanking region fragment with the plasmid *pUC-G418*, which is previously digested with *Bam*HI and *Kpn*I to generate the plasmid *pUC-G418-ast 5′ flanking*.
9. PCR amplify, an approximately 280 bp *ast* 3′ flanking region from the plasmid pFAST 725, and *Sac*I and *Eco*RI restriction sites are added to the 5′ and 3′ ends, respectively. The primers for the amplification of the *ast* 3′ flanking region are the "3′ flanking region forward primer" and the "3′ flanking region reverse primer."
10. Clone the PCR product (*ast* 3′ flanking region) into the TOPO Blunt PCR vector.
11. Excise the *ast* 3′ flanking region fragment by *Sac*I and *Eco*RI double digestion.
12. Ligate the *ast* 3′ flanking region fragment with the plasmid *pUC-G418-ast 5′ flanking* which is previously digested with *Sac*I and *Eco*RI to generate the final plasmid *pUC-G418-ast 5′-3′ flanking*.
13. Verify all inserts by restriction enzyme double digestion and sequencing (see Note 10).

3.12. Generation of an ast Knockout in X. dendrorhous by Double Cross Homologous Recombination

1. Double digest the transfer vector *pUC-G418-ast 5′-3′ flanking* by *Bam*HI and *Eco*RI.
2. Purify the linearized 2.67 kb fragment containing the G418 expression cassette plus the *ast* 5′ and 3′ flanking regions.
3. Transform at least 1 μg of the fragment into *X. dendrorhous* CBS 6938 competence cells by electroporation using a BioRAD GenePulser Xcell Electroporation System (see Subheading 3.10).
4. Plate the cell suspension (100 μL/plate) on YPD agar plates containing 40 μg/mL G418 sulfate and incubate the plates at 21°C for 4 days.
5. Select colonies growing on the G418 plates based on color to perform further verification.

3.13. Verification of the ast Knockout

PCR analysis is used to confirm the insertion of the G418 expression cassette and absence of the *ast* gene in *X. dendrorhous* CBS 6938 (Fig. 3).

1. The method to extract the genomic DNA from knockout mutants described in Subheading 3.9.
2. Verify the insertion of the G418 expression cassette using primer E (G418 5′ forward primer) and F (G418 3′ reverse primer) to amplify the entire G418 expression cassette.

3. Verify that the G418 expression cassette is inserted into the original location and replaces the *ast* gene using a pair of primers A and G (5′ location checking prime), which is based on the genome sequence located upstream and outside of the *ast* 5′ flanking region and the G418 inside 3′ primer that coincided with a sequence within the G418 gene. The second pair of primers H and D consisted of a 3′ flanking region checking primer that coincided with the genome sequence locate downstream and outside of the *ast* 3′ flanking region and the G418 inside 5′ primer that coincided with a sequence within the G418 gene.
4. Verify the removal of the *ast* gene; two pairs of primers (A+B and C+D) are used including a 5′ location checking primer and partial *ast* reverse primer which coincided with a sequence within the *ast* gene. The second pair of primers consisted of the 3′ flanking region checking primer and the partial *ast* forward primer which coincided with a sequence within the *ast* gene. All the PCR products are verified by sequencing.

3.14. Generation of an ast Repair Strains in X. dendrorhous by Double Crossover Homologous Recombination

The strategy to generate the *ast* repair strains is similar to that of the knockout (see Subheading 3.12) (Fig. 2). A different antibiotic selection cassette other than G418 needs to be chosen for selection, e.g., hygromycin.

4. Notes

1. Stock samples are stored in 50% glycerol at –80°C.
2. Plasmid pFAST 72, which contains *ast* cDNA and G418 expression cassette, and a *gapdh* promoter and terminator from *X. dendrorhous* was kindly provided by NIPPON ROCHE KK.
3. Yeast cells will not grow above 25°C, recommended grow temperature is between 20 and 22°C.
4. Following conjugation between a cell and its bud, a slender cylindrical holobasidium forms, which is generally 2–3 μm in diameter and 30–165 μm (usually 70–80 μm) in length (Fig. 4). Other morphological forms of basidia and basidiospores have also been observed. Rarely does conjugation occur between independent cells.
5. UV light could damage skin and eyes, wear appropriate personal protection equipment like gloves and goggle when UV light is used.

6. NTG is a mutagen. It needs to be treated with hydrochloric acid before being discarded.
7. Solvents used in HPLC are hazardous. Wear appropriate personal protection equipment including lab coat, gloves, and goggles; also handle the solvents in a chemical hood.
8. Genomic DNA is very fragile, do not vortex the DNA which will break it. Carefully mix the DNA by turning tube up and down using hand.
9. Transformation cuvette and competence cells need to be put in ice at least 10 min before transformation.
10. Ethidium bromide is highly mutagenic. Wear double gloves and lab coat whenever handing it.

Acknowledgment

This research was conducted by diligent students and scientists in our laboratories.

References

1. Phaff HJ, Miller MW, Yoneyama M, Soneda M (1972) A comparative study of the yeast florae associated with trees on the Japanese islands and on the west coast of North America. In: Terui G (ed) Fermentation technology today: proceedings of fourth international fermentation symposium. Society of Fermentation Technology, Osaka, pp 759–774
2. Miller MW, Yoneyama M, Soneda M (1976) *Phaffia*, a new yeast species in the in the Deuteromycotina (Blastomycetes). Int J Syst Bacteriol 26:88–91
3. Fell JW, Johnson EA (2011) *Phaffia* M. W. Miller, Yoneyama & Soneda (1976). In: Kurtzman CP, Fell JW, Boekhout T (ed) The yeasts. A taxonomic study, 5th edn. Elsevier, Amsterdam, p. 1853–1855
4. Fell JW, Johnson EA, Scorzetti G (2011) *Xanthophyllomyces* Golubev (1995). In: Kurtzman CP, Fell JW, Boekhout T (eds) The yeasts. A taxonomic study, 5th edn. Elsevier, Amsterdam, pp 1555–1595
5. Andrewes AG, Phaff HJ, Starr MP (1976) Carotenoids of *Phaffia rhodozyma*, a red-pigmented fermenting yeast. Phytochemistry 15:1003–1007
6. Johnson EA (2003) *Phaffia rhodozyma*: a colorful odyssey. Int Microbiol 6:169–174
7. Johnson EA, Schroeder WA (1995) Microbial carotenoids. Adv Biochem Eng Biotechnol 53:119–178
8. Golubev WI (1995) Perfect state of *Rhodomyces dendrorhous*. Yeast 11:101–110
9. Retamales P, Hermosilla G, Leon R, Martínez C, Jimenez A, Cifuentes V (2002) Development of the sexual reproductive cycle of *Xanthophyllomyces dendrorhous*. J Microbiol Methods 48:87–93
10. An G-H, Schuman DB, Johnson EA (1989) Isolation of *Phaffia rhodozyma* mutants with increased astaxanthin content. Appl Environ Microbiol 55:116–124
11. An GH, Bielich J, Auerbach R, Johnson EA (1991) Isolation and characterization of carotenoid hyperproducing mutants of yeast by flow cytometry and cell sorting. Biotechnology (NY) 9:70–73
12. Echavarri-Erasun C, Johnson E (2002) Fungal carotenoids. In: Khachatourians G, Arora D (eds) Applied mycology and biotechnology. Elsevier Science, Amsterdam, pp 45–85

Chapter 17

DNA Assembler Method for Construction of Zeaxanthin-Producing Strains of *Saccharomyces cerevisiae*

Zengyi Shao, Yunzi Luo, and Huimin Zhao

Abstract

DNA assembler enables design and rapid construction of biochemical pathways in a one-step fashion by exploitation of the in vivo homologous recombination mechanism in *Saccharomyces cerevisiae*. It has many applications in pathway engineering, metabolic engineering, combinatorial biology, and synthetic biology. Here we use the zeaxanthin biosynthetic pathway as an example to describe the key steps in the construction of pathways containing multiple genes using the DNA assembler approach. Methods for the construction of the clones, *S. cerevisiae* transformation, and zeaxanthin production and detection are shown.

Key words: DNA assembler, In vivo homologous recombination, Pathway engineering, Synthetic biology, Metabolic engineering, Zeaxanthin biosynthesis

1. Introduction

Methods that enable design and rapid construction of biochemical pathways will be invaluable in pathway engineering, metabolic engineering, combinatorial biology, and synthetic biology (1–6). In all these studies, the conventional multistep, sequential cloning method, including primer design, PCR amplification, restriction digestion, in vitro ligation and transformation, is typically involved and multiple plasmids are often required (7, 8). This method is not only time-consuming and inefficient but also relies on unique restriction sites that become limited for large recombinant DNA molecules.

Thanks to its high efficiency and ease to work with; in vivo homologous recombination in yeast has been widely used for gene cloning, plasmid construction, and library creation (9–12). Recently, we developed a new method, called "DNA assembler," which enables design and rapid construction of large biochemical pathways in a one-step fashion by exploitation of the in vivo

José-Luis Barredo (ed.), *Microbial Carotenoids From Fungi: Methods and Protocols*, Methods in Molecular Biology, vol. 898,
DOI 10.1007/978-1-61779-918-1_17,

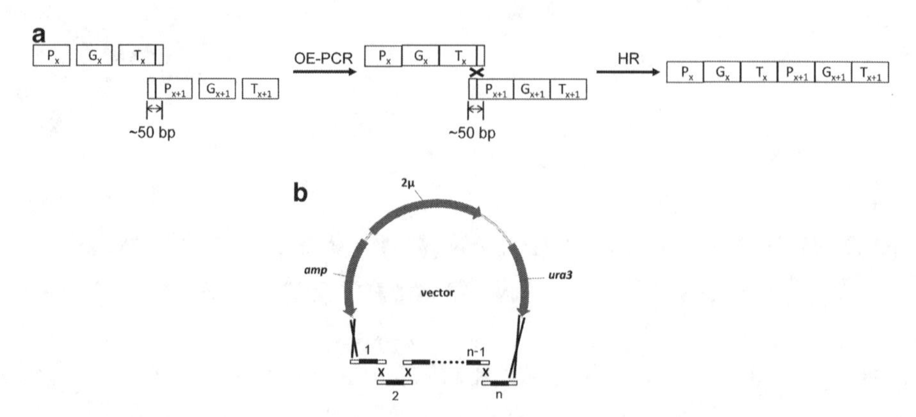

Fig. 1. (**a**) Preparation of each gene expression cassette using OE-PCR. Promoters (P_x, P_{x+1}), genes (G_x, G_{x+1}), and terminators (T_x, T_{x+1}) are individually PCR-amplified and joined together through OE-PCR. The resulting two cassettes are fused through the in vivo homologous recombination (HR) process. To generate an overlap of approximately 50 bp, the reverse primer used to amplify T_x contains a sequence of the first 20–25 nucleotides of P_{x+1}, and the forward primer used to amplify P_{x+1} contains a sequence of the last 20–25 nucleotides of T_x. (**b**) One-step method for assembly of a biochemical pathway using in vivo homologous recombination in *S. cerevisiae.*

homologous recombination mechanism in *S. cerevisiae* (13). This method is highly efficient and circumvents the potential problems associated with the conventional cloning methods, representing a versatile approach for the construction of biochemical pathways for synthetic biology, metabolic engineering, and functional genomics studies. Here we use the zeaxanthin biosynthetic pathway as an example to illustrate the experimental procedures.

As shown in Fig. 1, for each individual gene in the zeaxanthin pathway, an expression cassette including a promoter, a structural gene, and a terminator is PCR amplified and assembled using overlap extension PCR (OE-PCR) (14). The 5′ end of the first gene expression cassette is designed to overlap with a vector, while the 3′ end is designed to overlap with the second cassette. Each successive cassette is designed to overlap with the two flanking ones, and the 3′ end of the last cassette overlaps with the vector. All overlaps are designed to be at least 50 bp for efficient in vivo homologous recombination (see Note 1). The resulting multiple expression cassettes are cotransformed into *S. cerevisiae* with the linearized vector through electroporation, which allows the entire pathway to be assembled into a vector. Restriction digestion is subsequently used to verify the correctly assembled pathway, after which the cells carrying the correct construct are checked for zeaxanthin production.

2. Materials

Prepare all solutions using ultrapure water, prepared by purifying deionized water to attain a sensitivity of 18.2 mΩ cm at 25°C. Prepare and store all reagents at room temperature unless indicated otherwise.

2.1. DNA Preparation

1. pRS416 (New England Biolabs, Beverly, MA, USA).
2. pRS416m: pRS416 with a *hisG* sequence and a *Delta2* (15) sequence that flank the multiple cloning site and serves as the vector for assembly of the zeaxanthin pathway (Fig. 2a) (see Note 2).
3. pCAR-ΔCrtX: It contains the genes *crtE*, *crtB*, *crtI*, *crtY*, and *crtZ* from *Erwinia uredovora* for zeaxanthin biosynthesis (Prof. E.T. Wurtzel, City University of New York, NY, USA) (16–18).
4. 0.5 M Ethylenediaminetetraacetic acid (EDTA) solution pH 8.0: For a 500 mL of stock solution of 0.5 M EDTA, weigh out 93.05 g of EDTA disodium salt (MW = 372.2) and dissolve it in 400 mL of deionized water. Adjust to pH 8.0 with NaOH and correct the final volume to 500 mL. EDTA will not dissolve completely in water unless the pH is adjusted to about 8.0.
5. Concentrated stock solution of TAE (50×): Weigh 242 g of Tris base (MW = 121.14) and dissolve it in approximately 750 mL of deionized water. Carefully add 57.1 mL of glacial

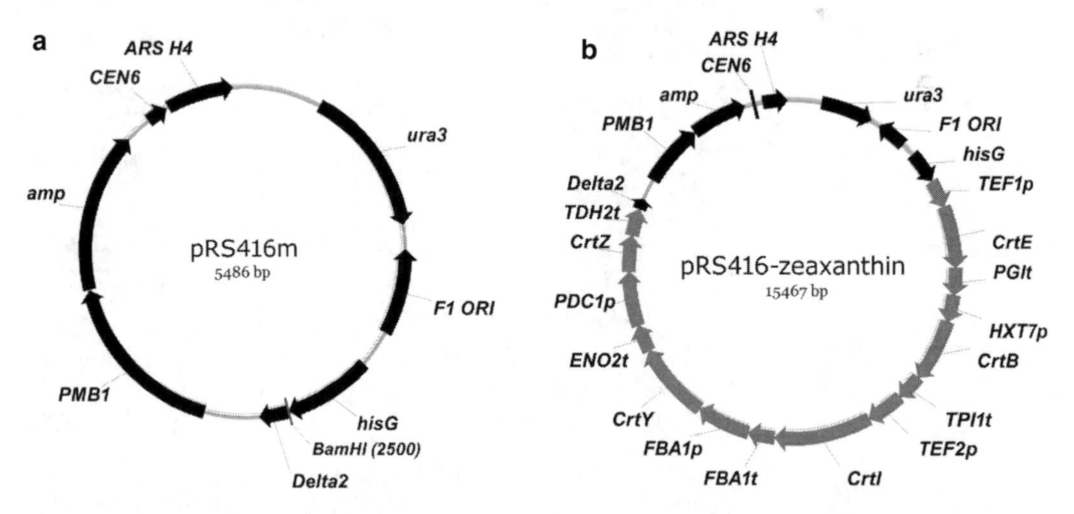

Fig. 2. (**a**) The vector map of pRS416m. (**b**) The vector map of the construct pRS416m-zeaxanthin. *CEN6*: centromere, *ARS H4*: automatic replication sequence, *ura3*: selection marker in *S. cerevisiae*, *amp*: selection marker in *Escherichia coli*, *hisG* and *Delta2*: two regions flanking the zeaxanthin biosynthetic pathway, *PMB1*: *E. coli* origin of replication, *F1 ORI*: this region is contained in the original pRS416 vector, but is not required for the construction of the plasmid containing the zeaxanthin biosynthetic pathway.

acid and 100 mL of 0.5 M EDTA, and adjust the solution to a final volume of 1 L. This stock solution can be stored at room temperature. The pH of this buffer is not adjusted and should be about 8.5.

6. Working solution of TAE buffer (1×): Dilute the stock solution by 50-fold with deionized water. Final solute concentrations are 40 mM Tris acetate and 1 mM EDTA.
7. 0.7% Agarose gel in 1× TAE buffer: Add 0.7 g of agarose into 100 mL of 1× TAE buffer and microwave until agarose is completely melted. Cool the solution to approximately 70–80°C. Add 5 μL of ethidium bromide into the solution and mix well. Pour 25–30 mL of solution onto an agarose gel rack with appropriate 2- or 8-well combs.
8. QIAquick Gel Extraction Kit (QIAGEN, Valencia, CA, USA).
9. QIAprep Miniprep Kit (QIAGEN, Valencia, CA, USA).
10. DNA polymerase: Any polymerase with high fidelity can be used.
11. Failsafe PCR 2× PreMix G containing dNTP and PCR reaction buffer (EPICENTRE Biotechnologies, Madison, WI, USA).
12. *Bam*HI restriction enzyme.
13. 3 M Sodium acetate pH 5.0: Weigh 12.3 g of sodium acetate (MW = 82.03) and dissolve it in 50 mL of deionized water. Adjust to pH 5.0 by HCl.
14. 10 mg/mL Glycogen: Dissolve 10 mg of glycogen in 1 mL of deionized water.
15. NanoDrop2000c: Used to measure the concentration of DNA and check the OD_{600} of the cells (Thermo Scientific, Wilmington, DE, USA).
16. Benchtop centrifuges to separate cells and supernatant.
17. Molecular imager gel doc: Used to check DNA on the agarose gel (Bio-Rad, Hercules, CA, USA).

2.2. Transformation

1. *S. cerevisiae* YSG50 (*MAT*α, *ade*2-1, *ade*3Δ22, *ura*3-1, *his*3-11, 15, *trp*1-1, *leu*2-3, 112, and *can*1-100): Used as the host for DNA assembly (see Note 3).
2. YPAD medium: Dissolve 6 g of yeast extract, 12 g of peptone, 12 g of dextrose, and 60 mg of adenine hemisulphate in 600 mL of deionized water. Autoclave at 121°C for 15 min.
3. Synthetic complete dropout medium lacking uracil (SC-Ura): Dissolve 3 g of ammonium sulfate, 1 g of yeast nitrogen source without ammonium sulfate and amino acids, 0.5 g of complete synthetic medium minus uracil (CSM-Ura; MP Biomedicals, Solon, OH, USA), 26 mg of adenine hemisulphate, and 12 g

of dextrose in 600 mL of deionized water, and adjust pH to 5.6 by NaOH. Autoclave at 121°C for 15 min.

4. SC-Ura–agar: SC-Ura and 20 g/L g of agar.
5. 1 M Sorbitol solution: Dissolve 91.1 g of sorbitol (MW = 182.17) in 400 mL of deionized water and adjust to a final volume of 500 mL. Sterilize the solution by filtering it through a filter with a pore size of 0.22 μm.
6. Gene pulser II and pulse controller plus: Used to transform plasmids into *S. cerevisiae* and *E. coli* through electroporation (Bio-Rad, Hercules, CA, USA).

2.3. Verification of the Clones

1. Zymoprep II yeast plasmid miniprep (Zymo Research, Orange, CA, USA).
2. *Sca*I restriction enzyme (see Note 4).
3. *Psi*I restriction enzyme (see Note 4).
4. 1 M Glucose solution: Dissolve 90 g of D-glucose in 400 mL of deionized water and adjust to a final volume of 500 mL. Filter-sterilize it.
5. SOC: Add 20 g of Bacto-tryptone, 5 g of yeast extract, 0.5 g of NaCl, 186.4 mg of KCl into 980 mL of deionized water. Adjust pH to 7.0 with NaOH. Autoclave at 121°C for 15 min. After the solution cools down to 70–80°C, add 20 mL of sterile 1 M glucose.
6. 100 mg/mL Ampicillin stock solution: Dissolve 1 g of ampicillin powder in 10 mL of deionized water and filter-sterilize it.
7. LB broth: Add 10 g of Bacto-tryptone, 5 g of yeast extract, 10 g of NaCl into 1 L of deionized water. Autoclave at 121°C for 15 min.
8. LB agar: LB and 20 g/L agar.
9. LB-Amp^{+} agar plates: Autoclave LB agar and when the solution cools down to 70–80°C, add 1 mL of 100 mg/mL ampicillin to 1 L of LB agar. Pour 20–25 mL into each Petri dish.

2.4. Detection of Zeaxanthin

1. 0.1% Trifluoroacetic acid (TFA) buffer: Add 1 mL of neat TFA into 999 mL of deionized water.
2. 10 μg/mL Zeaxanthin standard solution: Dissolve 1 mg of zeaxanthin (Sigma, St Louis, MO, USA) in 100 mL of methanol.
3. French pressure cell press: Used to lyse the yeast cells.
4. Rotavapor R-205: Used to evaporate the solvent.
5. High performance liquid chromatography (HPLC) equipment.
6. ZORBAX SB-C18 column (Agilent Technologies, Palo Alto, CA, USA).

3. Methods

3.1. DNA Preparation

1. Amplify the genes *crtE*, *crtB*, *crtI*, *crtY*, and *crtZ* from the plasmid pCAR-ΔCrtX and amplify the corresponding promoter and terminator from the genomic DNA of *S. cerevisiae* using the primers listed in Table 1. Set up the reaction mixtures as

Table 1
The primers used in assembling the zeaxanthin pathway in pRS416m

Name	Sequence
hisG-f	GGCCAGTGAGCGCGCGTAATACGACTCACTATAGGCGCGCCTGCGT GAAGTCGAAG
hisG-r	GGAGTAGAAACATTTTGAAGCTATTTCCAGTCAATCAGGGTATTG
TEF1p-f	CTTCAATACCCTGATTGACTGGAAATAGCTTCAAAATGTTTCTACTC
TEF1p-r	GAACGTGTTTTTTTGCGCAGACCGTCATTTTGTAATTAAAACTTAGATTAG
CrtE-f	CTAATCTAAGTTTTAATTACAAAATGACGGTCTGCGCAAAAAAACACGTTC
CrtE-r	GGTATATATTTAAGAGCGATTTGTTTAACTGACGGCAGCG
PGIt-f	CGCTGCCGTCAGTTAAACAAATCGCTCTTAAATATATACC
PGIt-r	CCGAAATTGTTCCTACGAGAAGTGGTATACTGGAGGCTTCATGAGTTATG
HXT7p-f	CATAACTCATGAAGCCTCCAGTATACCACTTCTCGTAGGAACAATTTCGG
HXT7p-r	GAGTAACGACGGATTATTCATTTTTTGATTAAAATTAAAAAAAC
CrtB-f	GTTTTTTTAATTTTAATCAAAAAATGAATAATCCGTCGTTACTC
CrtB-r	GATAATATTTTTATATAATTATATTAATCCTAGAGCGGGCGCTGCCAGAG
TPI1t-f	CTCTGGCAGCGCCCGCTCTAGGATTAATATAATTATATAAAAATATTATC
TPI1t-r	CTATATGTAAGTATACGGCCCTATATAACAGTTGAAATTTGG
TEF2p-f	CCAAATTTCAACTGTTATATAGGGCCGTATACTTACATATAG
TEF2p-r	CACCAATTACCGTAGTTGGTTTCATGTTTAGTTAATTATAGTTCGTTG
CrtI-f	CAACGAACTATAATTAACTAAACATGAAACCAACTACGGTAATTGGTG
CrtI-r	CTCATTAAAAAACTATATCAATTAATTTGAATTAACTCATATCAGATCC TCCAGCATC
FBA1t-f	GATGCTGGAGGATCTGATATGAGTTAATTCAAATTAATTGATATAGTTTTT TAATGAG
FBA1t-r	GTTCAAGCCAGCGGTGCCAGTTGGAGTAAGCTACTATGAAAGACTTTAC
FBA1p-f	GTAAAGTCTTTCATAGTAGCTTACTCCAACTGGCACCGCTGGCTTGAAC
FBA1p-r	CAGATCATAATGCGGTTGCATTTTGAATATGTATTACTTGGTTATGG
CrtY-f	CCATAACCAAGTAATACATATTCAAAATGCAACCGCATTATGATCTG

(continued)

Table 1 (continued)

Name	Sequence
CrtY-r	CTAATAATTCTTAGTTAAAAGCACTTTAACGATGAGTCGTCATAATGG
ENO2t-f	CCATTATGACGACTCATCGTTAAAGTGCTTTTAACTAAGAATTATTAG
ENO2t-r	GGAACATATGCTCACCCAGTCGCATGAGGTATCATCTCCATCTCCCATATG
PDC1p-f	CATATGGGAGATGGAGATGATACCTCATGCGACTGGGTGAGCATATGTTCC
PDC1p-r	GGCATTCCAAATCCACAACATTTTGATTGATTTGACTGTG
CrtZ-f	CACAGTCAAATCAATCAAAATGTTGTGGATTTGGAATGCC
CrtZ-r	CATTAAAGTAACTTAAGGAGTTAAATTTACTTCCCGGATGCGGGCTC
TDH2t-f	GAGCCCGCATCCGGGAAGTAAATTTAACTCCTTAAGTTACTTTAATG
TDH2t-r	GATCCGTTAGACGTTTCAGCTTCCAGCGAAAAGCCAATTAGTGTGATAC
Delta2-f	GTATCACACTAATTGGCTTTTCGCTGGAAGCTGAAACGTCTAACGGATC
Delta2-r	TTACGCCAAGCGCGCAATTAACCCTCACTAAAGGCGCGCCGAGA ACTTCTAGTATATTC

follows: 50 μL of FailSafe PCR 2× PreMix G, 2.5 μL of forward primer (20 pmol/μL), 2.5 μL of reverse primer (20 pmol/μL), 1 μL of template (10–50 ng of *S. cerevisiae* genomic DNA or the plasmid pCAR-ΔCrtX), 1 μL of DNA polymerase, and 43 μL of ddH_2O in a total volume of 100 μL.

2. PCR condition: Fully denature at 98°C for 30 s, followed by 25 cycles of 98°C for 10 s, 58°C for 30 s, and 72°C for 1 min, with a final extension at 72°C for 10 min.
3. Load the 100 μL of PCR products onto 0.7% agarose gels and perform electrophoresis at 120 V for 20 min.
4. Gel-purify PCR products using the QIAquick Gel Extraction Kit.
5. Check the concentrations of the purified products using NanoDrop.
6. Perform OE-PCR to generate each gene expression cassette (see Note 5). Set up the first-step reaction mixture as follows: 10 μL of FailSafe PCR 2× PreMix G, 100 ng of promoter fragment, 100 ng of *crt* gene fragment, 100 ng of terminator fragment, and 0.2 μL of DNA polymerase. Add ddH_2O to a final volume of 20 μL.
7. Reaction condition: Fully denature at 98°C for 30 s, followed by 10 cycles of 98°C for 10 s, 58°C for 30 s, and 72°C for 1 min, with a final extension at 72°C for 10 min.

8. Set up the second-step reaction mixture as follows: 50 μL of FailSafe PCR 2× PreMix G, 10 μL of first-step reaction mixture, 1 μL of DNA polymerase, 2.5 μL of forward primer (20 pmol/μL), 2.5 μL of reverse primer (20 pmol/μL), and 34 μL ddH_2O in a total volume of 100 μL.
9. Reaction condition: Fully denature at 98°C for 30 s, followed by 25 cycles of 98°C for 10 s, 58°C for 30 s and 72°C for 1 min, with a final extension at 72°C for 10 min.
10. Digest pRS416m by *Bam*HI at 37°C for 3 h. Digestion condition: 5 μL of 10× buffer, 0.5 μL of 100× BSA, 3 μg of pRS416m, and 30 U of *Bam*HI. Add ddH_2O to a final volume of 50 μL.
11. Load the PCR and digestion products onto 0.7% agarose gels and perform electrophoresis at 120 V for 20–30 min.
12. Gel-purify the PCR and digestion products using the QIAquick Gel Extraction Kit.
13. Check the concentrations of the purified products using NanoDrop.
14. Take 200–300 ng of each fragment, mix in a tube, and calculate the final volume.
15. Add 10% v/v 3 M sodium acetate and 2% v/v 10 mg/mL glycogen (e.g., if you have 100 μL of mixture, add 10 μL of sodium acetate and 2 μL of glycogen) and mix well.
16. Add 2× v/v 100% ethanol (e.g., if the final volume is about 110 μL, add 220 μL ethanol) and mix well.
17. Store the DNA mixture at −80°C for at least an hour.
18. Centrifuge at 4°C, 16,100 × *g* for 20 min. Usually the precipitated DNA can be seen at the bottom of the tube.
19. Remove the supernatant completely (do not touch the DNA).
20. Add 500 μL of 70% ethanol to wash the DNA pellet and centrifuge at room temperature, 16,100 × *g* for 3 min.
21. Remove the ethanol completely and air dry the pellet for 1–2 min (do not overdry it).
22. Resuspend the DNA pellet by 4 μL of ddH_2O. Now the DNA is ready for transformation (see Note 6).

3.2. Transformation

1. Inoculate a single colony of YSG50 into 3 mL of YPAD medium and grow overnight in a shaker at 30°C and 250 rpm.
2. Measure the OD_{600} of the seed culture and inoculate the appropriate amount to 50 mL of fresh YPAD medium to obtain an OD_{600} of 0.2 (e.g., if the overnight culture has an OD_{600} of 10, then add 1 mL into 50 mL of fresh YPAD medium).

3. Continue growing the 50 mL of culture for approximately 4 h to obtain an OD_{600} of 0.8 (see Note 7).
4. Spin down the yeast cells at 4°C, 3,220 × *g* for 10 min and remove the spent medium.
5. Use 50 mL of ice-cold ddH_2O to wash the cells once and centrifuge again.
6. Discard water, add 1 mL of ice-cold ddH_2O to resuspend the cells, and move them to a sterile Eppendorf tube.
7. Spin down the cells using a bench top centrifuge for 30 s at 4°C, 4,500 × *g*.
8. Remove water and use 1 mL of 1 M ice-cold sorbitol to wash the cells once (now the cells look slightly yellow). Centrifuge again and remove the sorbitol.
9. Resuspend the cells in 250–300 μL of chilled 1 M sorbitol and distribute them into 50 μL of aliquots.
10. Now each 50 μL of cells is ready for electroporation (see Note 8). Mix the 4 μL of DNA with 50 μL of yeast cells and put the mixture into a chilled electroporation cuvette.
11. Electroporate the cells at 1.5 kV, and quickly add 1 mL of pre-warmed (30°C) YPAD medium to resuspend cells (see Note 9).
12. Grow in a shaker at 30°C, 250 rpm for 1 h.
13. Spin down the cells in a sterile tube at 16,100 × *g* for 30 s and remove the YPAD medium.
14. Use 1 mL of room temperature sorbitol solution to wash the cells two to three times and finally resuspend the cells in 1 mL sorbitol.
15. Spread 100 μL of resuspended cells onto SC-Ura plates.
16. Incubate the plates at 30°C for 2–3 days until colonies appear.

3.3. Verification of the Correctly Assembled Zeaxanthin Pathway

1. Randomly pick ten colonies from the SC-Ura plate and inoculate each colony into 1.5 mL of SC-Ura liquid medium. Grow at 30°C for 1.5 days (see Note 10).
2. Purify yeast plasmid DNA from each 1.5 mL of culture using the Zymoprep II kit.
3. Mix 2 μL of isolated plasmid with 50 μL of *E. coli* BW25141 cells and put the mixture into a chilled electroporation cuvette (see Note 11).
4. Electroporate the cells at 2.5 kV, and quickly add 1 mL of SOC medium to resuspend cells (see Note 9).
5. Grow in a shaker at 37°C, 250 rpm for 1 h.
6. Spin down the cells, remove 800 μL of SOC medium, resuspend the pellet with the remaining 200 μL of SOC medium, and spread the cells on a LB-Amp^+ plate.

7. Incubate the plates at 37°C for 16–18 h until colonies appear (see Note 12).
8. Inoculate a single colony from each plate to 5 mL of LB supplemented with 100 μg/mL ampicillin and grow at 37°C for 12–16 h.
9. Purify *E. coli* plasmids from each 5 mL of culture using the QIAgen Miniprep kit.
10. Check the plasmid concentrations by NanoDrop.
11. Verify the correctly assembled pathway through two separate restriction digestion reactions (see Note 4).
12. Digestion condition by *Sca*I at 37°C for 3 h: 1.5 μL of 10× buffer, 0.15 μL of 100× BSA, 300 ng of plasmid, and 5 U of *Sca*I. Add ddH_2O to a final volume of 15 μL. Expected bands: 1,750, 2,131, 2,628, 3,223, 5,735 bp.
13. Digestion condition by *Psi*I at 37°C for 3 h: 1.5 μL of 10× buffer, 0.15 μL of 100× BSA, 300 ng of plasmid, and 5 U of *Psi*I. Add ddH_2O to a final volume of 15 μL. Expected bands: 215, 1,389, 1,689, 1,782, 2,425, 2,752, and 5,215 bp.

3.4. Detection of Zeaxanthin

1. Inoculate a single colony carrying the zeaxanthin biosynthetic pathway into 3 mL of SC-Ura liquid medium and grow at 30°C, 250 rpm for 1.5 days.
2. Inoculate 2.5 mL of seed culture into 250 mL of fresh SC-Ura medium and continue growing at 30°C, 250 rpm for 4 days.
3. Cells are collected by centrifugation at 3,220 × *g*, resuspended with 5 mL of acetone and lysed by French press at 10,000 psi.
4. Supernatants are collected after centrifugation at 16,100 × *g* for 3 min and evaporated to dryness using Rotavapor.
5. After resuspension in 0.5–1 mL of methanol, 100 μL of sample is loaded onto the Agilent ZORBAX SB-C18 column and analyzed at 450 nm by HPLC with a 0.5 mL/min flow rate as follows: buffer A: H_2O with 0.1% TFA, buffer B: 100% CH_3OH; 0–3 min, 60% CH_3OH; 3–15 min, linear gradient from 60% CH_3OH to 100% CH_3OH; 15–17 min, 100% CH_3OH; 17–20 min, linear gradient from 100% CH_3OH to 60% CH_3OH. Authentic zeaxanthin is used as standard which was eluted at 19.9 min.

4. Notes

1. If a larger biochemical pathway needs to be assembled, increasing the length of the overlaps between the adjacent fragments is necessary. For example, to assemble a pathway with a size of

~25 kb, a longer overlap (e.g., 125 bp) could ensure high assembly efficiency (>50%), while low efficiency (10–20%) is obtained if the length of the overlaps is only 50 bp (13).

2. The *hisG* and *Delta2* sequences are not essential for assembling a pathway. As long as the 5′ end of the first promoter and the 3′ end of the last terminator contain overlaps (at least 50 bp) with the vector backbone, the pathway can be assembled on a plasmid. Similarly, any linearized *S. cerevisiae–E. coli* shuttle vector containing an *ura3* gene as a selection marker can be used as the vector backbone.
3. *S. cerevisiae* YSG50 is used as the host for DNA assembly. However, any *S. cerevisiae* strain with a nonfunctional *ura3* gene can be used as a host.
4. In order to verify the correctly assembled constructs through restriction digestion, one or two enzymes that cut the expected construct multiple times are chosen. Usually, two to three groups of digestion need to be set up in order to ensure the correct assembly. For a plasmid with a size of 15–20 kb, such as pRS416-zeaxanthin, find one or two enzymes which cut the DNA molecule five to nine times. Try to avoid using enzyme digestion that will result in multiple fragments with similar sizes.
5. In the construction of the first gene expression cassette and the last gene cassette, the *hisG* sequence and the *Delta2* sequence are also included. Therefore, for these two reactions, four pieces of fragments are spliced together.
6. The fragment mixture can be maintained at –20°C for several months.
7. Normally, the doubling time for a *S. cerevisiae* laboratory strain is approximately 2 h.
8. Unlike *E. coli*, yeast competent cells need to be freshly prepared each time.
9. For an efficient electroporation, a time constant of 5.0–5.2 ms should be obtained.
10. Assembly efficiency is defined as the percentage of the correct clones among the transformants appearing on the plate. Usually, ten colonies are picked, and an average efficiency of 80% can be obtained for assembly of the zeaxanthin pathway.
11. *E. coli* strain BW25141 is used for plasmid enrichment and verification. However, any *E. coli* strain suitable for DNA cloning, such as DH5α and JM109, can be used.
12. The number of obtained *E. coli* transformants could vary from a few to several thousands. This is mainly due to the low quality of the isolated yeast plasmids. However, as long as colonies appear, experiments can proceed.

References

1. Hjersted JL, Henson MA, Mahadevan R (2007) Genome-scale analysis of *Saccharomyces cerevisiae* metabolism and ethanol production in fed-batch culture. Biotechnol Bioeng 97:1190–1204
2. Keasling JD (2008) Synthetic biology for synthetic chemistry. ACS Chem Biol 3:64–76
3. Menzella HG, Reid R, Carney JR, Chandran SS, Reisinger SJ, Patel KG, Hopwood DA, Santi DV (2005) Combinatorial polyketide biosynthesis by de novo design and rearrangement of modular polyketide synthase genes. Nat Biotechnol 23:1171–1176
4. Pitera DJ, Paddon CJ, Newman JD, Keasling JD (2007) Balancing a heterologous mevalonate pathway for improved isoprenoid production in *Escherichia coli*. Metab Eng 9: 193–207
5. Ro DK, Paradise EM, Ouellet M, Fisher KJ, Newman KL, Ndungu JM, Ho KA, Eachus RA, Ham TS, Kirby J, Chang MC, Withers ST, Shiba Y, Sarpong R, Keasling JD (2006) Production of the antimalarial drug precursor artemisinic acid in engineered yeast. Nature 440:940–943
6. Szczebara FM, Chandelier C, Villeret C, Masurel A, Bourot S, Duport C, Blanchard S, Groisillier A, Testet E, Costaglioli P, Cauet G, Degryse E, Balbuena D, Winter J, Achstetter T, Spagnoli R, Pompon D, Dumas B (2003) Total biosynthesis of hydrocortisone from a simple carbon source in yeast. Nat Biotechnol 21:143–149
7. Dejong JM, Liu Y, Bollon AP, Long RM, Jennewein S, Williams D, Croteau RB (2006) Genetic engineering of taxol biosynthetic genes in *Saccharomyces cerevisiae*. Biotechnol Bioeng 93:212–224
8. Yan Y, Kohli A, Koffas MA (2005) Biosynthesis of natural flavanones in *Saccharomyces cerevisiae*. Appl Environ Microbiol 71:5610–5613
9. Gunyuzlu PL, Hollis GF, Toyn JH (2001) Plasmid construction by linker-assisted homologous recombination in yeast. Biotechniques 31:1246–1250
10. Ma H, Kunes S, Schatz PJ, Botstein D (1987) Plasmid construction by homologous recombination in yeast. Gene 58:201–216
11. Oldenburg KR, Vo KT, Michaelis S, Paddon C (1997) Recombination-mediated PCR-directed plasmid construction *in vivo* in yeast. Nucleic Acids Res 25:451–452
12. Raymond CK, Pownder TA, Sexson SL (1999) General method for plasmid construction using homologous recombination. Biotechniques 26:134–141
13. Shao Z, Zhao H, Zhao H (2009) DNA assembler, an *in vivo* genetic method for rapid construction of biochemical pathways. Nucleic Acids Res 37:e16
14. Horton RM, Hunt HD, Ho SN, Pullen JK, Pease LR (1989) Engineering hybrid genes without the use of restriction enzymes: gene-splicing by overlap extension. Gene 77:61–68
15. Lee FW, Da Silva NA (1997) Sequential delta-integration for the regulated insertion of cloned genes in *Saccharomyces cerevisiae*. Biotechnol Prog 13:368–373
16. Chemler JA, Yan Y, Koffas MA (2006) Biosynthesis of isoprenoids, polyunsaturated fatty acids and flavonoids in *Saccharomyces cerevisiae*. Microb Cell Fact 5:20
17. Misawa N, Nakagawa M, Kobayashi K, Yamano S, Izawa Y, Nakamura K, Harashima K (1990) Elucidation of the *Erwinia uredovora* carotenoid biosynthetic pathway by functional analysis of gene products expressed in *Escherichia coli*. J Bacteriol 172:6704–6712
18. Misawa N, Shimada H (1997) Metabolic engineering for the production of carotenoids in non-carotenogenic bacteria and yeasts. J Biotechnol 59:169–181

Chapter 18

Neurosporaxanthin Production by *Neurospora* and *Fusarium*

Javier Avalos, Alfonso Prado-Cabrero, and Alejandro F. Estrada

Abstract

The orange pigmentation of the ascomycete fungi *Neurospora* and *Fusarium* is mainly due to the accumulation of neurosporaxanthin, a carboxylic apocarotenoid whose possible biotechnological applications have not been investigated. From the discovery of the first enzyme of the biosynthetic pathway in 1989, the prenyltransferase AL-3, to the recent identification of an aldehyde dehydrogenase responsible for the last biosynthetic step, all the enzymes and biochemical reactions needed for neurosporaxanthin biosynthesis in these fungi are already known. Depending on the culture conditions and/or genetic background, *Neurospora* and *Fusarium* may produce large quantities of this xanthophyll and minor amounts of other carotenoids. This chapter describes methods for the growth of *Neurospora crassa* and *Fusarium fujikuroi* for improved neurosporaxanthin production, the analysis of this xanthophyll, its separation from its carotenoid precursors, and its identification and quantification.

Key words: Apocarotenoids, Xanthophyll, Carotenoid overproduction, Light induction, TLC, HPLC

1. Introduction

Carotenoid production is a common trait in filamentous fungi (1, 2). Neurosporaxanthin, abbreviated hereafter NX, is a carboxylic apocarotenoid found in some fungi, such as the ascomycete genera *Neurospora* and *Fusarium*. The structure of this xanthophyll (Fig. 1), firstly described as an acidic pigment (3, 4), was elucidated by a combination of physical and chemical analyses (5). The knowledge on the fungal enzymes involved in NX biosynthesis started with the cloning of genes responsible for albino phenotypes in *Neurospora crassa*: *al-1* (6), *al-2* (7), and *al-3* (8). The encoded proteins are the prenyltransferase AL-3, that catalyzes the synthesis

José-Luis Barredo (ed.), *Microbial Carotenoids From Fungi: Methods and Protocols*, Methods in Molecular Biology, vol. 898,
DOI 10.1007/978-1-61779-918-1_18, © Springer Science+Business Media New York 2012

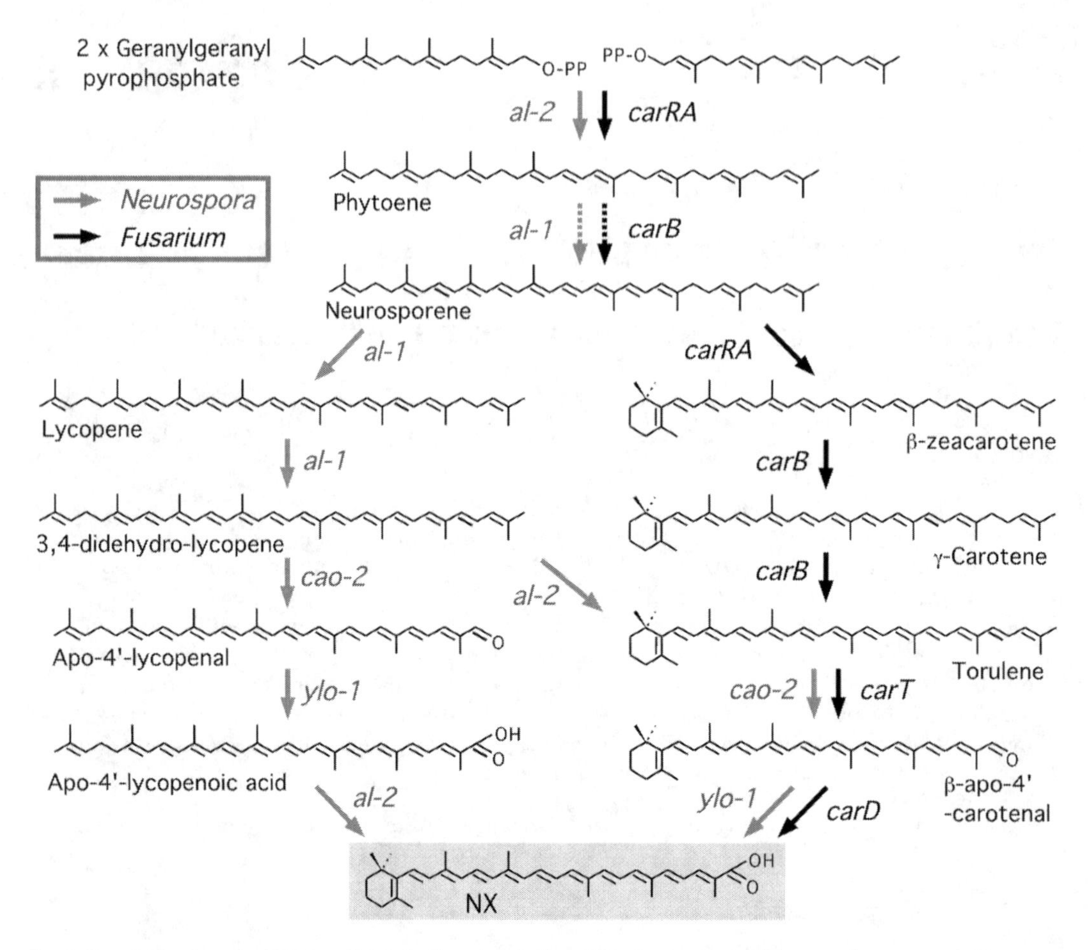

Fig. 1. Enzymatic steps for NX biosynthesis in *Fusarium fujikuroi* (*black arrows*) and *N. crassa* (*gray arrows*). The genes involved in each step are indicated. The *dotted arrows* represent several desaturation steps.

of geranylgeranyl pyrophosphate (GGPP) from farnesyl pyrophosphate (9), the bifunctional carotene cyclase/phytoene synthase AL-2, needed to build the colorless precursor phytoene and to introduce the beta cycles in later intermediates (10), and the dehydrogenase AL-1, responsible for all the desaturation steps in the pathway (11). These works facilitated the cloning of the orthologous genes in other carotenogenic fungi, including *Fusarium fujikuroi*, whose *al-3*, *al-2*, and *al-1* counterparts were called *ggs* (12), *carB* (13), and *carRA* (14), respectively. In *N. crassa* and *F. fujikuroi* these enzymes lead to the synthesis of torulene, a reddish NX precursor (Fig. 1). In *N. crassa*, depending on the culture conditions, these enzymatic activities may lead to the production of 3,4-didehydrolycopene (15). Torulene and 3,4-didehydrolycopene are substrates of a cleaving oxygenase, which produces a C35 apocarotenoid with an aldehyde group at the cleaved end.

The enzyme responsible for this key reaction is encoded by the gene *carT* in *F. fujikuroi* (16) and *cao-2* in *N. crassa* (17), and their mutants exhibit a reddish pigmentation and accumulate torulene. The last reaction for NX biosynthesis, the oxidation of the aldehyde group into the carboxylic one, is achieved in *N. crassa* by the aldehyde dehydrogenase YLO-1 (18), whose *F. fujikuroi* counterpart is called CarD.

The regulation of NX biosynthesis has been object of special attention in both fungi. Its most outstanding feature is the induction upon illumination (19–21), resulting from a rapid increase in the mRNA levels of the enzymatic genes (16, 22, 23). Large amounts of NX are found in carotenoid-overproducing mutants of *F. fujikuroi*, generically called *carS*, which exhibit a high carotenoid biosynthetic activity irrespective of light (19) and other culture conditions (24). The *carS* mutants are transcriptionally derepressed for the *car* genes in the dark (16, 22, 23) and exhibit a higher enzymatic activity (25). The genetic basis of this phenotype is currently under investigation by our research group. Similar *carS* mutants also exist in *F. oxysporum* (R. Rodríguez-Ortiz, unpublished), but their occurrence in other *Fusarium* species has not been described yet.

Here, we describe the basic methodology for NX production and analysis, from fungal incubations (see Subheadings 3.1 and 3.2), to NX extraction (see Subheadings 3.3 and 3.4), quantification (see Subheadings 3.5 and 3.6), and purification (see Subheadings 3.7 and 3.8). For high NX production in *F. fujikuroi* (up to several mg/g dry weight), we recommend to grow a *carS* strain in the dark in low nitrogen broth (24). No *carS*-like mutants have been described in *N. crassa*. However, NX overproduction is obtained with the standard wild-type strain in this fungus through illumination at low temperature, e.g., 8°C, of mycelia formerly grown in the dark at standard growth temperature (see ref. (18)). Under these conditions, the proportion and the total NX concentration are markedly enhanced.

2. Materials

2.1. Submerged Culture of *N. crassa* and *F. fujikuroi*

1. *N. crassa* Oak Ridge 74-OR23-1A (FGSC 987) (Fungal Genetics Stock Center, Kansas City, MI, USA) (26).
2. *F. fujikuroi* SG22 (or equivalent *carS* carotenoid overproducing mutant) (19, 24).
3. Vogel salts: Dissolve 125 g Na_3 citrate $2H_2O$ in 775 mL of distilled water, and dissolve afterwards 250 g KH_2PO_4, 100 g

NH_4NO_3, 10 g $MgSO_4 \cdot 7H_2O$, 5 g $CaCl_2 \cdot 2H_2O$ (formerly dissolved in 20 mL H_2O to avoid precipitation), 2.5 mL of biotine solution (5 mg biotine in 100 mL 50% ethanol), and 5 mL microelement solution (see ref. (27) for details).

4. Vogel's minimal broth: Dissolve 15 g of sucrose and 20 mL of Vogel salts in 965 mL distilled water (27).
5. Solution A: Add distilled water to 80 g glucose to complete 0.5 L, and dissolve.
6. Solution B: Dissolve 0.48 g of NH_4NO_3, 5 g of H_2KPO_4, 1 g of $MgSO_4 \cdot 7H_2O$, and 2 mL of ICI microelement solution (28) in 450 mL distilled water.
7. ICI microelement solution: Dissolve 100 mg of $FeSO_4 \cdot 7H_2O$, 15 mg of $CuSO_4 \cdot 5H_2O$, 161 mg of $ZnSO_4 \cdot 7H_2O$, 10 mg of $MnSO_4 \cdot 7H_2O$, and 10 mg of $(NH_4)_6Mo_7O_{24} \cdot 4H_2O$ in 100 mL distilled water (28).
8. Low nitrogen broth (10% ICI): Mix solutions A and B after autoclaving (28).

2.2. Carotenoid Analyses

1. Lyophilizer.
2. Porcelain mortar and pestle.
3. Centrifuge and rotor for 10-mL glass tubes.
4. Rotoevaporator.
5. Spectrophotometer.
6. Chromatography tank.
7. HPLC system equipped with a C30-reversed phase column.
8. On line HPLC photodiode array detector.
9. Mini-Beadbeater (Biospec, Bartlesville, OK, USA).
10. FastPrep® 24 device (MP Biomedicals, Solon, OH, USA).
11. Vacuum centrifuge.
12. HPLC solvent system A: Methanol/*tert*-butyl methyl ether (500:500, v/v).
13. HPLC solvent system B: Methanol/*tert*-butyl methyl ether/water (600:120:120, v/v/v).
14. KOH–EtOH solution: Dissolve 5 g KOH in 5 mL H_2O.
15. Silicagel 60 TLC plates (Merck, Darmstadt, Germany).
16. Sea sand (purified by acid).
17. Millex®-GV filter units 0.22 μm (Millipore, Billerica, MA, USA).

3. Methods

3.1. Submerged Culture of N. crassa

1. Grow the strain in a 500-mL Erlenmeyer flask with 200 mL of Vogel's minimal broth inoculated with 10^5 conidia, and shake at 150 rpm for 2 days in the dark at 30°C.
2. Incubate 1 day more at 8°C under white light (e.g., two standard fluorescent tubes at a distance of ca. 40 cm). Precool the flask at 8°C for 1 h in the dark before incubation under light.

3.2. Submerged Culture of F. fujikuroi

1. Grow the *carS* mutant in a 500-mL Erlenmeyer flask with 250 mL of low nitrogen broth inoculated with 10^6 conidia, and shake at 150 rpm for 15 days in the dark at 30°C.

3.3. Carotenoid Extraction by Cell Disruption in Mortar

NX is a polar carotenoid with different solubility properties than those of nonpolar precursor carotenoids. Extraction with a polar solvent is required for efficient NX extraction. Two methods are described depending on the breaking device: mortar or homogenization device.

1. Filter the mycelial sample from submerged cultures grown under appropriate conditions (see Note 1).
2. Freeze, lyophilize, and determine weight of approximately 0.1 g of dry samples (see Note 2).
3. Ground to a fine powder in a mortar with the help of 1–2 g of commercial sea sand and wash with acetone up to discoloration of the sample (see Note 3).
4. Transfer to acetone-resistant tubes, centrifuge to separate sand and mycelial debris, and transfer acetone supernatants to a rotoevaporation tube.
5. Dry in a vacuum rotoevaporator and conserve in a freezer (see Note 4).

3.4. Carotenoid Extraction by Cell Disruption in Homogenization Device

1. Proceed through steps 1 and 2 of Subheading 3.3.
2. Transfer about 0.5 mg mycelial sample to 2 mL screw tube with 0.5 g of sea sand. Use more than one tube if required.
3. Homogenize the sample in an automatic breaking device, such as the Mini-Beadbeater or the FastPrep® 24 device (see Note 5).
4. Centrifuge the tube and transfer the acetone supernatant to a rotoevaporation tube (see Note 6).
5. Repeat the process up to discoloration of the sample. Proceed as described in step 5 of Subheading 3.3.

3.5. Quantification of NX by the Spectrum Subtraction Protocol

Polarity of NX facilitates its chromatographic separation from neutral precursor carotenoids. If NX is the major carotenoid, we recommend this simple and reliable method to estimate its concentration.

1. Resuspend the sample in an appropriate volume of hexane (typically 10 mL) and measure absorption spectrum in a spectrophotometer (see Note 7). This is spectrum 1 (if possible, store in an e-file).
2. Transfer back the solution to the rotoevaporation tube and vacuum dry again the sample.
3. Add light petroleum 40–60° (abbreviated LP elsewhere in this chapter) to an appropriate amount of Al_2O_3 (grade II–III), in a small beaker (see Note 8).
4. Plug a 1-mL pipette tip with a small piece of soft cellulose paper and wash with 0.5 mL LP (see Note 9).
5. Add to the tip 1 mL LP and load with a Pasteur pipette up to 0.5 cm of Al_2O_3 from the beaker. Leave to flow most of the solvent (see Note 10).
6. While the solvent is flowing through the Al_2O_3 layer, resuspend the carotenoid sample in 200 μL of LP.
7. Place a collecting tube below the tip and add the resuspended carotenoid sample to the Al_2O_3 surface (see Note 11).
8. Let the sample to absorb completely and immediately add 1 mL of diethyl ether. Collect the sample (see Note 12).
9. Vacuum dry the collected carotenoid sample, resuspend in the same volume of hexane as in the step 1, and measure absorption spectrum in a spectrophotometer. This is spectrum 2.
10. NX spectrum is obtained upon subtraction of spectrum 2 from spectrum 1 (examples shown in Fig. 2) (see Note 13).
11. NX concentration (mg/g dry weight) is obtained from NX absorption through the formula (Absorption × dilution)/(171.5 × dry weight in grams) (see Note 14).

3.6. Quantification of NX by the Elution Protocol

If the proportion of NX is low (as a hint, less than 40%) or the total carotenoid content of the sample is too low (as a hint, less than 50 μg/g dry weight), this method is proposed.

1. Fill a 2-μm wide column with a 1 cm layer of Al_2O_3 (grade II–III prepared as in step 3 of Subheading 3.5) and wash with 3 mL LP.
2. Resuspend the carotenoid sample in 200 μL LP.
3. Place a collecting tube below the column. Once the added LP was absorbed, add the resuspended sample to the Al_2O_3 surface.

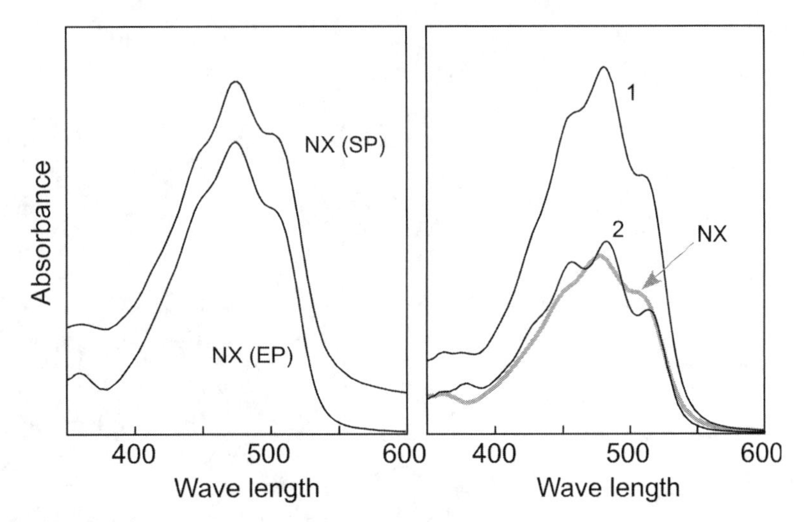

Fig. 2. *Left panel*: Examples of NX spectra obtained from *carS* mutants of *F. fujikuroi* by the subtraction protocol (SP) or by the elution protocol (EP). The analyses correspond to samples from the mutants SF4 (21) and SG22 (19), respectively. *Right panel*: Spectra 1 and 2 and NX subtraction spectrum (in *gray*) after application of the subtraction protocol to the mutant T17.1 of *F. fujikuroi* grown at 30°C under light (23), containing similar amounts of NX and neutral carotenoids.

4. Let it to absorb completely and immediately add 3 mL of diethyl ether (see Note 15).
5. Prepare 100 mL of KOH–EtOH solution. After cooling, mix with 90 mL ethanol 96° (see Note 16).
6. Once the diethyl ether is completely absorbed, change the collecting tube and add 8 mL of the KOH–EtOH solution. Let it to absorb completely and add again 8 mL of KOH–EtOH solution. The flow rate will decrease upon KOH–EtOH addition.
7. The eluted sample contains the NX fraction. To recover it, add 2 mL of LP and add slowly 8 mL 2 M HCl. Mix carefully as HCl is added. Confirm that NX passes to the top LP layer.
8. Collect the LP sample in a rotoevaporation tube and wash with further 2 mL LP samples up to total discoloration.
9. Vacuum dry the NX-containing LP fraction, and proceed as described in step 1 of Subheading 3.5. An example of NX spectrum obtained with this method is shown Fig. 2.
10. Calculate NX concentration from absorption spectrum at 477 nm as described in step 11 of Subheading 3.5.

3.7. Chromatographic Purification of NX by TLC

The different chemical properties of NX allow its easy separation by other chromatographic methods.

1. Cut an appropriate size of a Silicagel 60 TLC plate.
2. Resuspend the carotenoid sample in 200 μL diethyl ether and load carefully with a 200-μL pipette tip (see Note 17).

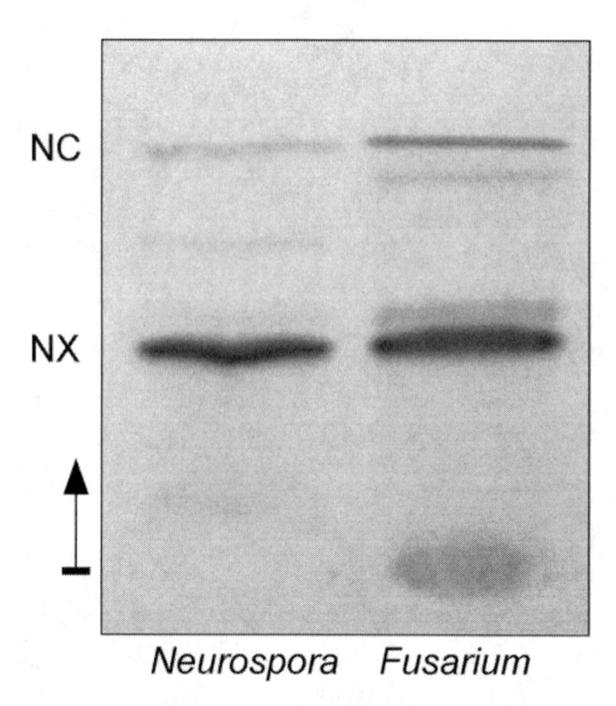

Fig. 3. Examples of TLC separations of NX from neutral carotenoids (NC). *Left lane*: Carotenoid sample from the *carS* mutant SF134 of *F. fujikuroi* grown for 12 days in the dark on minimal medium (mycelial sample was kindly provided by M. Macías, University of Sevilla, Spain). *Right lane*: Carotenoid sample from the wild-type Oak Ridge 74-OR23-1A of *N. crassa* grown for 48 h in the dark at 30°C and 24 h in the light at 8°C.

3. Place in chromatography tank with a bottom solvent layer of a mixture of 40 mL light petroleum, 10 mL diethyl ether, and 7 mL acetone.
4. Neutral carotenoids run in the front and NX appear in an intermediate position (Fig. 3) (see Note 18).
5. The NX band may be scraped out and resuspended in acetone for later spectrophotometric analysis (see Note 19).

3.8. Chromatographic Purification of NX by HPLC

1. Concentrate the sample in a small volume and clean the sample by centrifugation (see Note 20).
2. Inject in an HPLC system equipped with C30-reversed phase column (see Note 21).
3. Run the sample using the solvent systems A and B at a flow rate of 1 mL/min with a linear gradient from 100% B to 57% A 43% B within 25 min, followed by a linear gradient to 100% A with a flow rate of 1.5 mL within 7 min. Increase the flow to 2 mL/min of 100% A maintaining these conditions for 32 min.
4. Analyze chromatogram at appropriate wave length (Fig. 4).

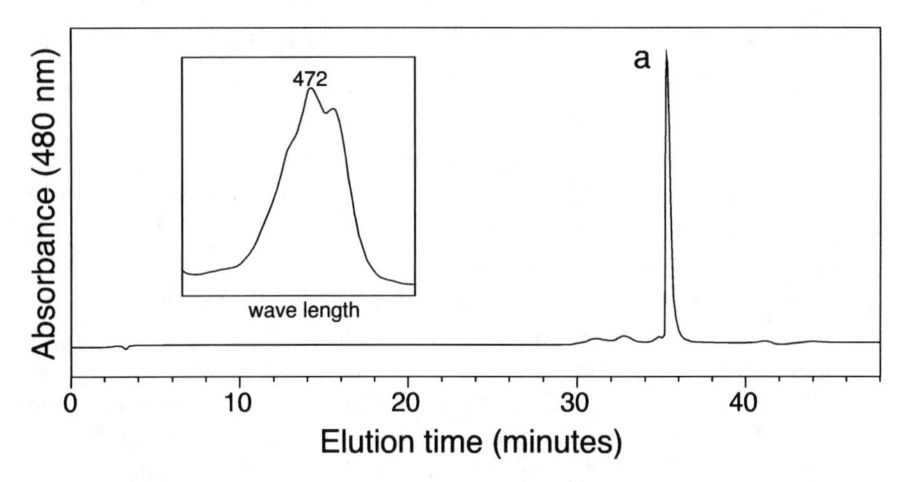

Fig. 4. Chromatogram of the HPLC separation of the carotenoids accumulated by the wild-type strain of *N. crassa* as described in Fig. 3 (adapted from ref. 18). The major peak is NX, whose spectrum is shown in the *inner box*.

4. Notes

1. For *Neurospora*, agar cultures are not recommended because of its abundant conidiation and difficult mycelia separation. Separation of mycelia from agar is more feasible in the case of *Fusarium*.
2. Store the weighed samples protected from moisture in the freezer.
3. Here and in further manipulation steps, avoid direct illumination of the carotenoid samples.
4. Carotenoids may be subject of oxidation. It is recommended to store the sample in an N-enriched atmosphere if N_2 supply is available.
5. Breaking pulses should not be longer than 30 s to avoid sample overheating. Cool tubes in ice between breaking cycles.
6. Acetone supernatants may also be transferred to microcentrifuge tubes and dried in a vacuum centrifuge.
7. Maximal absorption should not exceed the value of 1.
8. LP volume should be enough to cover the Al_2O_3 sample. 1–2 g Al_2O_3 is sufficient. Higher Al_2O_3 amounts may be needed for the elution protocol (see Subheading 3.6).
9. Use a Pasteur pipette to tighten the 3–4 mm plug, which should be small enough to be close to the tip end.
10. The Al_2O_3 layer should allow a flow of less than one drop of solvent per second.

11. Avoid the Al_2O_3 layer to get dry before sample addition.
12. Polarity of diethyl ether leads to the rapid elution of neutral carotenoids.
13. This step requires the availability of absorption data in an Excel-compatible computer file. If spectra data are not available, maximal NX absorption is obtained upon subtraction from the absorption at 477 nm of spectrum 1 of the absorption at the same wavelength of spectrum 2.
14. The specific extinction coefficient ($E_{1cm}^{1\%}$) for NX in hexane at 477 nm is 1,715 (29). This value corresponds to the absorption (extinction) at the indicated wave length of a 1% solution of the compound. Maximal absorption wavelength of NX may vary slightly between different spectrophotometers.
15. The eluted sample contains the neutral carotenoids, to be analyzed separately if desired.
16. The solution heats as KOH is dissolved. Proceed slowly to avoid overheating. Storing is not recommended. Prepare only the volume needed for the experiment.
17. Load at least at 1 cm distance from the lower edge to allow absorption of the solvent in the chromatography tank.
18. If highly polar pigments are present in the sample, such as bikaverins (30), they are retained in the origin.
19. Agitate the scraped sample in vortex to facilitate separation of NX from Al_2O_3; wash several times if required for total recovery.
20. The samples may also be cleaned through an appropriate filtering system (e.g., 0.22 μm Millex®-GV filter units).
21. Availability of a photodiode array detector is highly recommended for NX confirmation.

Acknowledgments

We thank the Spanish Government (projects BIO2006-01323 and BIO2009-11131) and Junta de Andalucía (project P07-CVI-02813) for financial support.

References

1. Sandmann G, Misawa N (2002) Fungal carotenoids. In: Osiewacz HD (ed) The Mycota X. Industrial applications. Springer, Berlin, pp 247–262
2. Avalos J, Cerdá-Olmedo E (2004) Fungal carotenoid production. In: Arora DK (ed) Handbook of fungal biotechnology, 2nd edn. Marcel Dekker, Inc, New York, pp 367–378

3. Zalokar M (1954) Studies on biosynthesis of carotenoids in *Neurospora crassa*. Arch Biochem Biophys 50:71–80
4. Zalokar M (1957) Isolation of an acidic pigment in *Neurospora*. Arch Biochem Biophys 70:568–571
5. Aasen AJ, Jensen SL (1965) Fungal carotenoids II. The structure of the carotenoid acid neurosporaxanthin. Acta Chem Scand 19:1843–1853
6. Schmidhauser TJ, Lauter FR, Russo VE, Yanofsky C (1990) Cloning, sequence, and photoregulation of *al-1*, a carotenoid biosynthetic gene of *Neurospora crassa*. Mol Cell Biol 10:5064–5070
7. Schmidhauser TJ, Lauter FR, Schumacher M, Zhou W, Russo VE, Yanofsky C (1994) Characterization of *al-2*, the phytoene synthase gene of *Neurospora crassa*. Cloning, sequence analysis, and photoregulation. J Biol Chem 269:12060–12066
8. Nelson MA, Morelli G, Carattoli A, Romano N, Macino G (1989) Molecular cloning of a *Neurospora crassa* carotenoid biosynthetic gene (*albino-3*) regulated by blue light and the products of the *white collar* genes. Mol Cell Biol 9:1271–1276
9. Sandmann G, Misawa N, Wiedemann M, Vittorioso P, Carattoli A, Morelli G, Macino G (1993) Functional identification of *al-3* from *Neurospora crassa* as the gene for geranylgeranyl pyrophosphate synthase by complementation with *crt* genes, in vitro characterization of the gene product and mutant analysis. J Photochem Photobiol B 18:245–251
10. Arrach N, Schmidhauser TJ, Avalos J (2002) Mutants of the carotene cyclase domain of *al-2* from *Neurospora crassa*. Mol Genet Genomics 266:914–921
11. Hausmann A, Sandmann G (2000) A single five-step desaturase is involved in the carotenoid biosynthesis pathway to beta-carotene and torulene in *Neurospora crassa*. Fungal Genet Biol 30:147–153
12. Mende K, Homann V, Tudzynski B (1997) The geranylgeranyl diphosphate synthase gene of *Gibberella fujikuroi*: isolation and expression. Mol Gen Genet 255:96–105
13. Fernández-Martín R, Cerdá-Olmedo E, Avalos J (2000) Homologous recombination and allele replacement in transformants of *Fusarium fujikuroi*. Mol Gen Genet 263:838–845
14. Linnemannstöns P, Prado MM, Fernández-Martín R, Tudzynski B, Avalos J (2002) A carotenoid biosynthesis gene cluster in *Fusarium fujikuroi*: the genes *carB* and *carRA*. Mol Genet Genomics 267:593–602
15. Estrada AF, Maier D, Scherzinger D, Avalos J, Al-Babili S (2008) Novel apocarotenoid intermediates in *Neurospora crassa* mutants imply a new biosynthetic reaction sequence leading to neurosporaxanthin formation. Fungal Genet Biol 45:1497–1505
16. Prado-Cabrero A, Estrada AF, Al-Babili S, Avalos J (2007) Identification and biochemical characterization of a novel carotenoid oxygenase: elucidation of the cleavage step in the *Fusarium* carotenoid pathway. Mol Microbiol 64:448–460
17. Saelices L, Youssar L, Holdermann I, Al-Babili S, Avalos J (2007) Identification of the gene responsible for torulene cleavage in the *Neurospora* carotenoid pathway. Mol Genet Genomics 278:527–537
18. Estrada AF, Youssar L, Scherzinger D, Al-Babili S, Avalos J (2008) The *ylo-1* gene encodes an aldehyde dehydrogenase responsible for the last reaction in the *Neurospora* carotenoid pathway. Mol Microbiol 69:1207–1220
19. Avalos J, Cerdá-Olmedo E (1987) Carotenoid mutants of *Gibberella fujikuroi*. Curr Genet 25:1837–1841
20. Avalos J, Schrott EL (1990) Photoinduction of carotenoid biosynthesis in *Gibberella fujikuroi*. FEMS Microbiol Lett 66:295–298
21. Prado-Cabrero A, Schaub P, Díaz-Sánchez V, Estrada AF, Al-Babili S, Avalos J (2009) Deviation of the neurosporaxanthin pathway towards β-carotene biosynthesis in *Fusarium fujikuroi* by a point mutation in the phytoene desaturase gene. FEBS J 276:4582–4597
22. Prado MM, Prado-Cabrero A, Fernández-Martín R, Avalos J (2004) A gene of the opsin family in the carotenoid gene cluster of *Fusarium fujikuroi*. Curr Genet 46:47–58
23. Thewes S, Prado-Cabrero A, Prado MM, Tudzynski B, Avalos J (2005) Characterization of a gene in the *car* cluster of *Fusarium fujikuroi* that codes for a protein of the carotenoid oxygenase family. Mol Genet Genomics 274:217–228
24. Rodríguez-Ortiz R, Limón MC, Avalos J (2009) Regulation of carotenogenesis and secondary metabolism by nitrogen in wild-type *Fusarium fujikuroi* and carotenoid-overproducing mutants. Appl Environ Microbiol 75:405–413
25. Avalos J, Mackenzie A, Nelki DS, Bramley PM (1988) Terpenoid biosynthesis in cell-extracts of wild type and mutant strains of *Gibberella fujikuroi*. Biochim Biophys Acta 966:257–265
26. McCluskey K (2003) The fungal genetics stock center: from molds to molecules. Adv Appl Microbiol 52:245–262

27. Davis RH, de Serres FJ (1970) Genetic and microbiological research techniques for *Neurospora crassa*. Methods Enzymol 17: 79–143
28. Geissman TA, Verbiscar AJ, Phinney BO, Cragg G (1966) Studies on the biosynthesis of gibberellins from (−)-kaurenoic acid in cultures of *Gibberella fujikuroi*. Phytochemistry 5: 933–947
29. Bindl E, Lang W, Rau W (1970) Untersuchungen über die lichtabhängige carotinoidsynthese. VI. Zeitlicher Verlauf der Synthese der einzelnen Carotinoide bei *Fusarium aquaeductuum* unter verschiedenen Induktionsbedingungen. Planta 94:156–174
30. Limón MC, Rodríguez-Ortiz R, Avalos J (2010) Bikaverin production and applications. Appl Microbiol Biotechnol 87:21–29

Chapter 19

Production of Torularhodin, Torulene, and β-Carotene by *Rhodotorula* Yeasts

Martín Moliné, Diego Libkind, and María van Broock

Abstract

Yeasts of the genera *Rhodotorula* are able to synthesize different pigments of high economic value like β-carotene, torulene, and torularhodin, and therefore represent a biotechnologically interesting group of yeasts. However, the low production rate of pigment in these microorganisms limits its industrial application. Here we describe some strategies to obtain hyperpigmented mutants of *Rhodotorula mucilaginosa* by means of ultraviolet-B radiation, the procedures for total carotenoids extraction and quantification, and a method for identification of each pigment.

Key words: Torularhodin, β-Carotene, Carotenoids, *Rhodotorula*, Mutants, UV-B radiation

1. Introduction

The genera *Rhodotorula* refers to a large group of asporogenous pigmented yeasts belonging to the Basidiomycota phylum. They have a wide occurrence in nature and have been isolated in almost all known environments, flowers, phylloplane, freshwater and seawater samples, glaciers, soil, etc. (1). A few species have also been described as causative agents of opportunistic mycoses (2).

Most *Rhodotorula* species produce different types of carotenoids pigments, four of which were identified in almost all species: torularhodin (3′, 4′-Didehydro-β, ψ-caroten-16′-oic acid), torulene (3′, 4′-Didehydro-β, ψ-carotene), γ-carotene (β, ψ-carotene), and β-carotene (β, β-carotene). Pigments are synthesized via the mevalonate pathway (3). Mevalonic acid is transformed in isopentenyl pyrophosphate units and then successively condensed to form phytoene. A successive transformation of phytoene leads to each of the aforementioned pigments (Fig. 1). The proportion of each carotenoid depends on the strain and culture conditions (4).

José-Luis Barredo (ed.), *Microbial Carotenoids From Fungi: Methods and Protocols*, Methods in Molecular Biology, vol. 898, DOI 10.1007/978-1-61779-918-1_19, © Springer Science+Business Media New York 2012

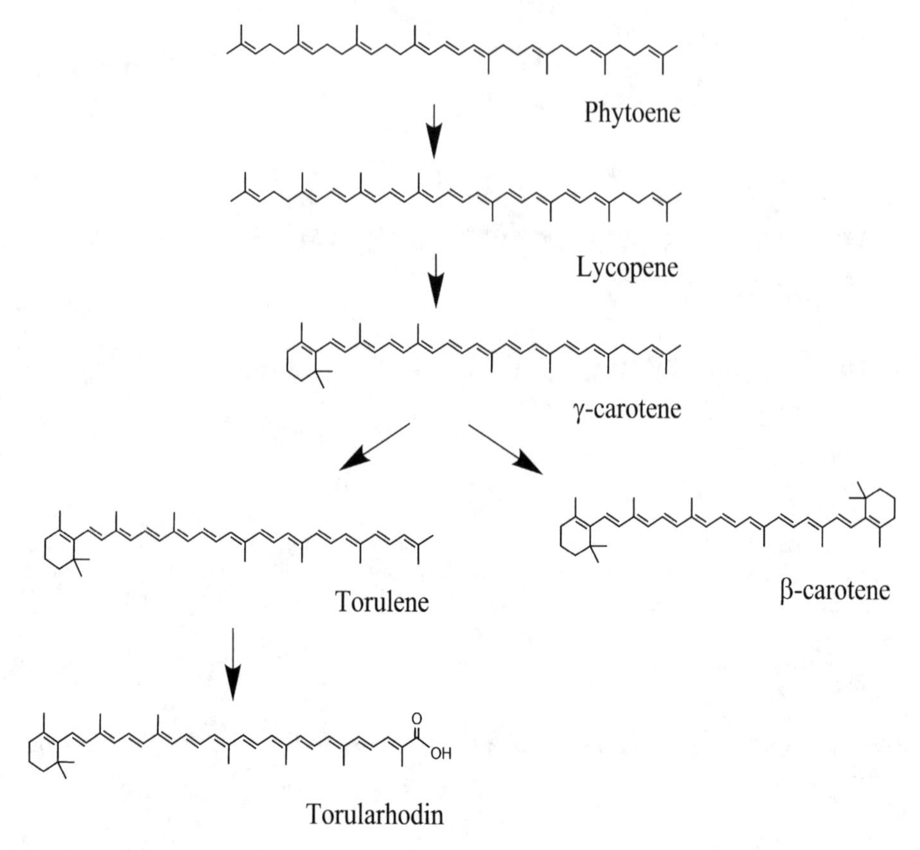

Fig. 1. Metabolic pathway of carotenoids biosynthesis in *Rhodotorula*.

Carotenoid production with *Rhodotorula* yeasts has advantages over other microorganisms such as algae and other fungi. Specific growth rate is high and large quantity of cell biomass is relatively easy to obtain at lab and pilot plant scale. Cells can be cultured in conventional bioreactors and biomass can be used directly as feed or as additive in pharmaceutical products. Besides, yeasts can adapt to different environmental conditions and grow under a wide variety of carbon and nitrogen sources (5).

Despite the biotechnological potential of *Rhodotorula* towards the production of carotenoid pigments, the low carotenoid content of natural strains limits its industrial exploitation. Therefore, obtaining hyperproducing mutants in order to develop an economic alternative for the production of carotenoids is needed. In this chapter we describe strategies we use to mutate *Rhodotorula mucilaginosa* in order to obtain hyperproducing strains or strains with other carotenoid composition. We also describe the methods for extraction, quantification and characterization of *Rhodotorula* pigments. Methods reported here can be properly adapted to other carotenogenic yeast species.

2. Materials

2.1. Mutation of R. mucilaginosa

1. *R. mucilaginosa* reference strain CBS 316^T (1).
2. Different *R. mucilaginosa* strains (wild or reference) (see Note 1).
3. MMS: 10 g/L glucose, 2 g/L $(NH_4)_2SO_4$, 2 g/L $KH_2(PO_4)$, 0.5 g/L $MgSO_4 \cdot 7H_2O$, and 1 g/L yeast extract. Mix, adjust to pH 5.0, and sterilize by autoclaving at 121°C for 15 min.
4. MMS agar: MMS and 15 g/L agar.
5. Diphenylamine stock solution: 10 mM diphenylamine in ethanol and filter through a 0.2-μm membrane.
6. MMS agar diphenylamine: Add 10–40 μL of diphenylamine stock solution to 10 mL of melt MMS agar medium, mix gently and plate (see Note 2).
7. UV-B lamp Spectroline XX-15B: Broadband (~290–380 nm; peak ~310 nm) medium intensity (15 W) (Spectronics Corporation, Westbury, NY, USA) (see Note 3).
8. Quartz tubes (20 mL).

2.2. Molecular Characterization of the Rhodotorula Mutant

1. YM agar: 3 g/L yeast extract, 3 g/L malt extract, 5 g/L peptone, 10 g/L dextrose, and 20 g/L agar. Adjust pH 4.5–5.0 and sterilize by autoclaving at 121°C for 15 min.
2. Genomic DNA extraction kit.
3. Lysing buffer: 50 mmol Tris, 250 mmol NaCl, 50 mmol EDTA, and 0.3% w/v SDS, pH 8.
4. $(GTG)_5$ primer: 5′-GTG GTG GTG GTG GTG-3′.
5. Polymerase chain reaction (PCR) reaction mix kit.
6. Thermocycler.
7. TBE buffer 10× stock solution: 108 g Tris, 55 g boric acid, and 100 mL 0.5 M EDTA; add water until 1 L.
8. UV transilluminator.

2.3. Carotenoid Extraction and Quantification

1. Liquid nitrogen.
2. Bench centrifuge.
3. UV spectrophotometer.

2.4. HPLC Pigments Characterization

1. Solvent A: Methanol:aqueous solution of 0.001 M tetrabutylammonium phosphate and 0.001 M propanoic acid (80:20).
2. Solvent B: Acetone:methanol (60:40).

3. Gemini C18 column: 5 μm, 250 × 4.6 mm (Phenomenex, Torrance, CA, USA).
4. HPLC with photodiode array detector.
5. β-carotene standard (Sigma, St. Louis, MO, USA).
6. Carotenoids extract from *R. mucilaginosa* CBS 316[T] (type strain).

3. Methods

3.1. Selection of R. mucilaginosa Hyperpigmented Mutants

There are numerous reports on the mutagenesis of several yeast species of the genus *Rhodotorula*, with the target of improving carotenoid accumulation or increasing the accumulation of β-carotene over other pigments (5–9). Mutagenic agents such as nitrosoguanidine, ethyl-methyl-sulfonate, methylene blue plus light or, UV-C have been effective for the generation of hyperpigmented mutants. However, these strategies usually have adverse effects over the growth rate, biomass yield, or even sugar assimilation capabilities (6–9). We recently established that accumulation of carotenoids is related with survival to UV-B radiation and color mutants frequently appear under this source of stress (10, 11). UV-B radiation can be used to obtain hyperpigmented yeast mutants at low mutagenic rate, without significant detrimental effects over active biotechnological traits.

1. To prepare fresh cultures of the *Rhodotorula* yeasts, grow the strains in 250-mL Erlenmeyer flasks, containing 50 mL of MMS medium, using 2.5 mL of fresh inoculum (48 h). Incubate the cultures at 20°C under agitation (200 rpm) for 48 h.
2. Centrifuge the culture (3,000 × *g*, 5 min at 20°C), discard the supernatant, and suspend the pellet in 5 mL of distilled water by vortexing for 1 min.
3. Repeat the procedure twice to eliminate the culture medium.
4. Suspend the pellet in distilled water and use a Neubauer chamber to estimate the number of cells per mL.
5. Prepare quartz tubes with 2×10^5 cells/mL in water.
6. Place each quartz tube under the UV-B lamp. Adjust the distance between the lamp and the tubes to get a final dose of 240 kJ m^2 (see Note 4).
7. Take 1 mL aliquots and make dilutions 1:10 and 1:100 with sterile distilled water.
8. Inoculate Petri dishes (Ø 9 cm) containing MMS agar with 100 μL of each cell dilution, and disperse the cells with a Drigalski spatula.

9. Incubate the plates at 20°C for 48 h in a dark chamber to avoid DNA photo repair processes (see Note 5).
10. Screen colonies and isolate those with different reddish color, including orange ones.
11. Determine specific growth rate at log phase and compare with parental strain values.
12. Colonies with increased carotenoid content and similar parental strain growth rate can be used in new mutagenic rounds. For strains with a carotenoid accumulation over 500 μM, use MMS-DPA agar plates in hyperpigmented mutant screening (see Note 2).

3.2. Molecular Characterization of the Rhodotorula Mutants

Once hyperpigmented yeast strains are obtained it is recommended to compare with parental strain in order to discard possible contamination. Yeast comparison can be done by MSP-PCR (Micro/Minisatellite-Primed PCR), a technique based on the amplification of repeated sequences in the genome with variable length. In *R. mucilaginosa* $(GTG)_5$ primer is used to reveal intraspecific variability among different strains (12). This primer is also useful to identify mutant and parental strains correspondence based on identical fingerprint patterns.

1. Prepare a 3–4 days YM agar culture of mutant and parental strains.
2. Extract genomic DNA with a DNA extraction kit (see Note 6).
3. Amplify DNA by PCR using the primer $(GTG)_5$ and the following mixture: reaction volume 25 μL, 1× PCR buffer, 2 mM dNTPs, 3.5 mM $MgCl_2$, 0.8 mM $(GTG)_5$ primer, 10–15 ng of genomic DNA, and 1 U Taq polymerase. Apply the following PCR program: initial denaturing step at 95°C for 5 min, followed by 40 cycles of 45 s at 93°C, 60 s at 50°C, and 60 s at 72°C, and a final extension step of 6 min at 72°C.
4. Separate DNA fragments by electrophoresis using agarose gels 1.4% (w/v) in 0.5× TBE buffer at 90 V for 3.5 h.
5. Stain with ethidium bromide or equivalent. Visualize and compare mutant and parental DNA banding patterns under a UV transilluminator.

3.3. Carotenoid Extraction and Quantification

Rhodotorula, cultured in MMS, produce torularhodin as a major carotenoid pigment, comprising ca. 60–80% of total carotenoids; β-carotene accounts for ca. 10–20% of total carotenoids, and torulene is usually present in minor proportions. Other carotenoid traces are γ-carotene and lycopene. Torularhodin is the pigment bearing the highest oxidation state and it is more polar than other similar pigments. This pigment appears first in the chromatogram followed by torulene and β-carotene (Fig. 2).

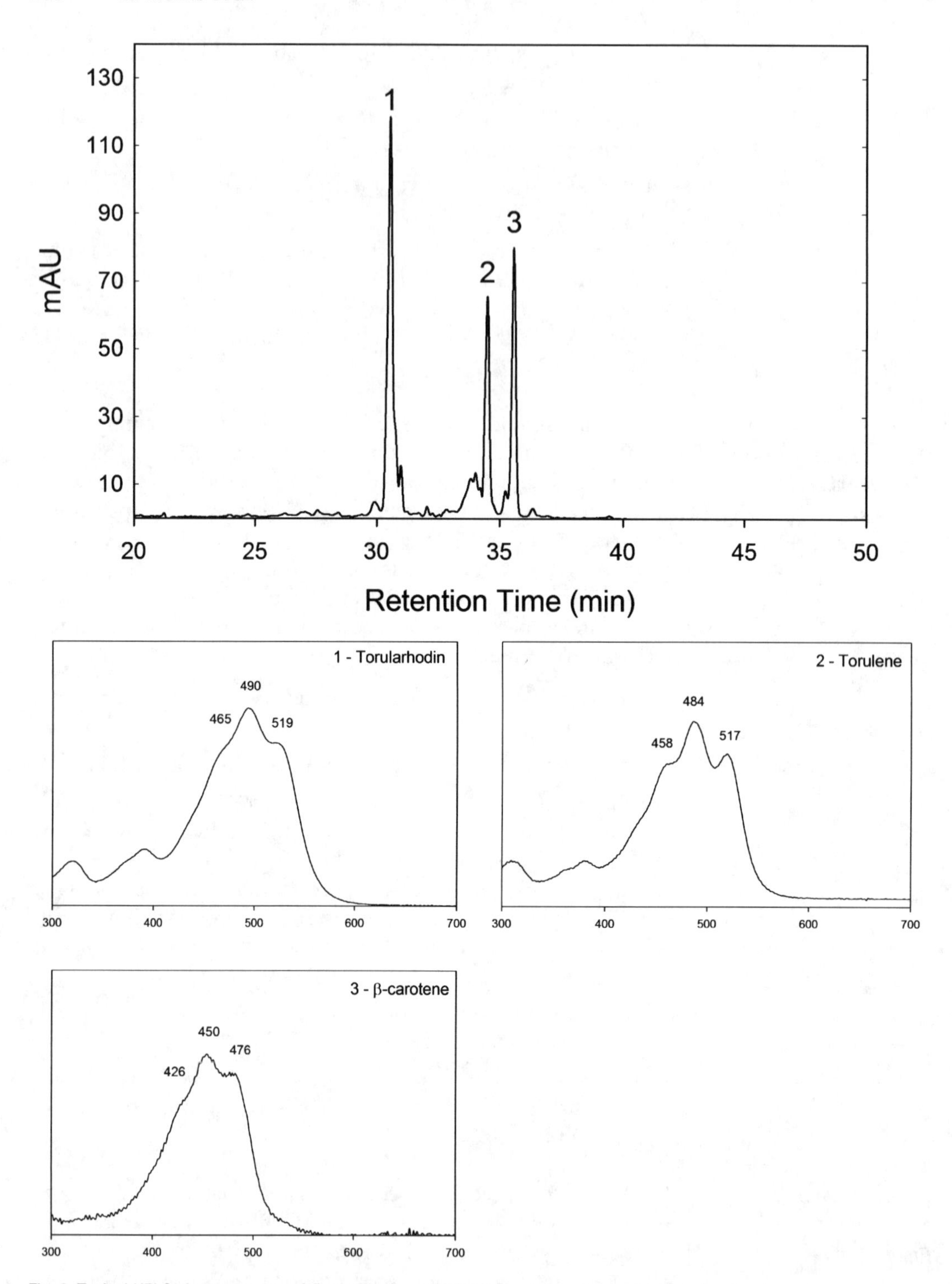

Fig. 2. Typical HPLC chromatogram of *R. mucilaginosa* CBS 316ᵀ and absorbance spectra of main pigments.

1. Culture the mutant strain in MMS agar.
2. Collect the cells by centrifugation (3,000 × *g*, 5 min), suspend the pellet in 4.5 mL of distilled water, and split it into four aliquots.
3. Pour 1 mL in each of two tubes for carotenoid extraction.
4. Pour 1 mL in each of other two pre-weighed tubes, dry at 100°C and record value at constant weight.
5. Add 1 mL of dimethyl sulfoxide to each carotenoid extraction tube and vortex the pellet for 1 min.
6. Incubate in a water bath at 55°C for 1 h.
7. After centrifugation (3,000 × *g*, 5 min), collect the supernatant in clean tubes and keep at −20°C.
8. Repeat steps 5–7.
9. Add 1 mL of acetone and extract pigments from the pellet by vortexing for 1 min.
10. Centrifuge at 3,000 × *g* for 5 min and pool the acetone fraction with dimethyl sulfoxide supernatants.
11. Repeat steps 9–10 until pellet becomes colorless.
12. Add 2 mL of petroleum ether (35–60°C) and 0.5 mL of a NaCl-saturated solution at 5°C to the tube containing supernatants, and mix in vortex for 15 s.
13. Centrifuge the tubes at 3,000 × *g* for 10 min at 5°C.
14. Collect the petroleum ether phase containing the pigments into a new tube.
15. Add 5 mL of distilled water at 5°C and mix in vortex for 30 s.
16. Centrifuge at 3,000 × *g* for 10 min and collect the supernatant.
17. Evaporate the petroleum ether under a continuous N_2 flux.
18. Add 1 mL of *n*-hexane or petroleum ether for spectrophotometric quantification or 100 μL of acetone for HPLC characterization (see Subheading 3.4).
19. Measure the absorbance of the sample at 485 nm (A_{485}).
20. Total carotenoid concentration can be estimated using the following simplified formula:

$$\text{Carotenoids}_t = \frac{A485 \times 1x10^4 (\mu\text{g / mL})}{2{,}680 \times \text{dry weight (g / mL)}}.$$

3.4. HPLC Pigments Characterization

1. Use the following assay conditions: start the elution with solvent A 100% and change gradually to solvent B during 20 min. Elute with solvent B for 20 min more using a flow rate of 0.8 mL/min.

2. Monitor the eluted fractions with a photodiode array detector. Identify carotenoids by retention time and compare with spectral absorbance of pure standards or with those obtained from *R. mucilaginosa* CBS 316^T strain (see Note 7).
3. Determine the amount of each pigment from the area of the peak detected at 450 nm using a calibration curve obtained with β-carotene standard. Determine the normalized quantity of carotenoid with respect to the dry weight of the pellet.

4. Notes

1. Each strain present differences in their mutation rate; it is recommended to include a high number of strains from different origins in order to select the most suitable one.
2. The concentration of diphenylamine should be used according to the accumulation of carotenoids in the mutant strain. Final concentration over 40 μM is toxic for yeasts, and concentrations lower than 10 μM present low effects on carotenoid concentration. Once you obtain a mutant carrying more than 500–600 μg/g of pigments, it is difficult to differentiate color changes in other hyperpigmented mutants with higher carotenoid concentrations. Diphenylamine inhibits carotenogenesis resulting in lower pigment accumulation and thus hyperpigmented mutants can be more easily detected. It is recommended to use the minimal diphenylamine concentration in order to get light pink colonies.
3. For the protocol developed in this chapter we use the model Spectroline XX-15B, although other models or brands can be adapted. The Lamp needs to be covered with an acetate filter. Absorbance of the acetate filters change during the irradiation and must be replaced after each use.
4. The Spectroline lamp placed at 20 cm from the tube for 240 min provides the necessary dose for mutation.
5. *R. mucilaginosa* survival is lower than 10%, according to the strain. Dilutions 1:10 and 1:100 produce an expected range of 200 to 20 colonies on each plate, although sometimes number of colonies can be inferior. Mutation frequency in the described conditions is low (0.1–0.04%), so it is necessary to use an adequate number of plates to get mutants. Plate incubation in the dark is mandatory since photo repair process will lower mutant generation.
6. An alternative DNA extraction procedure is described by Libkind et al. in this book (Chapter 12). Briefly, two loopfuls of YM agar grown cultures are suspended in Eppendorf tubes

containing 500 μL of lysing buffer and a 200 μL volume of 425–600 μm of glass beads. After vortexing for 3 min, tubes are incubated for 1 h at 65°C. After vortexing again for 3 min, the suspensions are centrifuged for 15 min at 4°C and 13,000 × *g*. Finally, the collected supernatant is diluted 1:750, and 5 μL is used directly for PCR studies.

7. Major pigments obtained in these conditions are, torularhodin, torulene, and β-carotene, in elution order. A reference curve with retention time and spectral absorbance of main pigments is presented in Fig. 2.

References

1. Fell JW, Statzell-Tallman A (1998) *Rhodotorula* F.C. Harrison. In: Kurtzman CP, Fell JW (eds) The yeasts, a taxonomic study. Elsevier Science Publishers, Amsterdam, pp 800–827
2. Libkind D, Sampaio JP (2010) *Rhodotorula*. In: Liu D (ed) Molecular detection of foodborne pathogens. CRC Press, New York, pp 603–618
3. Disch A, Rohmer M (1998) On the absence of the glyceraldehyde 3-phosphate/pyruvate pathway for isoprenoid biosynthesis in fungi and yeasts. FEMS Microbiol Lett 168:201–208
4. Buzzini P, Innocenti M, Turchetti B, Libkind D, van Broock M, Mulinacci N (2007) Carotenoid profiles of yeasts belonging to the genera *Rhodotorula, Rhodosporidium, Sporobolomyces,* and *Sporidiobolus.* Can J Microbiol 53:1024–1031
5. Bhosale PB (2001) Studies on yeast *Rhodotorula,* its carotenoids and their applications. Thesis, University of Pune, Pune
6. Bhosale P, Gadre RV (2001) Production of β-carotene by a *Rhodotorula glutinis* mutant in sea water medium. Bioresour Technol 76:53–55
7. Cong L, Chi Z, Li J, Wang X (2007) Enhanced carotenoid production by a mutant of the marine yeast *Rhodotorula* sp. hidai. J Ocean Univ China 6:66–71
8. Frengova GI, Simova ED, Beshkova DM (2004) Improvement of carotenoid-synthesizing yeast *Rhodotorula rubra* by chemical mutagenesis. Z Naturforsch C 59:99–103
9. Sakaki H, Nakanishi T, Satonaka K, Miki W, Fujita T, Komemushi S (2000) Properties of a high-torularhodin-producing mutant of *Rhodotorula glutinis* cultivated under oxidative stress. J Biosci Bioeng 89:203–205
10. Moliné M, Libkind D, Diéguez MC, van Broock M (2009) Photoprotective role of carotenoids in yeasts: response to UV-B of pigmented and naturally-occurring albino strains. J Photochem Photobiol B 95:156–161
11. Moliné M, Flores MR, Libkind D, Diéguez MC, Farías ME, van Broock M (2010) Photoprotection by carotenoid pigments in the yeast *Rhodotorula mucilaginosa*: the role of torularhodin. Photochem Photobiol Sci 9:1145–1151
12. Libkind D, Gadanho M, van Broock MR, Sampaio JP (2008) Studies on the heterogeneity of the carotenogenic yeast *Rhodotorula mucilaginosa* from Patagonia, Argentina. J Basic Microbiol 48:93–98

INDEX

José-Luis Barredo (ed.), *Microbial Carotenoids From Fungi: Methods and Protocols*, Methods in Molecular Biology, vol. 898, DOI 10.1007/978-1-61779-918-1, © Springer Science+Business Media New York 2012

D

E

O

P

R

S

Printed by Publishers' Graphics LLC